Disturbance and Ecosystems

Components of Response

Edited by
H. A. Mooney and M. Godron

Contributors
D. Auclair F. A. Bazzaz F. S. Chapin III R. T. T. Forman
M. Godron P.-H. Gouyon S. L. Gulmon G. Heim P. Jacquard
S. Jain M. Lamotte R. Lee R. Lumaret P. C. Miller
H. A. Mooney M. Rapp W. A. Reiners B. Saugier H. H. Shugart
G. Valdeyron Ph. Vernet P. M. Vitousek D. A. Weinstein
G. M. Woodwell

With 82 Figures

Springer-Verlag
Berlin Heidelberg New York Tokyo 1983

Prof. Harold A. Mooney, Department of Biological Sciences, Stanford University,
Stanford, CA 94305-2493, USA

Prof. M. Godron, Université des Sciences et Techniques du Languedoc, Laboratoire de
Systématique et d'Ecologie Méditerranéennes, Institut de Botanique,
Rue Auguste-Broussonnet, F-34000 Montpellier

ISBN-13: 978-3-642-69139-3 e-ISBN-13: 978-3-642-69137-9
DOI: 10.1007/978-3-642-69137-9

Library of Congress Cataloging in Publication Data. Main entry under title: Disturbance and ecosystems. (Ecological studies; v. 44) 1. Ecology. 2. Man – Influence on nature. 3. Pollution – Environmental aspects. I. Mooney, Harold A. II. Godron, Michel. III. Series. QH541.D57 1983 574.5 83-4771. ISBN 3-540-12454-3

Ecological Studies

Analysis and Synthesis

Volume 44

Preface

The earth's landscapes are being increasingly impacted by the activities of man. Unfortunately, we do not have a full understanding of the consequences of these disturbances on the earth's productive capacity. This problem was addressed by a group of French and U.S. ecologists who are specialists at levels of integration extending from genetics to the biosphere at a meeting at Stanford, California, sponsored by the National Science Foundation and the Centre National de la Recherche Scientifique. With a few important exceptions it was found at this meeting that most man-induced disturbances of ecosystems can be viewed as large-scale patterns of disturbances that have occurred, generally on a small scale, in ecosystems through evolutionary time.

Man has induced dramatic large-scale changes in the environment which must be viewed at the biosphere level. Acid deposition and CO_2 increase are two examples of the consequences of man's increased utilization of fossil fuels. It is a matter of considerable concern that we cannot yet fully predict the ecological consequences of these environmental changes. Such problems must be addressed at the international level, yet substantive mechanisms to do this are not available.

Ecologists are now viewing the dynamic processes operating at the level of whole landscapes. The consequences of the patterns of man's use of landscapes are just beginning to be understood. Networks connecting centers of man's activities often sever the flow of resources of populations between ecosystems. They further serve as pathways for invasions of weedy species. Concepts and tools to deal with landscape phenomena have only been developed recently – they are discussed in this book.

The impact of disturbance by large-scale natural catastrophies, such as fire and hurricanes, is similar to that caused by man's utilization of the landscape's natural productivity. Quantitative information is now available that permits us to predict: 1) the changing structure of forests through time after massive disturbance, 2) the changing status of the nutrient stores in ecosystems, and 3) the changes in water balance of whole ecosystems as they develop following disturbance. Knowledge of the events that immediately follow disturbance, when the link between the regulatory processes of the biota and their resources have been temporarily severed, is crucial to ecosystem management. Losses of system resources such as nitrates can be considerable if management practices discourage natural early successional species. These relationships, as outlined here, point to the value of studies of natural ecosystem processes as guides to resource management.

In a consideration of disturbance it was found that both size and frequency were of considerable importance to subsequent resource patterns and population structures. While concerning ourselves with large-scale patterns and processes, we appreciate that these patterns are the result of events occurring at microsites. The size of the disturbance site, the time of its occurrence, the proximity and the kind of potential colonizers are all important in determining subsequent patterns of population and resource development. The few studies available indicate this to be true. We need considerably greater documentation on how "patch size" influences recovery processes before we can make predictions. In particular we need to know the interrelationship between forest patches to determine how, in aggregate, they form a forest landscape.

The past decade has seen an explosive development in the fields of plant physiological ecology, plant demography and plant population genetics. The Stanford conference brought together experts in these subdisciplines to discuss these developments, with particular reference to the features of organisms that are characteristic of disturbance habitats. It became clear that these organisms have a commonality of features that adapt them to these habitats – that is, they possess a genetic–demographic–physiological syndrome which can be clearly circumscribed. They generally have distinctive nutrient and water-use and growth patterns. Further, they have characteristic reproductive and genetic patterns of variation. These features are obviously interconnected; however, they have traditionally been studied in isolation, and thus the nature of these interrelationships is poorly understood. It appears, though, that if we can understand the nature of resource exploitation by disturbance species where resources may be briefly abundant, we may begin to develop specific tools for managing habitats of differing frequencies and intensities of disturbance. It is important to gain a sound understanding in this area since – due to increased habitat disruptive activities by man – the earth's ecosystems are being simplified and in many cases are being increasingly occupied by "weedy" species. In order to control patterns of development subsequent to disruption, we must understand the ecology of the various successful invaders.

We hope the contents of this book provide an initial framework for a comprehensive view of the effects of disturbance on ecosystems and on their components.

Stanford University — H. A. Mooney
Centre d'Etudes Phytosociologiques et Ecologiques — M. Godron
July 1983

Contents

Contributors

AUCLAIR, DANIEL, Station de Recherches sur la Forêt et l'Environment, Institut National de la Recherche Agronomique, Olivet, France

BAZZAZ, FAKHRI A., Department of Botany, University of Illinois, Urbana, Illinois, USA

CHAPIN, F. STUART III, Institute of Arctic Biology, University of Alaska, Fairbanks, Alaska, USA

FORMAN, RICHARD T. T., Department of Biological Sciences-Botany, Rutgers University, New Brunswick, New Jersey, USA

GODRON, MICHEL, CNRS, Centre d'Etudes Phytosociologiques et Ecologiques L. Emberger, Route de Mende, B.P. 5051, Montpellier, France

GOUYON, PIERRE-HENRI, Institute National Agronomique, 16 rue Claude Bernard, Paris, France

GULMON, SHERRY, L., Department of Biological Sciences, Stanford University, Stanford, California, USA

HEIM, GEORGES, CNRS, Centre d'Etudes Phytosociologiques et Ecologiques L. Emberger, Route de Mende, B.P. 5051, Montpellier, France

JACQUARD, PIERRE, CNRS, Centre d'Etudes Phytosociologiques et Ecologiques L. Emberger, Route de Mende, B.P. 5051, Montpellier, France

JAIN, SUBODH, Department of Agronomy and Range Science, University of California, Davis, California, USA

LAMOTTE, MAXIME, Ecole Normale Supérieure, Laboratoire de Zoologie, 46 rue d'Ulm, Paris, France

LEE, RICHARD, Division of Forestry, West Virginia University, Morgantown, West Virginia, USA

LUMARET, R., CNRS, Centre d'Etudes Phytosociologiques et Ecologiques L. Emberger, Route de Mende, B.P. 5051, Montpellier, France

MILLER, PHILIP C., Systems Ecology Research Group, San Diego State University, San Diego, California, USA

MOONEY, HAROLD A., Department of Biological Sciences, Stanford University, Stanford, California, USA

RAPP, MAURICE, CNRS, Centre d'Etudes Phytosociologiques et Ecologiques L. Emberger, Route de Mende, B.P. 5051, Montpellier, France

REINERS, WILLIAM A., Department of Biological Sciences, Dartmouth College, Hanover, New Hampshire, USA

SAUGIER, BERNARD, Laboratoire d'Ecologie Végétale, Université de Paris-Sud, Bâtiment 431, Orsay, France

SHUGART, HERMAN H., Environmental Sciences Division, Oak Ridge National Laboratory, Oak Ridge, Tennessee, USA

VALDEYRON, G., CNRS, Centre d'Etudes Phytosociologiques et Ecologiques L. Emberger, Route de Mende, B.P. 5051, Montpellier, France

VERNET, PH., CNRS, Centre d'Etudes Phytosociologiques et Ecologiques L. Emberger, Route de Mende, B.P. 5051, Montpellier, France

VITOUSEK, PETER M., Department of Botany, University of North Carolina, Chapel Hill, North Carolina, USA

WEINSTEIN, DAVID A., Environmental Sciences Division, Oak Ridge National Laboratory, Oak Ridge, Tennessee, USA

WOODWELL, GEORGE M., The Ecosystems Center, Marine Biological Laboratory, Woods Hole, Massachusetts, USA

Section 1 Biosphere

1.1 The Blue Planet: of Wholes and Parts and Man

G. M. Woodwell

1.1.1 Introduction

It is axiomatic in ecology that the whole is greater than the sum of its parts. The axiom is almost an article of faith, not merely among ecologists, but also among the scientific community and held with only slightly less intensity among the public. The topic would be trifling in the larger scheme of things except that a useful axiom has been advanced beyond reality to engender an almost blind faith that the whole earth can continue to work as we have known it without its parts. Jim Lovelock has fed this dream recently with a fascinating small book, Gaia, in which he has advanced an hypothesis of homeostasis for the earth as a whole. The hypothesis may be sustained in evolutionary time; it holds little hope for our own time.

Meanwhile scientists live with a peculiar contradiction. We accept the fact that the earth is a biotic system, built and maintained by life itself, but also accept as reasonable the systematic destruction of the units of life, species and ecosystems, that sustain the biosphere as a place habitable by man. Here I examine some fundamental issues for man and the future, as well as details of the structure and function of nature.

What are the man-caused changes in the biosphere that are important? The subject is a monster that has been described in tedious books and I can but touch its outer platy hide. In touching it, however, I discover that the monster has some sensitive points, despite its size. There is guidance from a hand on its pulse and a look at its complexion. Sometimes a quick look at the whole shows limits and helps to keep a reasoned perspective in ecology. Some issues can be seen clearly in no other way.

Barring war or other catastrophe over the next 20 years, world population will probably grow from its present 4.58 billions to more than 6 billions. The importance of this more than 30% growth in less than two decades is difficult to exaggerate; it implies a surge in demands on all resources, including intensified tests of governments themselves, further demands on educational, economic and agricultural systems, as well as an increase in pressure on biotic and mineral resources. And the growth will not stop there. The series of problems for management of resources inherent in this growth is in itself enough to give pause to all thoughtful residents of the planet.

But there are additional transitions underway, some involving the biosphere itself. There is evidence that the world is already overpopulated to the point of destruction. Three transitions seem especially important although they are by no means the only transitions of significance. While they point to additional difficulties for us all in the next years, they also lend emphasis to the need for a new marriage of science and government, joined in the common objective of preserving choice and dignity for ourselves and our children in a world in which man is assuming dominance over more and more of nature.

1.1.2 The CO_2 Problem

The best documented and probably the most threatening transition apart from population growth itself and the erosion of social institutions, is the evidence that the CO_2 content of air is increasing worldwide. The evidence is unequivocal: a 22-year record from Mauna Loa in the Hawaiian Islands; parallel data from the South Pole and elsewhere. The annual increase has been about 1.5 ppm against a contemporary background of 336 ppm CO_2 by volume (Bacastow and Keeling 1973; Keeling and Bacastow 1977; Pearman 1980; Woodwell et al. 1973). The increase has been 3–4% annually until very recently, when according to a recent report from the U.S. National Oceanic and Atmospheric Administration (NOAA 1980), a rise in the price of energy seems to have reduced the CO_2 accumulation for 1979 to 1.2 ppm.

The source of the CO_2 that is accumulating in the atmosphere is the sum of human activities globally in burning fossil fuels and wood, and in harvesting and transforming forests for agriculture or other purposes. In the longer term of centuries, a new equilibrium will be established between atmosphere and oceans as the current pulse of CO_2 is slowly transferred into the very large oceanic reservoir. But for the next few centuries the dominant factors are the release of CO_2 from fossil fuels and the reduction of the biota by human action. As the world's remaining forests are destroyed, the fossil fuel contribution to the CO_2 burden will gain in relative and absolute importance.

The transition in the atmosphere due to the accumulation of CO_2 is profound. It has been shown best by the Mauna Loa record accumulated by C. D. Keeling and by NOAA (Fig. 1). That record has two important lessons for us: first, human activities are changing the globe fundamentally and at an accelerating rate; second, the sinusoidal, seasonal oscillation that is conspicuous in the Mauna Loa record and all records from the Temperate and Polar Zones reflects the metabolic activity of the terrestrial biota and provides ample proof of the importance of the biota in affecting, even controlling, the CO_2 content of the atmosphere. This point seems patent, but it is slow in seeping into common use among those who think about global issues. There is, nonetheless, no more powerful example of the importance of the biota in maintaining conditions on earth suitable for man. The importance should be clear from the magnitudes of the reservoirs of carbon associated with the biota and held within the atmosphere. The atmosphere is thought to contain about 700 billion metric tons (10^{15}g) of carbon as CO_2. The biota contains, according to various estimates, 400–1,200 billion tons (Ajtay et al. 1979; Rodin and Bazilevich 1967; Olson et al. 1978); 829 is a widely used contemporary estimate (Whittaker and Likens 1973). Organic matter in soils is another reservoir, not well known, but thought to contain 1,500–3,000 billion tons globally (Schlesinger 1977; Bohn 1976). The total carbon estimated as residing in the biota and terrestrial humus of

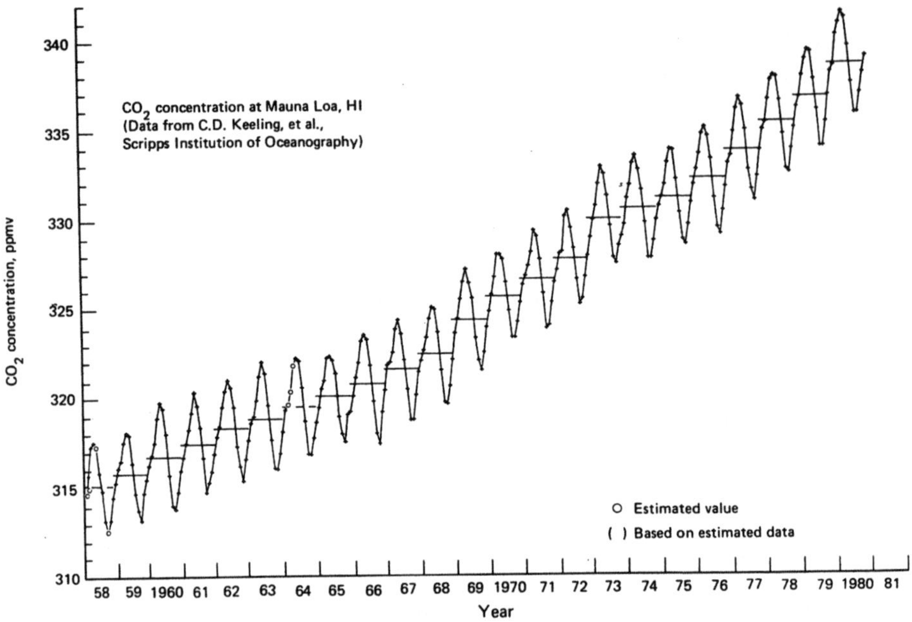

Fig. 1. Carbon dioxide concentration at Mauna Loa, Hawaiian Islands, since 1958 (MacCracken and Moses, 1982). The upward trend is due to the accumulation of CO_2 in the atmosphere from combustion of fossil fuels and from destruction of forests. The annual oscillation is due to the metabolic activity of forests in the northern hemisphere

the earth is a minimum of about 3 times that held in the atmosphere. A small change in that pool, say 0.1%, is enough to change the CO_2 content of the atmosphere by as much as 1 ppm. Current fossil fuel consumption is releasing about 5 billion metric tons of carbon as CO_2 annually (Rotty 1979). If we were to decide that the CO_2 increase should be controlled, we would look, first, toward restraint in the use of fossil fuels; second, we would seek the re-establishment of the forests of the earth. There is no other solution.

The potential effect of the increase in CO_2 in the atmosphere is very much more uncertain than the fact of the increase. Most climatologists who have examined the problem believe that a doubling in the atmospheric burden, expected to occur early in the next century, possibly by 2030, will bring a warming of the earth that will average 2–4 °C (Manabe and Wetherald 1980; Manabe and Stouffer 1979). The warming will be differential, however; being greater at the poles than in the equatorial zones. A temperature increase at the poles of 10–15 °C is thought possible. Such an increase will lead to a rise in sea level of 15–20 feet if the West Antarctic Ice Sheet disappears or slips into the ocean over the next century or so. The rise will inundate coastal areas worldwide and will certainly present a series of crises to a crowded human population [MacDonald 1978; Woodwell et al. 1979; AAAS 1980; Council on Environmental Quality (CEQ) 1981].

This physical disturbance will be preceded by a series of transitions in climate that will move climatic zones generally poleward, displace agriculture, and cause the displacement of forested zones. It is important to remember that in this transition forests can be destroyed rapidly by a change in conditions, but are rebuilt slowly. The development of forests is further impeded if development is dependent on the rapid migration of species. Such migrations normally occur slowly, are accompanied by considerable genetic sorting, the development of ecotypes, and ultimately by speciation. While those processes are obviously not required to obtain a stand of trees, they are an intrinsic part of establishing a new forest community in a different place. The forest community has the advantage of self-regulation and development, a significant advantage in a world short of energy.

The CO_2 problem is but one of several disruptions of global chemical cycles now recognized as man-caused. Others include nitrogen, sulphur, and phosphorus. The carbon cycle, however, is the best known and alone provides ample evidence of the magnitude of the human intrusion on biotic processes globally.

1.1.3 Toxification

A corollary of the central axiom is that the evolutionary processes that have made the whole have also defined a community chemistry, an environment whose chemistry has been defined by the biota itself. How precise this definition may be is elusive because measurements destroy the precision of the coupling. We are left to infer the basic relationship, as A. C. Redfield (1958) did when he drew attention to the constancy of atomic ratios between the biota and waters of the oceans. Nonetheless, progress in interpreting evolution leaves no other conclusion, so we are not surprised to discover that important chemical messages are sent between organisms by pheromones at concentrations in the range of 1 part in 10^{10} or 10^{12}. Similarly, we expect to find changes in the structure of communities if we shift the availability of nutrient elements or mobilize heavy metals, release novel, biotically active compounds or change the acidity of rain. All the evidence we have suggests that the biota is sensitive to extraordinarily small concentrations of chemical substances, that a variety of mechanisms exist for discriminating among substances and increasing or decreasing their concentrations, and that the response of a community to these types of disturbance is much more complex than can be indicated by any such single criterion as net primary production or an index of diversity.

These considerations make a mockery of those convenient crutches of the practical ecologist and the reasonable man charged with controlling toxins or wastes: thresholds and assimilative capacities. They also make a mockery of the assertions that systematic oceanic disposal of sewage or other, more noxious, industrial wastes is benign. We may choose to indulge in it, but we do so with the clear knowledge that there are biotic effects, although unmeasured and possibly unmeasurable and that persistent toxins will almost certainly return to haunt us.

The problems with toxins are larger than most wish to acknowledge. Fortunately, few are known to be global, but these few are important. They are radioactivity, persistent pesticides and virtually any other long-lived substance that enters air as small particles. Such substances can be transported worldwide in 21–26 days in mid-latitudes. The pollution of Lake Michigan with chlorinated hydrocarbons at concentrations that prevent the reproduction of fish and render the Coho salmon unfit for food is not reversible. Neither is the contamination of blue fish on the East Coast of North America with PCB's. Yet the arguments for oceanic disposal of waters, nationally and internationally, continue. The oceans are sufficiently closely connected that wastes are not isolated. What is required of the scientific and political communities is, first, sufficient information to define the issues and, second, international agreement on systems for management of toxins to eliminate any worldwide contamination.

Experience is abundant; so is precedent in management. Radioactivity from bomb tests in the atmosphere was recognized as a sufficiently significant problem to warrant an international treaty banning such tests. In a somewhat different procedure most of the technologically advanced nations of the Temperate Zone moved to ban the use of DDT and other chlorinated hydrocarbon pesticides once these compounds were found to have a global distribution, to affect a wide variety of organisms and to accumulate in man. These two steps showed that there is both the need and the political will to abandon segments of technological progress that reach beyond reason and control.

Despite these examples there is a growing number of instances of changes in global or regional chemistry due either to toxins introduced by human activity or to shifts in the availability of major nutrients. The results are similar: gradual changes in the structure of nature. These changes are usually not consistent with human interests in preserving the variety of choices offered by life in a biologically rich and stable environment (Woodwell 1967).

1.1.4 The Effects: Biotic Impoverishment

This book examines the structure of man-dominated as opposed to natural ecosystems. The first step must be an examination of the effect of man on natural ecosystems. Objective, persuasive data are rare.

The clearest example for me comes, even now, from an experiment started in 1961 at Brookhaven National Laboratory where we examined experimentally the effects of ionizing radiation on a forest. The most important aspect of the design was the selection of the community, a late-successional stand of oak-pine forest. The forest was selected on the basis of considerable analysis as likely to yield the greatest amount of information about the response of natural communities in general to this type of disturbance. The information gleaned from the work reached far beyond the effects of ionizing radiation, however.

A single Cs^{137} source of gamma radiation was established in the center of the forest. The source was large enough to produce a gradient of effects from virtual sterilization of the landscape at exposures in excess of 1,000 R/day to no conspicuous effects at exposures of less than 0.1 R/day. After 9 months' exposure over one winter, the pattern of response had been established and was modified subsequently only in detail. The pattern is of special interest because it is not peculiar to the effects of ionizing radiation but a general pattern, segments of which are to be found surrounding virtually any disturbance of a forest anywhere (Woodwell 1970).

The effects were conspicuous as a series of zones of vegetation. Close to the source of radiation no higher plants survived although certain crustose lichens survived, in this zone where exposures were extreme. At slightly lower exposures a frulticose lichen, *Cladonia cristatella,* formed an extensive ground cover after several years.

At lower exposure rates, farther from the source, there was an almost pure stand of the sedge, *Carex pensylvanica,* a common herb of the forest.

At a still greater distance and lower exposure rate, there was a zone made up of the shrubs of the forest – lowbush blueberry *(Vaccinium angustifolium)*, a common upright-growing blueberry *(Vaccinium heterophyllum)*, and the huckleberry *(Gaylusaccia baccata)*.

At lower exposures the pines *(Pinus rigida)* were eliminated, leaving the oaks in addition to the more resistant species of the inner zones. The forest appeared normal in the sense that there are other areas in the oak pine forest where no pine occurs and their absence might be considered as within the normal variation intrinsic in the forest.

At yet lower exposures, those below about 5 R/day, the pitch pine survived and the forest appeared intact.

Time brought adventives to the gradient; these survived in the inner zones, as part of the sedge zone and at its inner boundary. The adventives were the sweet fern, *Myrica asplenifolia, Phytolacca,* a spiny *Rubus,* and the fireweed, *Erechtites,* all hardy plants, commonly found in heavily disturbed, impoverished sites, and also resistant to ionizing radiation. The reason for the radiation resistance is surely one of the most remarkable facts of nature, but beyond our consideration here. The important point is the pattern, a transition from intact oak-pine forest through an oak forest stage that is in fact well within the variation expected in the normal forest, through a series of stages that are less common in nature but are far from unique in the mosaic that covers the region inhabited by the oak-pine forest. I think of these zones as stages of impoverishment set forth in an unusually clear manner.

A similar transition toward impoverishment occurs downwind from the Sudbury, Ontario smelters in the mixed spruce-hardwoods forests of southern central Ontario. A parallel series of changes occurs in the structure of aquatic communities under eutrophication or pollution. Details of these transitions are beyond the reach of this discussion. The pattern, however, is clear enough, familiar to all on reflection from personal experience: persistent shrub, field or barren land in a forested zone; the *Cladophora* communities of polluted estuaries; the filamentous blue-green algal communities of polluted lakes and streams. The patterns are those of impoverishment, shortened food webs composed of hardy species of limited utility to man (Woodwell 1970).

If we were to look for a general law, it would be that disturbance favors communities dominated by small-bodied, rapidly reproducing, hardy species: these are the groups from which we normally recognize pests. No one argues that human life in such a world is simpler, cheaper, richer, more comfortable, more desirable, or represents the heritage we seek for our children.

1.1.5 Conclusions

The transition from "unmanaged" to "managed" ecosystems is far more pervasive and important than is apparent at first glance. Natural, unmanaged ecosystems have been the dominant influence on the biosphere throughout most of the history of the earth. Their continued dominance is now clearly in question. The fact that a sudden change in the CO_2 content of the atmosphere is occurring over decades is ample evidence that the fundamentally biotic controls on the biosphere are weakening; dominance is shifting to man.

We ask what the implications may be. The answers are always less clear than we might wish, but human life is likely to be more, not less difficult, as the burdens for joint action in the common interest increase. The CO_2 problem is again a good example. The warming of the earth anticipated for the next decades can be expected to accelerate respiration globally more than it will accelerate photosynthesis. Zones that are now forested will lose their forests more rapidly through climate change than new forests can develop elsewhere. The effect will be a further net release of carbon as CO_2 into the atmosphere, aggravating the CO_2 problem and accentuating the changes in climate. The solution is a drastic one: restriction in the use of fossil fuels and careful management of forests globally to control or prevent the further accumulation of carbon dioxide in the atmosphere. The cost of intensified use of the biosphere is intensified management.

The CO_2 problem is but one of several. There are similar imbalances with nitrogen, sulphur, phosphorus, each of which, separately, and all of which together pose a series of challenges for man in management of both industrial and biotic resources. The challenge includes the hazards of toxification, hazards that may appear trifling when only natural communities are involved until we recognize the implications of progressive biotic impoverishment and the simple fact that there is no way that man can be insulated in his intensively managed ecosystems from continuous interaction with the rest of the world.

The litany of problems is virtually limitless; the solutions seem clear enough, although beyond reach at the moment. The solution lies in recognition of the importance of preserving the dominance of the biota globally. That requires only that the managed systems of the earth mimic in their function, in all of their exchanges with the rest of the world, the unmanaged systems they replace. We know enough to start. And this symposium is as good an instrument as any.

Résumé

Ce premier article examine quelques-unes des principales conséquences des actions humaines *à l'échelle du globe terrestre*.

La plus importante de ces conséquences est vraisemblablement l'augmentation de la concentration en CO_2 de la basse atmosphère (de 315 ppm à 336 ppm, en 22 ans), qui risque de conduire à un réchauffement de l'air à une fusion partielle des glaces polaires, et à une augmentation du niveau des mers au moins égale à 5–7 m. En outre, le réchauffement de la basse atmosphère mettrait en danger les forêts elles-mêmes, et accentuerait de ce fait l'augmentation du taux de gaz carbonique.

Une autre série de conséquences est l'augmentation de la production de produits toxiques (soit en raison de leurs propriétés radio-actives, soit en raison de leurs propriétés chimiques). L'auteur souligne que certains produits peuvent avoir une influence même si leur concentration est seulement égale à $1/10^{10}$ ou $1/10^{12}$.

La troisième série d'effects est un appauvrissement biologique: l'action de l'homme entraîne une régression, assez exactement inverse de la progression suivie naturellement par les communautés au cours de la succession qui va du sol nu à la forêt. Il serait utile de mieux comprendre pourquoi certaines espéces sont plus résistantes que d'autres aux accidents causés par l'homme, d'autant plus que les caractères de ces espèces résistantes sont aussi ceux que l'on attribue à celles que l'on considère comme des fléaux (petite taille, reproduction rapide, faible sensibilité aux traitement éradicants).

La conclusion de l'auteur est que les écosystèmes artificiels, qui deviennent de plus en plus dominants, sont de plus en plus difficilement contrôlables: nous brisons des systèmes de régulation qui se sont établis en plus d'un milliard d'années. Les systèmes qui résultent des actions humaines risquent d'être auto-accélérateurs (rétro-actions positives) au lieu d'être dotés de régulations (rétro-actions négatives).

References

AAAS (1980) Environmental and societal consequences of a possible CO_2-induced climate change. Am Assoc Adv Sci Meet 1980, San Francisco

Ajtay GL, Ketner P, Duvigneaud P (1979) Terrestrial primary production and phytomass. In: Bolin B, Degens ET, Kemps S, Ketner P (eds) SCOPE 13 – The global carbon cycle. SCOPE – Int Counc Sci Unions. Wiley, New York, pp 129–182

Bacastow R, Keeling CD (1973) Atmospheric carbon dioxide and radiocarbon in the natural carbon cycle: II. Changes from A.D. 1,700 to 2,070 as deduced from a geochemical model. In: Woodwell GM, Pecan EV (eds) Carbon and the biosphere. Tech Inf Center, Off Inf Serv. US At Energy Comm, Springfield, VI, pp 86–135

Bohn HL (1976) Estimate of organic carbon in world soils. Soil Sci Soc Am J 40:468–470

Council on Environmental Quality (1981) Global energy futures and the carbon dioxide problem. Superintendent of Documents. US Gov Print Off, Washington DC, 92 pp – ix

Keeling CD, Bacastow R (1977) Impact of industrial gases on climate. In: Energy and climate Nat Acad Sci Rep, Washington DC, pp 72–75

Lovelock JE (1979) Gaia, a new look at life on earth. Oxford University Press, Oxford New York

MacCraken MC and Moses H (1982) The first detection of carbon dioxide effects: Workshop summary. In: Department of Energy Report, *Carbon Dioxide Effect Research and Assessment Program.* DOE/CONF-8106 214:3–44, Washington D.C.

MacDonald GJF (1978) An overview of the impact of carbon dioxide on climate. Mitre Corp, M78–79, McLean VA

Machta L (1979) Air Resources Laboratories. Nat Oceanic Atmos Administ, Washington DC (fide GO Barney, Global 2000 Report to the President. Counc Environ Qual, Washington DC 1980)

Manabe S, Stouffer RJ (1979) A CO_2-climate sensitivity study with a mathematical model of the global climate. Nature (London) 282:491–493

Manabe S, Wetherald RT (1980) On the distribution of climate change resulting from an increase in CO_2-content of the atmosphere. J Atmos Sci 37:99–118

National Oceanic Atmosphere Administration (1980) Oral presentation before Department of Energy. Meeting on CO_2, April 1980, Washington DC

Olson JS, Pfuderer HA, Chan YH (1978) Changes in the global carbon cycle and the biosphere. ORNL/ETS-109, Oak Ridge Nat Lab, Oak Ridge Tenn, pp 169

Pearman GI (1980) Global atmospheric carbon dioxide measurements: A review of methodologies, existing programmes and available data. WHO Tech Rep, Geneva, Switzerland

Redfield AC (1958) The biological control of chemical factors in the environment. Am Sci 46:205–222

Rodin LE, Bazilevich NI (1967) Production and mineral cycling in terrestrial vegetation. English translation: Fogg GE (ed). Oliver and Boyd, Edinburgh London

Rotty R (1979) Uncertainties associated with global effects of atmospheric carbon dioxide. ORAU/IEA-79-60 (O) Oak Ridge Assoc Univ. Inst Energy Anal, Oak Ridge Tenn

Schlesinger WH (1977) Carbon balance in terrestrial detritus. Annu Rev Ecol Syst 8:51–81

Whittaker RH, Likens GE (1973) Carbon in the biota. In: Woodwell GM, Pecan EV (eds) Carbon and the biosphere. AEC Symp Ser 30. NTIS US Dep Commerce, Springfield Va, pp 281–302

Woodwell GM (1967) Toxic substances and ecological cycles. Sci Am 216(3):24–31

Woodwell GM (1970) The effects of pollution on the structure and physiology of ecosystems. Science 4:29–33

Woodwell GM, Houghton RA, Tempel NR (1973) Atmospheric CO_2 at Brookhaven, Long Island, New York: Patterns of variation up to 125 meters. J Geophys Res 78:932–940

Woodwell GM, Revelle R, MacDonald GJ, Keeling CD (1979) The carbon dioxide problem: Implications for policy in the management of energy and other resources. Bull At Sci 35(8):56–57

Section 2 Landscape

2.1 Landscape Modification and Changing Ecological Characteristics

M. Godron and R. T. T. Forman

We have two objectives in this chapter. First, we wish to use some of our most encompassing or fundamental ecological concepts to pinpoint emergent patterns when comparing the natural landscape with the major types of human-modified landscapes. Second, we focus more specifically on how the structural characteristics of landscapes change along a gradient of increasing human modification.

2.1.1 Ecosystems and Landscapes

As background, we must briefly digress to distinguish ecosystems from landscapes. The ecosystem concept is used widely, but often ambiguously, in the world's literature, because (a) for some ecologists it has only a functional meaning, (b) for others it also has a spatial connotation, which includes any level of scale, e.g., from a rabbit pellet to the biosphere, and (c) for others the spatial aspect is included, but additionally, relative homogeneity must characterize the ecosystem (Johnson and French 1981).

Tansley used the term ecosystem without giving it a spatial significance. He writes in his 1935 paper, "... ecosystems, as we may call them, are of the most various kinds and sizes. They form one category of the multitudinous physical systems of the universe, which range from the universe as a whole down to the atom." Similarly, Fosberg (1967) stated that an ecosystem is "... the sum total of vegetation, animal, and physical environment, in whatever size segment of the world." Odum (1971) writes, "Ecosystems may be conceived of and studied in various sizes."

Perhaps the inherent ambiguities in the ecosystem concept would have been avoided if we had universally adopted the earlier term "biocenosis" [Mobius (1877), as cited by Kormondy (1965)], whose definition was a model of clarity: a biocenosis exists, "... where the sum of a community of living species and individuals, being mutally limited and selected under the average external conditions of life, have, by means of transmission, continued in possession of a certain definite territory." The concept of community indicates relative homogeneity in the distribution of organisms (e.g., Greig-Smith 1964; Daubenmire 1968). Mobius' definition shows that the biocenosis includes the action of the physical environment plus the action of competition and selection of organisms in a community.

A similar term, biogeocenosis, is also commonly used. The most widely cited definition by Soviet ecologists is ".. an assembly on a specific area of land surface of homogeneous natural elements (atmosphere, mineral strata, vegetation, animals and microorganisms, soils and hydrological conditions), with its own specific interrelationships among these components and a definite type of inter-change of materials and energy among themselves and with other natural phenomena and representing an internally-contradictory dialectical unity, being in constant movement and development" (Sukachev and Dylis 1964), as cited by Johnson and French (1981). Biogeocenosis appears to be essentially the same as the biocenosis, in including the concepts of structure and dynamics, specific spatial limits, and homogeneity. Therefore, in view of the definitions of the authors cited above, a biocenosis or biogeocenosis is an ecosystem at the community level.

This seems to us to be a possible, but awkward, state of affairs. One additional factor seems useful. An ecologist studying an ecosystem almost always selects an area with relative homogeneity in the distribution of organisms. Thus, for example, the structure and dynamics of a woodlot or a field are studied, but rarely would one study the woodlot with an adjacent field and hedgerow as a single ecosystem. This is because methodologically one wishes to characterize the ecosystem with a limited number of samples, and such heterogeneity requires much more sampling. In addition, results from a heterogeneous area are less easily compared or extrapolated to other areas, since the spatial configuration of the heterogeneity may be critical in the results. Indeed, practically-speaking, ecosystems studied almost always are selected with the homogeneity criterion in mind. We therefore recommend that the ecosystem concept be considered synonymous with the biocenosis and biogeocenosis concept, namely including the characteristics of structure and dynamics, specific spatial limits, and relative homogeneity, and use it ourselves in this sense.

The landscape, then, contrasts clearly with the ecosystem, since the landscape is almost always highly heterogeneous, including such landscape elements as fields, woods, marshes, villages, and corridors (McHarg 1969; Rackham 1976; Baudry and Baudry-Burel 1978). While the dictionary says the landscape is the area which may be seen by the eye, this definition is too imprecise scientifically, since the eye sees different areas depending upon characteristics of the eye, location of a person, and atmospheric conditions. Rather, we suggest that a landscape is a kilometers-wide area where a cluster of interacting stands or ecosystems is repeated in similar form (Troll 1968; Forman and Godron 1981). In a suburban landscape, for example, a cluster of landscape elements such as a residential area, a patch of woods, a commercial area, and an open grassy area would be found throughout most of the landscape, whereas a different cluster of ecosystems would be found in an adjoining agricultural landscape or a sandy forested plain. The boundary of a landscape encloses the area subject to a common overall disturbance regime (including both human and natural disturbances), and normally a common geomorphology.

2.1.2 Some Ecological Attributes for Comparing Landscapes

Here we compare natural and human-modified landscapes by examining the attributes which change along a gradient of increasing human modification. Many ecological attributes could be used (Odum 1969, 1971) but we will examine ten: horizontal structure, stability, thermodynamic characterization, chorology, minimal area (or grain), nutrient cycling, net production, tactics, phylogeny, and type of resistance.

TYPES OF LANDSCAPES AND ECOLOGICAL ATTRIBUTES	NATURAL LANDSCAPE	MANAGED GRASSLAND OR FOREST	CROPLAND	URBAN
1. Horizontal structure	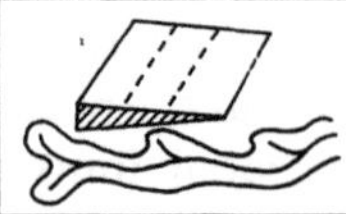		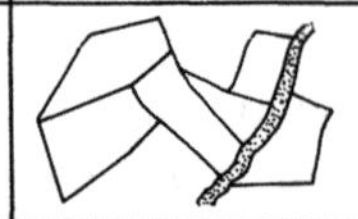	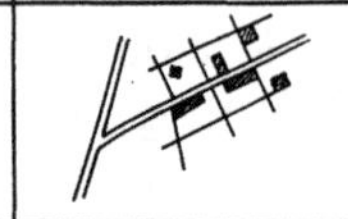
2. Stability	Highest metastability (with high potential energy)	High metastability	Intermediate metastability	Low metastability (with low potential energy)
3. Thermodynamic characterization	Relatively closed systems	Partly open systems	Open systems	Very open systems
4. Chorology	Viscosity	Fluidity	Introduction of exotic species	Cosmopolitanism
5. Minimal area (or grain)	Proportional to the size of the plants	Bimodal	Irregular	Unknown
6. Nutrient cycling	Relatively rapid	Rapid	Forced	Polluting
7. Net landscape production	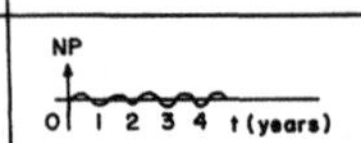	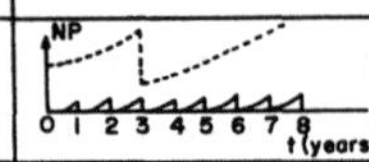	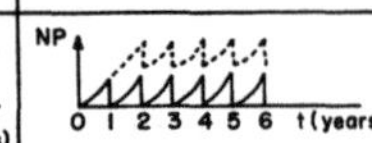	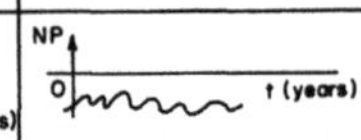
8. Tactics	k >> r	k ≈ r	r > k	Very variable
9. Phylogeny	Anagenesis > Cladogenesis	Some remarkable co-evolution	Loss of polymorphism	Unknown
10. Type of resistance	Senescence resistant	Accident resistant	Pesticide resistant	Pollution resistant

Fig. 1. Hypothesized changes in major ecological attributes along a gradient of landscapes increasingly modified by human activities

2.1.2.1 Horizontal Structure

In natural landscapes subjected to minimal human disturbance, the horizontal structure which appears, for example, on aerial photographs at the scale of 1/10,000, generally has little contrast. Usually ecotones separating communities are frequent, and spatial continua are present or common. The primary breaks in structure come from natural disturbances and from geomorphology, particularly dendritic patterns produced by erosion. With increasing landscape modification by human activities (moving to the right in Fig. 1), strong contrasts in the form of sharp boundaries between ecosystems become more numerous. Equally striking is the proliferation of straight lines in the landscape, composed of both boundaries between ecosystems, and corridors crossing ecosystems (Lebeau 1969). Thus a predominant human imprint on landscape structure is an increase in contrast as well as a linearization and rectangularization process.

2.1.2.2 Stability

For brevity, we will utilize the classical characteristics of stability used in mechanics, yet attempt to be cognizant of the added dimension of evolution. Living organisms often start from highly stable states whose "neg-entropy" is low, and

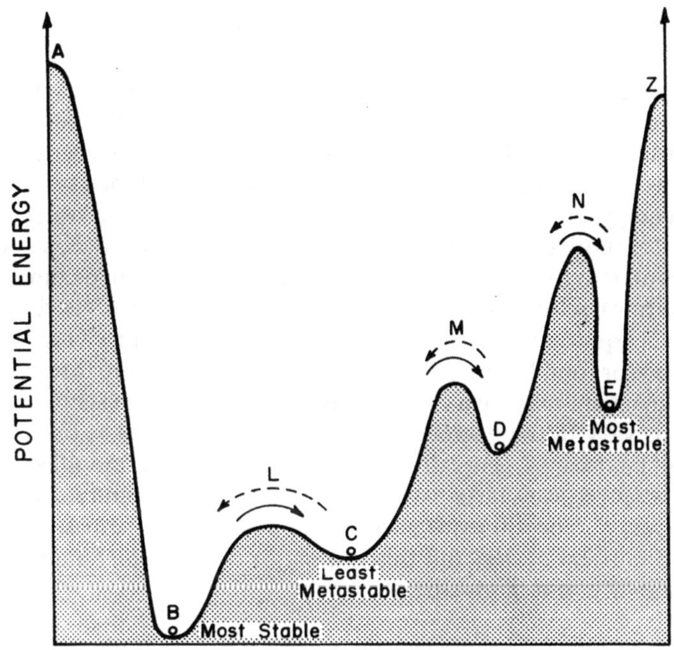

Fig. 2. Stability states in terms of potential energy. Point *B* it the most stable. Points *C*, *D*, and *E* are metastable. Points *A*, *L*, *M*, *N*, and *Z* are unstable. The transitions from *B* to *C*, and to *E* are difficult, but increase the degree of metastability (such a progression corresponds to a series of successional stages which begins with the most stable state, moves rapidly to the least stable state, and thereafter gradually increases its metastability). The transitions from *E* to *D*, to *C* and to *B* are less difficult. (They correspond to a degradation in ecological systems)

may be represented as state B in Fig. 2. Organisms build locally metastable structures which are more complex and have less entropy, and may be represented by points C, D or E in Fig. 2. Hence, natural successional stages from bare soil to forest proceed toward states of increasing metastability.

When this stability model is applied to the landscape (Fig. 2), we see the natural landscape has considerable potential energy in the form of biomass, and thus must be high on the graph. Yet, the landscape is also rather stable. Consequently, the natural landscape must be analogous to the metastable point E or a hypothetical F on the graph. With increasing landscape modification, both potential energy and metastability decrease, such that the town or suburban landscape might be analogous to the low metastability point C. Note that any landscape may be devastated by a disturbance to point B, which has little potential energy, but has maximum stability. If a modified landscape such as point C is disturbed, recovery may be rapid (Horn 1974). In contrast, recovery to the pre-disturbance level by a natural landscape may be slow since so much potential energy was lost. Therefore, we see both potential energy (or neg-entropy) and metastability are both useful landscape descriptors, and that they are curvilinearly related to one another.

2.1.2.3 Thermodynamic Characterization

When considering the thermodynamic characterization of ecosystems, it must be remembered that all systems may be classified into three types: isolated, closed, and open (Bruhat 1962). Isolated systems have no input of energy or matter. Closed systems have an input of energy, but not of matter. Open systems have inputs of both energy and matter. It is often emphasized that living systems are open, which is true for individual plants, animals or people, considered as systems. But ecosystems containing both autotrophs and heterotrophs, while commonly open (to solar radiation), may function quite well closed, with matter recycling. For example, a closed aquarium may stay in equilibrium without receiving any external mineral nutrient or food for an indefinite period. An old mature forest with its surrounding atmosphere may also function as a nearly closed system.

Ecosystems vary in the degree of thermodynamic openness, but for our purpose, natural landscapes generally are more closed than modified landscapes (Fig. 1). The latter receive a large external supply of nutrients and fossil energy, and give off "products". The influence of people almost always increases the openness of energy and nutrient cycles, and this is a major way in which humans stress or modify the stability of a landscape.

2.1.2.4 Chorology

Chorology, which focuses on the processes and results of species dispersal (Good 1962), may also be used to characterize a landscape. In natural landscapes, the dispersion of reproductive structures is rather "viscous", and the spectrum of dispersal types is dominated by barochores (heavy fruits) and zoochores (mainly disseminated by frugivory) (Van der Pijl 1969).

Managed forests and grasslands produce more "fluid" chorologic systems dominated, for example, by anemochory (wind dispersal) and by long-range zoochory. In contrast, along a successional sequence, dispersal mechanisms commonly change from predominately anemochory to barochory and short-range zoochory.

Cultivation is generally characterized by the planting and accidental introduction of non-native (exotic) species. At the end of the modification gradient (Fig. 1) the most human-influenced landscapes provide special opportunities for cosmopolitan species.

2.1.2.5 Minimal Area (or Grain)

In temperate and cooler regions, there is a minimal area in which almost all species of a community are found. This classic concept (Jaccard 1902; Braun-Blanquet 1964) may be related to chorology as well as to horizontal structure. However, it appears useful to characterize minimal area separately, since we know that it applies at a fine scale as well as at a coarse scale.

In natural forests, the minimal area may be so great that the limits of the community cannot easily be observed (Emberger 1954). In other natural communities

We now take a closer look at a few of the points just outlined. We begin by considering the critical concept of disturbance, and in so doing, will add two more important levels of landscape modification to the four utilized in Fig. 1. Finally, we will elucidate the expected patterns of variation of landscape elements, such as patches, corridors and matrix, as a function of level of landscape modification.

2.1.3 Disturbance Regimes

Landscapes are defined by a common geomorphologic origin and a common disturbance regime. Geomorphologic processes of deposition, rock formation, uplift, erosion, aeolian action, and glaciation produce more or less distinct physiographic units, such as coastal plains, parallel ridge and valley terrain, muskeg lands, large river deltas, calcareous plains, and the like. In the absence of human activities these units contain various natural disturbances. Thus, fire may be frequent in some, flooding effects, salt deposition, hurricanes, or insect plagues in others. These observations suggest that landscapes have both different physiographies and different natural disturbance regimes.

Disturbance regime is the sum of types, frequencies, and intensities of disturbance through time in the landscape. Disturbance, in turn, is considered as something that causes a community or ecosystem characteristic, such as species diversity, nutrient output, biomass, vertical and horizontal structure, etc., to exceed or drop below its common (homeostatic) range of variation.

As noted earlier, a landscape is an ecological mosaic of specific ecosystems (Forman 1979). Among the ecosystem patches, several community or vegetation types are normally present. These result from the presence of soil catenas and variations in soil moisture, elevation, and aspect.

Human-modified landscapes are changed by new disturbances introduced by people. Where are people concentrated and where are their disturbances concentrated? People are not distributed evenly or randomly among landscapes, but instead are concentrated in certain landscapes and sparse in others. For example, people settle near coasts and rivers where water is available for drinking, agriculture, transportation, and protection. Agricultural activity (e.g., plowing and applying pesticides), as a disturbance, is concentrated in landscapes which are relatively flat, fertile, warm, and moist. Similarly, disturbances such as tree cutting, air pollution, fire suppression, and constructing buildings and roads are concentrated in different landscapes. In essence, human disturbance regimes differ by landscape, and are superimposed on natural disturbance regimes and geomorphologies, which also differ sharply. The resulting landscapes vary markedly, and the boundary between landscapes is relatively distinct, at least in vegetation structure.

We have mentioned the primary characteristics which differentiate landscapes. But what are the types of landscapes produced? We will not approach the task of a classification of landscapes, and further, will ignore particular disturbance regimes and geomorphologic differences. Rather we wish to focus on the types of modified landscapes produced by disturbance regimes. Six broad patterns are suggested, which differ in the level of human modification.

(1) Natural vegetation (grassland, rain forest, desert, etc.) is produced by a natural disturbance regime in the absence of significant human effects. (2) The second level along the gradient of landscape modification is illustrated by areas recently colonized by people, where scattered clearings in the natural vegetation are formed for habitations and adjacent plantings. Significant effects on the natural vegetation are localized near the clearings. (3) At the third level the majority of the landscape is similar to natural vegetation, but is managed for grazing, cutting, etc., such as rangeland and timber tracts. This pattern will be referred to as managed vegetation, since important differences in species, energy and nutrient fluxes from natural vegetation exist (Franzreb 1978; Southwood 1960; Vitousek and Reiners 1975). Planted ecosystems such as crops or tree plantations are also found commonly in this landscape. (4) At the fourth level, most of the land is planted, although considerable managed vegetation may remain. Much cropland is in this category. (5) At the fifth level, habitation areas such as towns and suburban areas are common along with planted vegetation. Managed vegetation is generally in isolated patches. This is a heterogeneous town and country mix of agriculture, urbanization, and managed vegetation. (6) At the end of the scale, an insignificant amount of managed or planted vegetation remains in the urbanized or industrialized matrix.

This rough scale of increasing landscape modification by people will be used to examine the changing patterns of several characteristics of landscape structure.

2.1.4 Effect on Landscape Patch Structure

The structure of a landscape is the spatial distribution of energy, mineral nutrients and species in relation to the patches, corridors, and matrix of the landscape (Forman and Godron 1981). Patches are plant and animal communities surrounded by an area with a dissimilar community structure or composition, and corridors have the added characteristic of being long and narrow. Both are typically embedded in a matrix. Here we will focus on patches, their origin, size, shape, number, and configuration. How does the increasing level of landscape modification by people affect each of these patch characteristics?

2.1.4.1 Patch Origins

Natural vegetation may contain gradients (Curtis 1959; Whittaker 1967) where there are gradual environmental changes, and the patches present may or may not have distinct boundaries. Yet, quite distinct patch types (Forman and Godron 1981) in natural vegetation are also normally produced through a variety of mechanisms (Daubenmire 1968; Heinselman 1973; Forman et al. 1976; Forman and Boerner 1981). Spot disturbance patches result from disturbance of small areas. Environmental resource patches result from the heterogeneous spatial distribution of environmental resources. Remnant patches result from disturbance surrounding small areas.

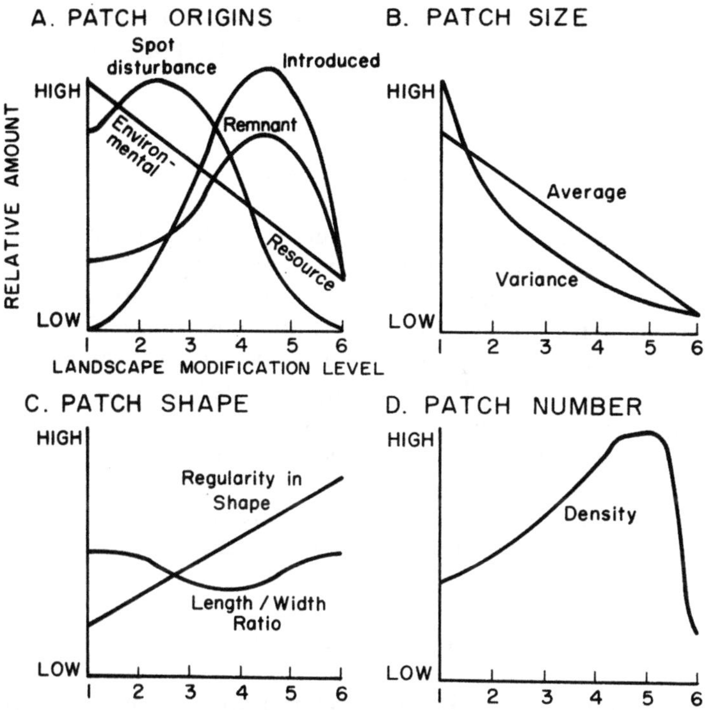

Fig. 3. Effects of landscape modification on patch characteristics. Landscape modification levels refer to landscapes dominated by: (1) Natural Vegetation; (2) Natural Vegetation with Scattered Clearings the colonization stage); (3) Managed Vegetation (such as managed grassland for livestock or forest for logging); (4) Planted Vegetation (such as cropland); (5) Town or Suburban Areas (a town and country stage with a mixture of residential, managed and planted patches); (6) Urban or Heavy Industry. See text for further description

The density of environmental resource patches decreases steadily along the modification gradient (Fig. 3 A), as introduced agricultural vegetation and urbanization spread, thus obliterating the natural vegetation and its component patches.

Spot disturbance patches increase in the colonization level 2 (Heinselman 1973; Lorimer 1977), and in the case of forest cutting in level 3 (Fig. 3 A). Beyond this level, spot disturbance patches decrease as cropland and urbanization increase.

Remnant patches, which change little in number until most of the landscape is agriculture, peak in the agricultural and town levels, before dropping in the last level (Curtis 1956).

Introduced patches, such as pine plantations or bean fields, are dominated by species introduced by people. Such patches, which are absent in natural vegetation, first appear in the colonization level 2, predominate in the cropland level 4, and decrease with urbanization (Fig. 3 A). A bimodal distribution would be produced where the cropland landscape is essentially all extensive fields of one or two crops, resulting in a relatively non-patchy homogeneous matrix.

In short, with increasing landscape modification, there are major shifts in the proportions of patches with different causal mechanisms (and consequently species dynamics). The overall trend is from predominantly environmental resource patches to spot disturbance patches to introduced patches and, finally, to remnant patches.

2.1.4.2 Patch Size

In natural vegetation influenced minimally by human activities, patches are highly variable in size, with the average patch size probably quite large (Pickett and Thompson 1978) (Fig. 3 B). These are predominantly environmental resource and spot disturbance patches, such as produced by fire, insect outbreaks or hurricanes.

With colonization, the average patch size and its variance would appear to decrease (Forman and Boerner 1981), since people generally initially colonize an area for similar objectives. In addition, the presence of people would tend to introduce barriers to the spread of some disturbances such as fire (Wacker 1979). Though in some landscapes patch sizes might be large in levels 3 and 4, in general, we suggest the progressive utilization of the land for agriculture and urbanization would tend to progressively subdivide the area into smaller patches (Lebeau 1969).

2.1.4.3 Patch Shape

In natural vegetation the predominant environmental resource and spot disturbance patches are highly irregular in shape, where factors such as fire, weather, erosion, and other geological processes are major controls on the extensiveness of patches (Fig. 3 C). The progressive modification of the landscape by both agriculture and urbanization tends to eliminate narrow peninsula-like projections, straighten patch edges, and result in polygons, particularly rectangles (Curtis 1956; Lebeau 1969).

The average length to width ratio of patches may be somewhat lower in the middle levels of modification than at either extreme. Thus, in natural vegetation erosion patterns and sedimentary strata on the one hand, and disturbances controlled by weather and migration on the other, would tend to be rather long and narrow. Settlement patterns tend to eliminate long patches. At levels 5 and 6 where patches are small and being further subdivided, narrow patches become common again, resulting from the tendency to maintain frontages on roads and bodies of water when subdividing property (Lebeau 1969).

2.1.4.4 Patch Numbers and Configuration

In natural vegetation there is probably on average a modest density of patches (in addition to continua), with a large average size (Fig. 3 D). This density increases progressively to a maximum in the heterogeneous level 5 and then decreases sharply as vegetation patches are eliminated in urban areas (Curtis 1956). The spatial configuration of patches to one another is also highly important ecologically, but it is uncertain what the trends are along the gradient of landscape modification.

2.1.4.5 Summary for Patch Trends

We see major changes in the basic patch structure of landscapes as we progressively modify the environment. On average, in natural vegetation, patches are suggested to be large, narrow, irregular in shape, few in number, and caused mainly by environmental resource distributions and spot disturbances. By contrast, in suburban or town-dominated landscapes, patches are suggested to be on average, small, narrow, rectangular in shape, numerous, and caused mainly by plantings or remnants of managed vegetation.

2.1.5 Effect on Linkage Characteristics of the Landscape

While the number, size, shape, origin, and configuration of patches provide the primary structure of a landscape, other structural characteristics are evident, which often play key roles in the dynamics across a landscape. We recognize four types of corridors (Forman and Godron 1981); (1) line corridors, which are narrow strips of edge habitat, such as paths, hedgerows and roadsides (Rackham 1976); (2) strip corridors, with a width sufficient for the ready movement of species characteristic of patch interiors [for example, a wide power line corridor permitting movement of open country species though a forest (Anderson et al. 1977)]; (3) stream corridors, which may function as one of the previous two, but which additionally control stream bank erosion, siltation and stream nutrient levels; and (4) networks, which are formed by the intersection of corridors, this usually resulting in the presence of loops, as well as subdividing the matrix into many patches. Finally habitations, including the adjacent yard and buildings, are nodes of people and introduced species which radiate into the surrounding landscape elements.

As disturbance regimes induced by people increase in intensity, we may expect significant changes in these four landscape structures. Because of their key roles in landscape function, however, the changes along the human modification gradient are probably more marked ecologically than the patch changes analyzed above.

2.1.5.1 Line Corridors

With relatively few line corridors in natural vegetation, the paths and roadsides introduced in the colonization level 2 and the managed vegetation level 3 increase the number of corridors (Fig. 4A). Line corridors such as hedgerows, irrigation ditches, and roadsides increase further in the cropland level 4 (Pollard et al. 1974; Les Bocages 1976) and reach a peak in the heterogeneous town and country level 5. As the line corridors associated with agriculture and roadside vegetation decrease in the urban level 6 (Schmid 1975), line corridor density drops.

2.1.5.2 Strip Corridors

Strip corridors, excluding stream corridors, appear to be mainly a product of human modification of the landscape. Although generally of limited abundance,

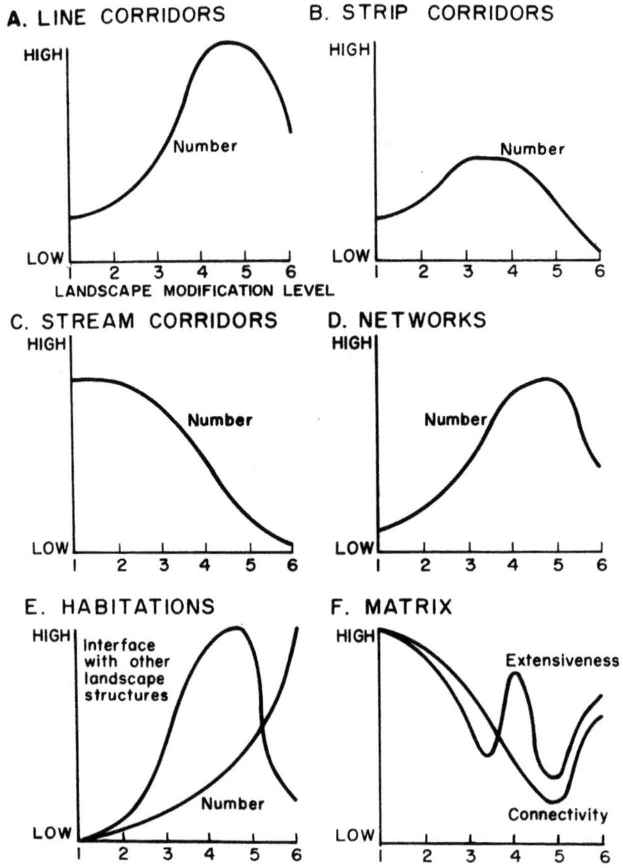

Fig. 4. Effect of landscape modification on linkage characteristics of the landscape. See legend for Fig. 3

they are predominantly in the intermediate levels of modification (Fig. 4B). Most are remnants of natural vegetation left by managing blocks of land on each side, though some are disturbance-caused, such as cutting for powerline corridors. At higher modification levels, the corridors are lost by being cut into patches or merged with the matrix.

2.1.5.3 Stream Corridors

The stream corridors in natural vegetation are differentiated by being tolerant of high soil moisture levels and flooding. Modification levels 2 and 3 decrease the stream corridors only slightly, whereas stream corridors are broken into patches and progressively eliminated through the urbanization process of levels 5 and 6 (Fig. 4C).

2.1.5.4 Networks

The pattern for networks (Fig. 4 D), and the mechanisms producing it, appear to be similar to those for line corridors.

2.1.5.5 Habitations

The density of habitations, including their immediate surroundings, increases sharply with landscape modification (Fig. 4 E). Also of importance, however, is the perimeter of habitations or the interface with the surrounding landscape structures such as matrix, patches, and corridors. This appears to be high in the agricultural level 4 and town and country level 5. Habitations become progressively more adjacent to one another in urban areas, thus decreasing the total interface with other landscape components. The addition of the two habitation curves for number and interface (Fig. 4 E) may be a useful indication of the effect of habitations on the landscape along this modification gradient.

2.1.5.6 Matrix

The matrix changes from natural vegetation in the first two levels, to cropland in level 4, to urban in level 6. Therefore, three peaks are evident (Fig. 4 F), separated by a dip between levels 3 and 4 where managed vegetation is being removed, and other dip at the town and country level where the balance among several types of areas is rapidly changing in the landscape. The connectivity of the matrix is largely a function of the amount of network present, which tends to subdivide the matrix and decrease connectivity (Pollard et al. 1974; Les Bocages 1976). Hence, the connectivity curve is the converse of the network curve, and is high at levels 1 and 6, but dips at the heterogeneous town and country level 4.

2.1.5.7 Summary of Trends for Linkage Characteristics

Most of the landscape structures other than patches are primarily characteristic of human-modified landscapes. Line corridors, strip corridors, networks and habitations all increase sharply with human disturbance regimes, and most decrease somewhat at the highest level, the urban landscape. Stream corridors and matrix characteristics, in contrast, are abundant in natural vegetation and generally decrease with increased landscape modification.

2.1.6 Conclusions

Landscape modification by humans is, at first glance, a hopelessly complex horizontal axis against which to seek trends. It is even considered independent of biome type and human culture. Yet, when each of the structural characteristics of landscapes is separately examined against this axis, patterns emerge. These patterns are clearly hypotheses, and need to be tested for their accuracy in description and their generality in application.

Further work will clarify landscape types and modification patterns (McHarg 1969; Rackham 1976; Baudry and Baudry-Burel 1978) as well as trends in functional landscape characteristics. For example, how do the flux of introduced species, the nutrient inputs and outputs, herbivory, or advection between two adjacent landscape elements (Forman 1981) change as the landscape is progressively modified? How do the flow rates of energy, mineral nutrients and species across an entire landscape (Pollard et al. 1974) change as that landscape is progressively modified by human disturbance regimes?

This overview of a very general, but little-studied, problem is imprecise on many points at this time. Nevertheless, these diverse considerations can be and must be connected if we wish to utilize the resources of our planet wisely and understand the human niche thereon.

Résumé

Pour comprendre les modifications des *paysages*, il faut commencer par savoir ce qu'est un paysage: un paysage est généralement un *ensemble* de biogéocénoses (ou d'écosystèmes relativement homogènes tels que des champs, des prés, des bois, des villages et villes, etc.), comprenant une matrice englobante, des taches et des corridors.

Pour accroître notre compréhension des implications écologiques des modifications du paysage liées aux activités humaines, nous mettons en relation les concepts fondamentaux et les caractéristiques structurelles spécifiques des paysages avec un gradient de modification croissante, allant d'un paysage naturel, où l'influence humaine est minime, à une forêt aménagée ou des prairies permanentes, puis aux cultures, pour aboutir à un territoire urbanisé.

Les attributs conceptuels des paysages qui dépendent le plus nettement des actions humaines sont la structure horizontale, la stabilité, le degré d'ouverture (au sens thermodynamique), les modes de dispersion des diaspores, l'aire minimale des biocénoses, les cycles d'éléments minéraux, la production nette, les tactiques des espèces, leurs modes de phylogénèse, et leurs types de résistance (Fig. 1).

Les perturbations dues à l'homme sont très diverses; elles peuvent être ordonnées par intensité croissante, et leurs effets sur les caractères des paysages sont rarement linéaires.

Parmi les caractéristiques de la structure (origine, taille, forme et nombre de taches, corridors en ligne, en bande ou le long des cours d'eau, réseaux, habitations, matrice), quelques-unes ont une réponse linéaire aux modifications du paysage, mais la plupart ont des réponses en pics négatifs ou positifs aux niveaux intermédiaires. Ces tendances générales, décelables dans les modifications des paysages, doivent être confrontées à la diversité des cas particuliers, mais elles fournissent des indices utilisables aussi bien pour un aménagement rationnel de nos ressources, que pour la compréhension de la niche écologique occupée par l'Homme.

Acknowledgements. We thank Jacques Baudry for aid, Harold A. Mooney for perceptive comments on the ideas presented, and the National Science Foundation for grant DEB-80-04653 in support of a portion of this work.

References

Anderson SH, Mann K, Shugart HH Jr (1977) The effect of transmission line corridors on bird populations. Am Midl Nat 97:216–221

Baudry J, Baudry-Burel F (1978) Contribution à la connaissance écologique du basin versant de la Rance. These, Univ Rennes, Rennes, France, 213 pp

Braun-Blanquet J (1964) Pflanzensoziologie. Springer, Wien New York, 865 P

Bruhat G (1962) Thermodynamique. Masson, Paris, 822 pp

Curtis JT (1956) The modification of mid-latitude grasslands and forests by man. In: Thomas WL (ed) Man's role in changing the face of the earth. Univ Chicago Press, Chicago Ill, pp 721–736

Curtis JT (1959) The vegetation of Wisconsin. Univ Wisconsin Press, Madison Wis, 657 pp

Daubenmire R (1968) Plant communities: a textbook of plant synecology. Harper & Row, New York, 300 pp

Emberger L (1954) Observations sur la fréquence en forêt dense équatoriale. Vegetatio 7:169–176

Forman RTT (1979) The Pine Barrens of New Jersey: an ecological mosaic. In: Forman RTT (ed) Pine Barrens: ecosystem and landscape. Academic Press, London New York, pp 569–585

Forman RTT (1981) Interactions among landscape elements: a core of landscape ecology. In: Perspectives in landscape ecology. Proc Int Congr Landscape Ecol 1981, Veldhoven. Pudoc Publ, Wageningen, The Netherlands, pp 35–48

Forman RTT, Boerner REJ (1981) Fire frequency and the Pine Barrens of New Jersey. Bull Torrey Bot Club 108:34–50

Forman RTT, Godron M (1981) Patches and structural components for a landscape ecology. Bio Science 31:733–739

Forman RTT, Galli AE, Leck CF (1976) Forest size and avian diversity in New Jersey woodlots with some land use implications. Oecologia 26:1–8

Fosberg R (1967) A classification of vegetation for general purposes. A guide to the check sheet for IBP areas. IBP Handbook 4:73–120

Franzreb KE (1978) Tree species used by birds in logged and unlogged mixed-coniferous forest. Wilson Bull 90:221–238

Futuyma D (1979) Evolutionary biology. Sinauer, Sunderland, 565 pp

Godron M (1966) Une application de la theorie de l'information à l'étude de l'homogéneité et de la structure de la végétation. Cent Etud Phytosociol Ecol, Montpellier, France, 67 pp

Good R (1962) The geography of the flowering plants. Longmans, London, 518 pp

Greig-Smith P (1964) Quantitative plant ecology. Butterworths, London, 256 pp

Heinselman ML (1973) Fire in the virgin forests of the Boundary Waters Canoe Area, Minnesota. Quat Res (NY) 3:329–382

Horn HS (1974) The ecology of secondary succession. Annu Rev Ecol Syst 5:25–37

Jaccard P (1902) Lois de distribution florale dans la zone alpine. Bull Soc Vaudoise Sci Nat 144:69–130

Johnson WC, French NR (1981) Soviet Union. In: Kormondy EJ, McCormick JF (eds) Handbook of contemporary developments in world ecology. Greenwood Press, Westport Conn, pp 343–383

Kershaw K (1964) Quantitative and dynamic ecology. Edward Arnold, London, 183 pp

Kormondy E (1965) Readings in ecology. Prentice-Hall, Englewood Cliffs NJ, 219 pp

Lebeau R (1969) Les grands types de structures agraires dans le monde. Masson, Paris, 120 pp

Les Bocages: Histoire, Ecologie, Economie (1976) Univ Rennes, Rennes, France 586 pp

Levins R (1968) Evolution in changing environments. Princeton Univ Press, Princeton NJ, 120 pp

Littre E (1898) Dictionnaire de medecine. Baillère, Paris, 1894 pp

Lorimer CG (1977) The presettlement forest and natural disturbance cycle of northeastern Maine. Ecology 58:139–148

McHarg IL (1969) Design with nature. Doubleday and Co, Garden City NY, 198 pp

Odum EP (1969) The strategy of ecosystem development. Science 164:262–270

Odum EP (1971) Fundamentals of ecology. Saunders, Philadelphia PA, 754 pp

Pickett STA, Thompson JN (1978) Patch dynamics and the design of nature reserves. Biol Conserv 13:27–37

Pijl L Van der (1969) Principles of dispersal in higher plants. Springer, Berlin Heidelberg New York, 153 pp

Pollard E, Hooper MD, Moore NW (1974) Hedges. Collins, London, 256 pp

Rackham O (1976) Trees and woodland in the British landscape. Dent, London, 204 pp

Schmid JA (1975) Urban vegetation: a review and Chicago case study. Dep Geogr, Univ Chicago, Chicago Ill, 266 pp

Southwood TRE (1960) The number of species of insects associated with various trees. J Anim Ecol 30:1–8
Tansley A (1935) The use and abuse of vegetational concepts and terms. Ecology 16:284–307
Trabaud L (1970) Quelques valeurs et observations sur la phyto-dynamique des surfaces incendiées dans le Bas-Languedoc (Premiers resultats). Nat Monspel Ser Bot 21:231–242
Troll C (1968) Landschaftsökologie. In: Tuxen R (ed) Pflanzensoziologie und Landschaftsökologie. Junk, Den Haag, P 1–21
Vitousek PM, Reiners WA (1975) Ecosystem succession and nutrient retention: a hypothesis. Bio Science 25:376–381
Wacker PO (1979) Human exploitation of the New Jersey Pine Barrens before 1900. In: Forman RTT (ed) Pine Barrens: Ecosystem and landscape. Academic Press, London New York, pp 3–23
Whittaker RH (1967) Gradient analysis of vegetation. Biol Rev 49:207–264

2.2 Ecological Modeling of Landscape Dynamics

D.A. WEINSTEIN and H.H. SHUGART

2.2.1 Introduction

To determine man's role in changing natural landscapes, one must first develop methods for measuring landscape attributes, patterns, and rates of change. Classically, a large body of methodologies has been developed for classifying and ordering patterns in landscapes using fairly sophisticated mathematical procedures (a sampling of such procedures is given in Whittaker 1978a, b). However, most of these methodologies have little analytical power in terms of noting changes in the fundamental dynamics of landscape systems.

This chapter will discuss methods for studying the dynamics of landscapes based on the use of mathematical models of landscape elements. These methods do not compete with ordination and classification methods for landscape studies. In fact, information obtained from properly conducted statistical analyses of landscape patterns provides valuable background information for some of these methods. The utility of these methods is restricted by one fairly fundamental assumption – they are predicated on the idea that one can reasonably divide a landscape into discrete, smaller elements.[1] Accepting this, one can then simulate the dynamics of a landscape by building models that keep track of the changes in landscape elements over time.

Landscape elements can be defined in several different ways. Their definition could be largely arbitrary. Their scale could be determined by sampling considerations (e.g., a LANDSAT satellite senses an area circa 1 ha for each datum). The scale might be determined by the natural pattern of disturbance in a forested landscape (Shugart and West 1980). Ideally, definitions of landscape elements should be reasonably homogeneous, such that one would expect the pattern of change of a landscape element of a given category to operate under the same rules (although not necessarily the same pattern) of change as another element of the same category. The categorization of landscape elements (in the form of sample data) has been the primary objective of much of the phytosociological research for over the past three decades.

The patterns of change of landscape elements can be simulated continuously in time or at discrete points of time. The dynamics of a landscape that was described by a regular resurvey of the landscape elements might best be considered to operate in discrete time, namely the time lapse between resurveys.

1 If one is unwilling to make this particular assumption, an alternative formulation would use partial differential equations. While this has been done with some success in certain spatially dynamic problems (e.g., Smith 1980) and for water bodies in environmental impact studies involving distributions of waste heat or toxic chemicals in rivers and estuaries (Okubo 1980), these applications require much more effort than those discussed below

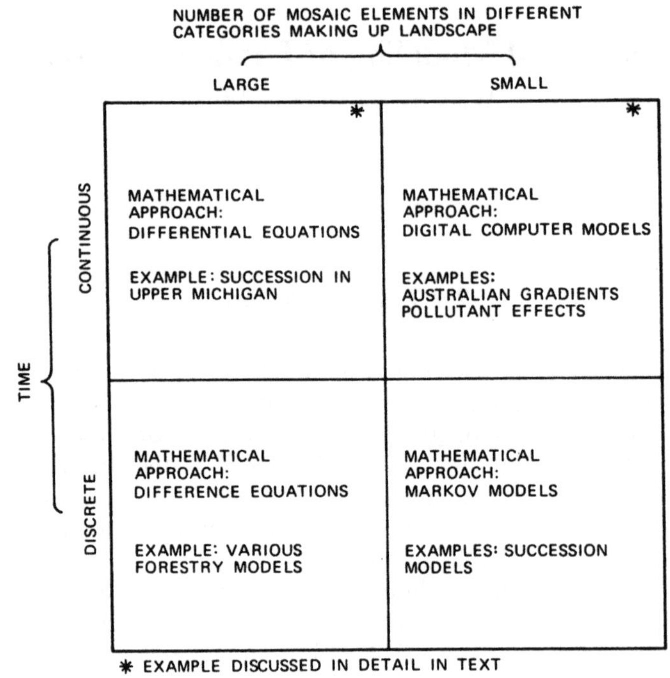

Fig. 1. Modeling approaches used to simulate the dynamics of landscape systems

In some cases, the number of landscape elements in a given category can be considered large enough to allow change in the tabulation of landscape elements to be considered a continuous variable. In other cases, the number of elements comprising a landscape could be so small that each element would have to be considered explicitly in projecting the dynamics of that landscape. The time and element units utilized determines what modeling approaches should be used in a given application (Fig. 1). Examples of these approaches as used in our own research will be developed and efforts in landscape modeling will be reviewed.

2.2.2 Approaches to Modeling Landscape Dynamics

Figure 2 diagrams changes in categorizations of forest landscape elements continuously through time. Although the changes in forest cover over time and space are relatively continuous, the cover is characterized as giving a finite number of landscape elements in four cover-state categories. In this case, cover type is defined by the densities (or mass) of various tree species found during succession. The trajectory that a given landscape element will take through this abstract N-dimensional species space is indicated by one of the arrows (Fig. 2). Each trajectory is the specific response of a particular landscape element, which may be guided in its path through the species space by local environmental conditions and by various stochastic events. However, the set of landscape elements can be said to have an

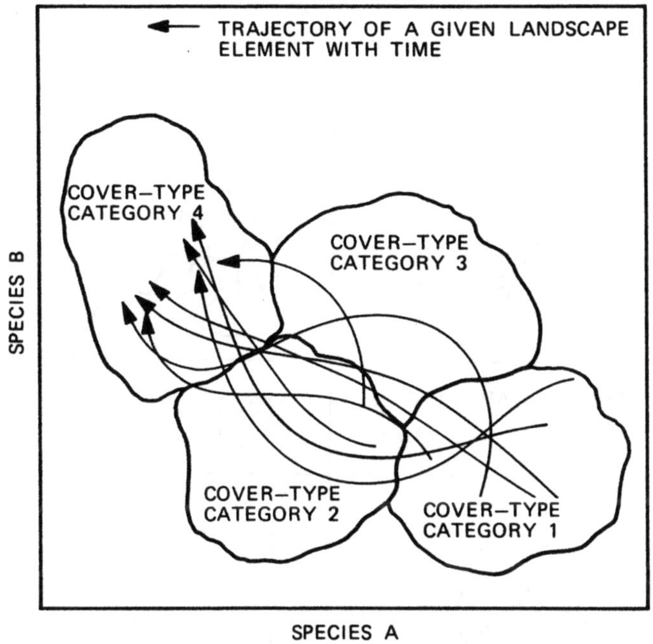

Fig. 2. Abstraction representing forest succession in a large region

expected average response to the heterogeneous complex of environmental conditions over the entire landscape. This response yields an average rate of change with which land elements in one cover-type category will change to another cover-type category. The rate of change of acreage in any cover type at any time can be expressed as a model using as model parameters the rates at which land is being converted to that cover type and the rates at which land is being converted to some other type. Based on this conceptualization of a landscape as being composed of different elements moving from one category to another through time, one can use several different modeling approaches (Fig. 1), depending on the discrete or continuous nature of time and on whether the identity of each landscape element is essential to the problem (this is usually related to the number of elements). The basic differences in these modeling approaches are shown in Fig. 3.

If time is continuous and the number of elements is large, then an appropriate model paradigm is to use differential equations (Fig. 3a). In this case the change in the number of elements in a given category is the balance between elements moving in and out of the category due to succession. These rates of change are the parameters of the model. A detailed example of the derivation and use of such a model for a large region will be provided below. If time is discrete (as in periodic sampling) then a difference equation approach may be more appropriate (Fig. 3b).

If time is discrete and the elements can be thought of as having different probabilities of being in one category or another at different times, then a Markov modeling approach may be used (Fig. 3d). In fact, Markov models (and related ap-

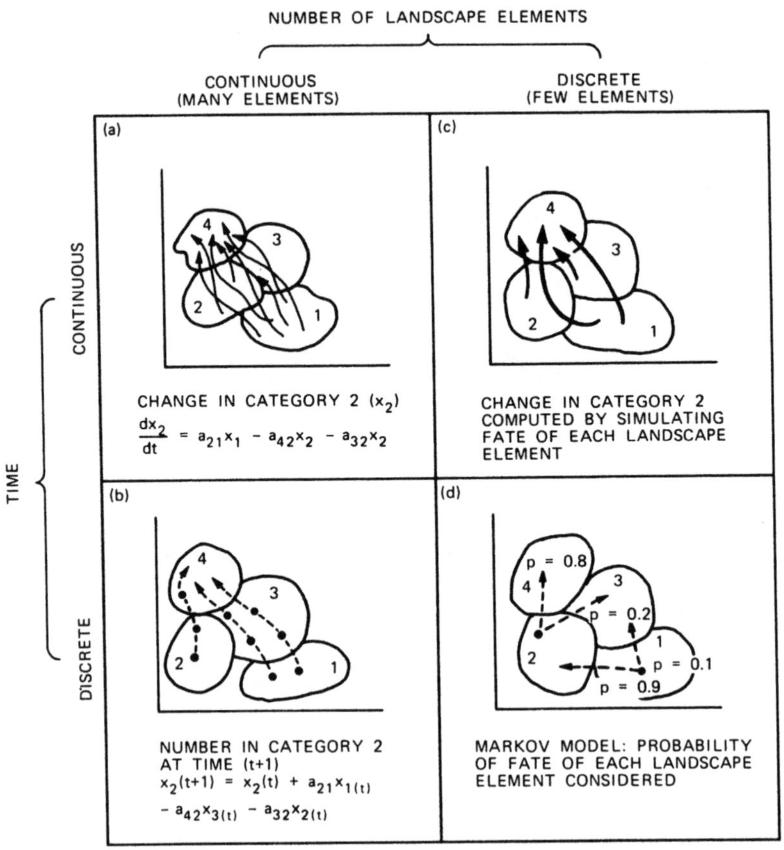

Fig. 3 a–d. Conceptualization and mathematical models of landscape dynamics

proaches) have been successfully used as landscape modeling paradigms in both applied (Hool 1966) and basic (Waggoner and Stephens 1970; Horn 1976; Wilkins 1977) contexts. One good example, which includes discussions of parameter estimates and which is explicitly oriented toward landscapes, is found in the succession modeling approach used by Kessell and his colleagues (e.g., Kessell 1976, 1977; Cattelino et al. 1979).

In practice, there are two data-determined modeling paradigms used in succession modeling. All of the approaches discussed thus far (Fig. 3 a, b, d) are somewhat related; three different types of models can be formed from the same sorts of data sets, depending on which assumptions seem appropriate, and therefore constitute the first modeling paradigm. A second, mathematically distinct paradigm is used in cases in which a few landscape elements change continuously over time. The approach for this second class has thus far been to define the dynamics of each landscape element with fairly detailed models (Fig. 3 c). An example of an application of each paradigm class will be discussed below.

2.2.3 A Differential Equation Model of Landscape Change

The development of a system of differential equations (Shugart et al. 1973) for a region involves the calculation of the rates of conversion from one cover type to another. The rate of transfer of acreage between cover types can be derived in several ways. For cover stands with no direct pathway between them, a rate constant equal to zero can be assumed. For cover states in which the time necessary for transfer of a given percentage of acreage to another cover state is known, the rate constant can be calculated as the inverse of this time period. This assumes that the loss of acreage follows a negative exponential curve. For cover types in which detailed knowledge of their dynamics is limited, the rates can be derived using autecological information and assumptions about species life spans. There is, for example, a direct relationship between the life span of white pine (light-intolerant habit) and the time it takes another species to succeed it as dominant. Whenever there are several successional paths, however, a thorough knowledge of properties of a region is necessary for allocating among these links. A more complex discussion of the construction of specific coefficients of the differential equations used in this model type is contained in Shugart et al. (1973).

Figure 4 shows the complexity of successional cover types and patterns of change on three different site conditions in a landscape in the Western Great Lakes region (compiled from Curtis 1959). On xeric sites (Fig. 4) the successional progression is from intolerant oak to mixed-oak-dominated stands (Curtis 1959). Tamarack (*Larix laricina*) dominates pioneer communities on the wettest sites, many of which are sphagnum sedge mats. These communities generally succeed to cover states dominated by either black spruce (*Picea mariana*) or firmer organic substrates, or nothern white cedar (*Thuja occidentalis*) on partly to well-decomposed woody peats. This later cover type will eventually become dominated by hemlock (*Tsuga canadensis*), yellow birch (*Betula alleghaniensis*), and black ash (*Fraxinus nigra*).

Succession on mesic sites (Curtis 1959) is much more complex (Fig. 4). Pioneer forests may be dominated by either aspen (*Populus tremuloides*), pin cherry (*Prunus pensylvanica*), or jack pine (*Pinus banksiana*). An aspen association (*Populus tremuloides, Populus grandidentata, Betula papyrifera, Prunus pensylvanica*) is by far the most predominant cover-state on disturbed sites. This type will succeed to either a white spruce (*Picea glauca*) and balsam fir (*Abies balsamea*) association, a mixed northern hardwood (*Acer saccharum, Fagus grandifolia, Tilia americana, Populus* spp., *Betula* spp., *Quercus* spp.), or to a cover type dominated by white pine (*Pinus strobus*). Tertiary mature communities in undisturbed locations become composed largely of hemlock or sugar maple. Jack pine will be succeeded by the more shade-tolerant and longer lived red pine (*Pinus resinosa*) on drier sites. These sites will eventually be occupied by white-pine-dominated stands. Pin cherry will be the initial colonizer of clayey soils or burns.

Specifics of the complex successional patterns of this region (see Curtis 1959) and details of the construction of the coefficients of the differential equations (see Shugart et al. 1973) are available elsewhere. The method is used here to illustrate the simulation of the dynamics of such a landscape system.

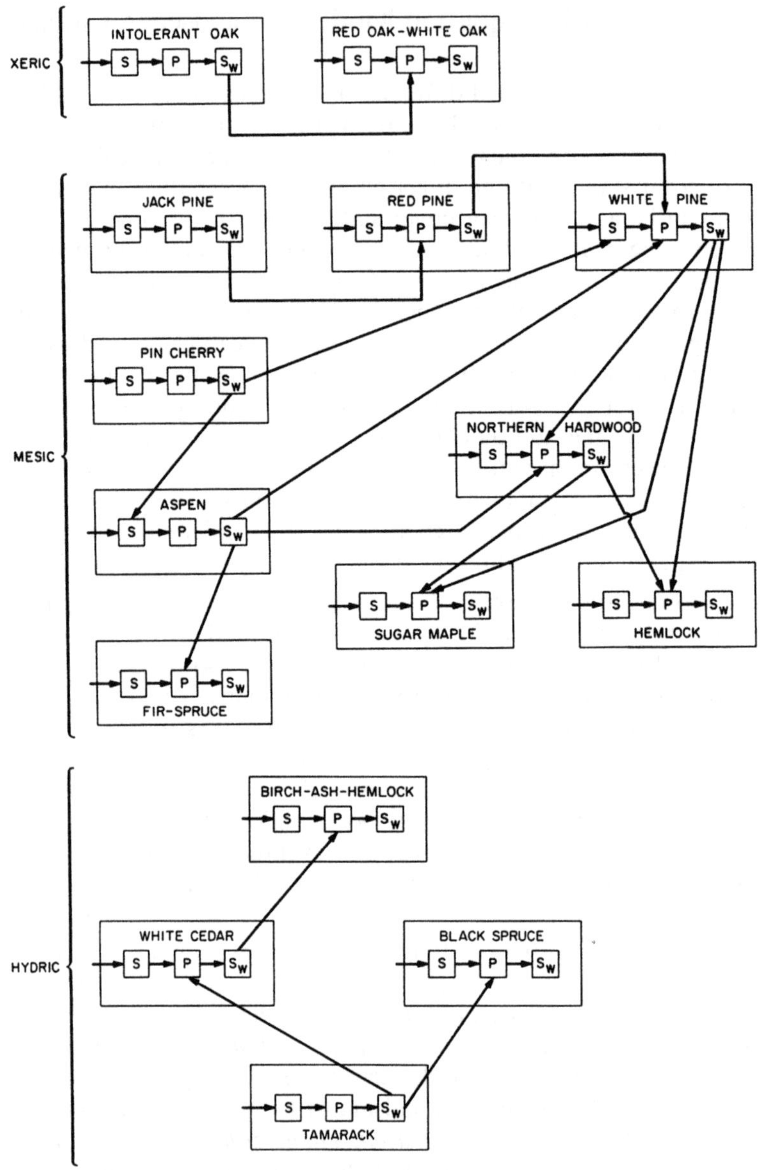

Fig. 4. Successional pattern on three different site conditions in a landscape in the Western Great Lakes region (S = small saplings, P = pole size trees, S_w = saw timber)

Figure 5 gives the results of a 250-year simulation of the mesic succession portion of this system, beginning from present-day conditions. It expresses the proportions at any given time of total acreage in the region in the various cover-type categories.

In the absence of disturbance and its associated feedback mechanisms, pioneer forest types would show a rapid exponential decrease of the area covered by the

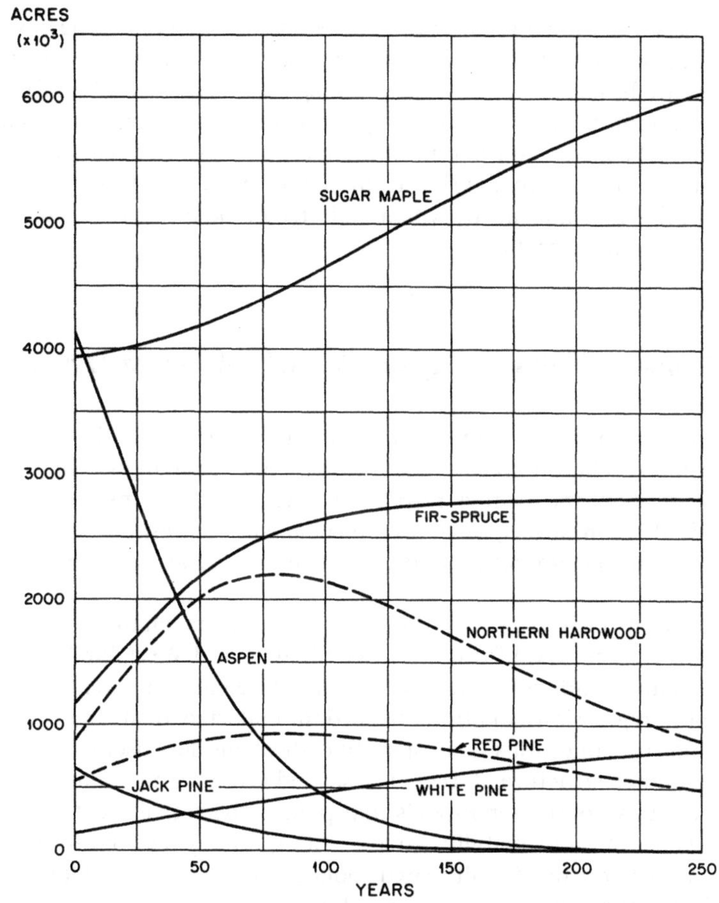

Fig. 5. Simulation results from a model of the mesic succession portion of the Great Lakes landscape

aspen-birch type by a factor of 10 within 100 years. Communities dominated by intermediately light-tolerant species (e.g., the northern hardwood and red pine types) initially increase in acreage in the region and then begin a slow decline. This dynamic response reflects a rapid input of landscape elements from pioneer community types and a slower output to later successional types. Terminal forest types, such as fir-spruce or sugar maple, show an increase in number of landscape elements through the simulation and eventually approach maximum. Qualitatively, these results are not unexpected given our initial information on the dynamics of each cover type. However, the model provides a capability of quantitative assessment of the rate at which these changes occur when several conditions, each changing, are operating simultaneously. This would not otherwise be possible, particularly given the regional scale of the problem.

The rates of change used in these models are intended to summarize the average output of many biological mechanisms operating simultaneously. The problem with this approach is that the model results are not sensitive to changes that may occur during the simulation, and alter the rates of conversion from one cover type to another. That is, the results are entirely confined by the initial set of data on which the rates are based. If, for example, the region was cleared of all forest, the return of some communities would be delayed due to a lack of tree propagules on large areas; a model based on the assumption that the seed source remains constant would not simulate the recovery. What is missing in this type of model is a simulation of the specific mechanisms by which each tree in the forest responds biologically to environmental alterations.

2.2.4 Complex Digital Computer Models of Forest Dynamics

Complex digital computer forest models simulate independently the behavior of each important mechanism in the system. Rates of change of system properties based on a summary of the average response of these mechanisms are not as constrained by initial assumptions, since each mechanism can respond to changes in the environment during simulation and can create nonlinear behavior in the system.

Three types of models in general use incorporate the mechanisms of response at the landscape element level for forests (see Shugart and West 1980 for review). Yield models simulate individual trees on a landscape element in great detail. They typically employ regression equations from data collected from the response of trees to specific site conditions. Foresters find these models useful in predicting bole form and branch structure of trees. The second group of modeling approaches at this level contains population models that predict the recruitment of trees into new age classes based on probability tables. Both model types tend to be site specific and for this reason are difficult to extrapolate to the dynamics of large regions.

The third type of model consists of mixed-age, mixed-species stand models (also called gap replacement models). Tree establishment, growth, and mortality simulations are based on general silvicultural information. This information includes: (1) site requirements for germination, (2) palatability of seedlings for browsers, (3) sprouting potential, (4) shade tolerance, (5) germination and growth temperature requirements, (6) inherent growth potential, (7) longevity, and (8) sensitivity to crowding stress. Since the information is consistent over entire species' ranges, the model is capable of simulating the response of these tree species over large regions. Figure 6 is a diagrammatic representation of an example of the model type formulated for a Southern Appalachian Mountain region (Shugart and West 1977).

In gap replacement models, the maximum growth of each tree on a plot (often 1/12 ha) is computed from the inherent radial and height increment potential and the longevity of the tree species. This maximum growth is modified by the response of the tree to environmental conditions such as light availability (determined from the leaf area of overtopping trees), annual temperature and growing degree days, spatial crowding, and root competition. These environmental factors are considered to be homogeneously distributed over the plot (for more detail see Shugart and West 1977). The establishment of seeds and sprouts is based on the availability

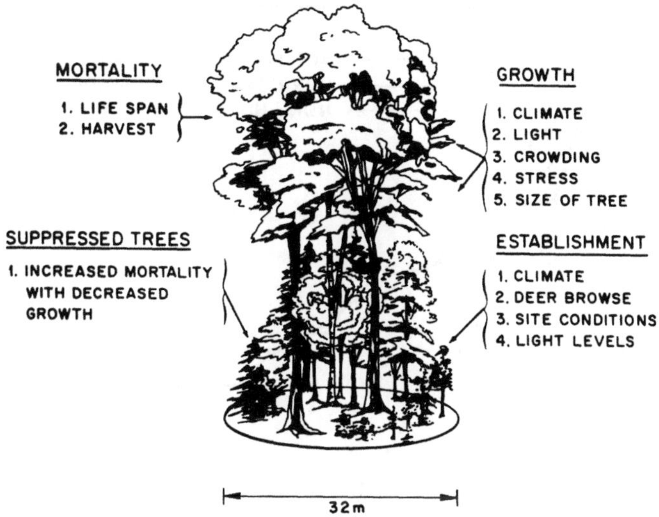

Fig. 6. Diagramatic representation of the southern Appalachian gap succession model

of light, proper temperatures, substrate requirements for germination, and sprouting capabilities of dying trees. Mortality is a function of the growth and expected maximum age of the tree. Each growth response to environmental conditions is described as a probability function. Therefore the model is stochastic, meaning that the ultimate output is a mean outcome of 100 simulations of a given set of initial stand conditions. This takes into consideration wide variation in the response of any individual tree. Further, provided that the mean and the variance of the biological response of an individual tree to a natural disturbance or to management techniques are known, the model can simulate the subsequent effect on the tree community. This type of model, then, has the required mechanistic quality.

Models of this type have been constructed for many different forests around the globe. One validation test has been run for each application of each model. For example, the Southern Appalachian Forest model successfully mirrored the recovery of the forest following chestnut blight. Examples of the validity of this type of model will now be examined.

2.2.5 Evaluating the Impact of Regional Scale Problems on Localized Stands

How can these models be used for assessing regional scale influences of man-induced disturbance of natural systems? First, these gap replacement models can be used to evaluate the impact of regional scale problems on small localized stands. This evaluation can be extrapolated to large areas in regions where the heterogeneity of the forest does not increase greatly when the spatial scale under consideration

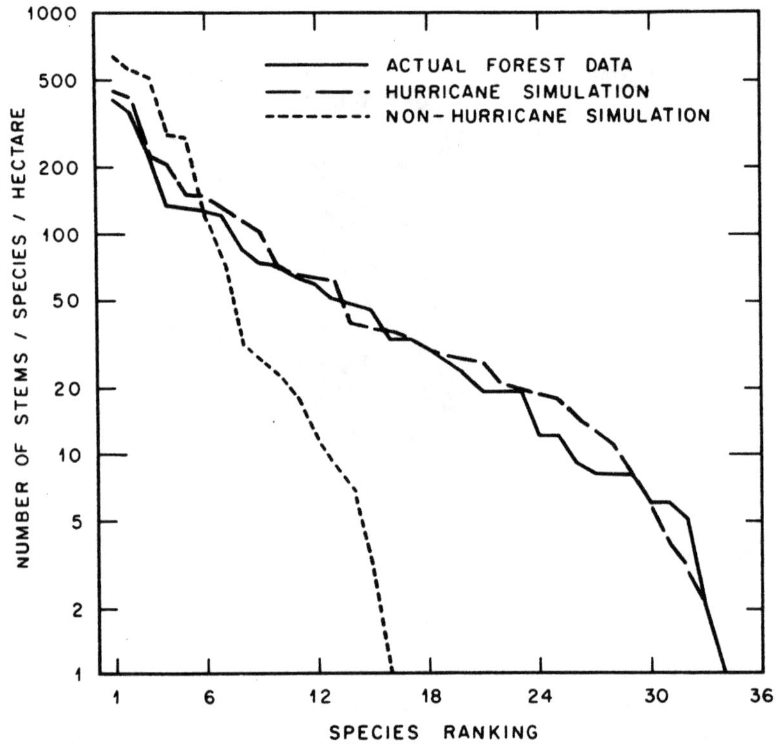

Fig. 7. Dominance diversity diagram of the Puerto Rican tropical rain forest and of a simulation of this forest

is expanded from localized plot to large area. An example of this application comes from Doyle (1981) in his simulation of the effect of hurricanes on Puerto Rican tropical rain forests using a Puerto Rican rain forest succession gap model. Figure 7 shows a dominance diversity graph of species in that forest. Numbers of individuals per species are plotted against species rank. These ranks are arrayed along the abscissa, from those species with the largest number of individual species to those with the fewest. In order that the dominance diversity curve produced by the simulation should match the field data, the natural hurricane frequency and associated tree mortality caused by hurricanes had to be included in model runs. The result provides convincing evidence of the historical importance of hurricanes in the development of the tropical rain forest. Since hurricane activity and subsequent devastation is typical throughout a large region, this result would lead us to suspect similar patterns in other forests of that area. More importantly, the result suggests that the system may be well adapted to a disturbance frequency such as that of the hurricane recurrence and therefore would show the highest resilience and recovery from man-induced disturbances that mimicked this frequency.

A second example in this category involves evaluating the impacts of the man-induced disturbance of SO_2 on localized stands. In the eastern United States, the high density of multiple point sources and the high frequency of air stagnation events have led to elevated levels of pollutant oxidants over large regions. Abundant evidence exists documenting the harmful effects of those pollutants on individual trees (Ziegler 1973; Mudd and Kozlowski 1975; Heck and Brandt 1977). The extension of this information to the forest community scale involving multiple stresses over multiple-year time scales has been hampered by the complexity of intracommunity competition dynamics.

An investigation into this problem was conducted by West et al. (1980). The Southern Appalachian Forest succession gap replacement model, which considers 32 tree species, was used to study the effects of air pollution stress on the growth and development of eastern forests. Species were classified with respect to their relative sensitivity to the stress on growth and development induced by SO_2 pollution. Species-specific growth reductions based on their sensitivity ranking were incorporated into the model to reflect the expected behavior of individual trees observed under field conditions. Although no two species exhibit identical tolerance levels to this stress, the range of behavior was simplified to three classes of responses: resistant, intermediate, and sensitive. Species were placed in one of these classes based on their relative susceptibility to visible foliar injury. The sensitivity classification was based on 10 years of field survey data of vegetation near a coal-fired electric plant (McLaughlin and Lee 1974) and an extensive summary of field and laboratory data on the susceptibility of woody plants to SO_2 and photochemical oxidants reported by Davis and Wilhour (1976). The basic assumption of this approach was that the relative sensitivity of species to reductions in physiological function under chronic stress regimes would parallel their relative sensitivity to visible foliar injury resulting from the higher concentrations generally reported in the literature.

Responses of a simulated forest to air pollution were examined for varying levels of stress and for the forest ages at which the stress was initiated. An initial simulation imposed a maximum growth stress of 20% on the six most sensitive species. The ten intermediate species received a 10% stress, and the sixteen resistant species were unstressed. Stress was begun at year 1, when trees were still in the seeding stage.

The results of these simulations are summarized in Fig. 8. Increases in the biomass of four major species are compared with and without simulated air pollution stress. These species show representative responses of the three tolerance classes: yellow poplar (intermediate), white oak (resistant), black oak (intermediate), and black cherry (sensitive). The collective "other" species category and total stand biomass are shown as well.

Yellow poplar, an early successional species with a rapid growth rate, shows rapid positive response to stress. The enhancement of the total biomass of yellow poplar on the plot, despite a 10% growth stress, may be attributed to the relatively greater effect of the stress on other species competing at early stages of the developing forest stand. Black oak, on the other hand, with the same level of applied growth stress (10%), shows biomass reductions in excess of 10%. Reductions of this magnitude may be attributed primarily to intrinsically slower growth and response, suggesting the inability of black oak to compete with other species such as yellow and white oak under the additional stress imposed by air pollution. White oak, a tolerant but late successional species, shows a positive response in the sim-

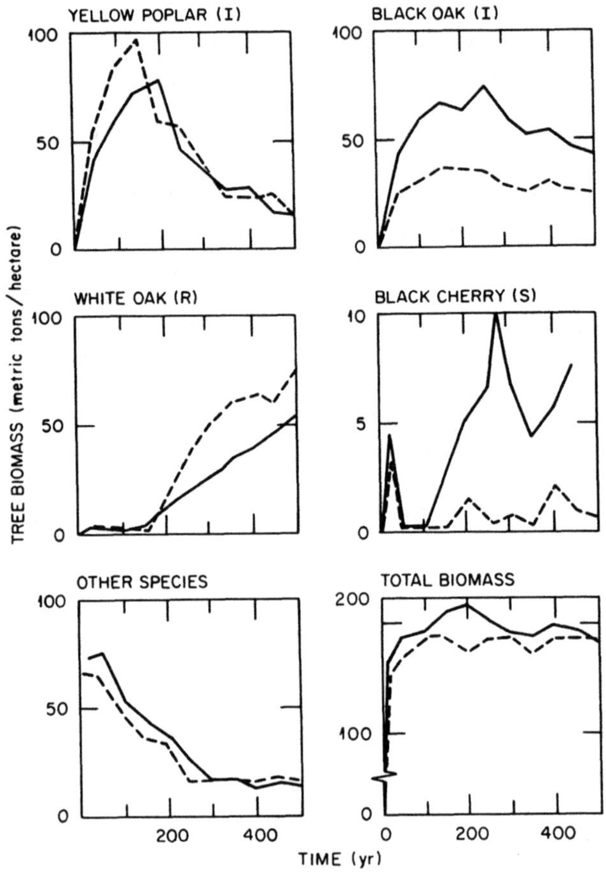

Fig. 8. Species and stand dynamics of a forest with and without continuous exposure to air pollution stress (— unaffected, --- affected) (after West et al. 1980)

ulation, which, however, becomes evident only as the stand matures and begins to stabilize. The effects of a 20% growth stress on black cherry, a sensitive species, are quite dramatic, particularly on stands older than 100 years; the result was almost total elimination. Responses of other species and of the whole stand indicate an overall growth suppression by the simulated stress. These results strongly suggest that the response of trees to air pollution may be quite different under the competitive conditions in a forest stand than would be expected from experiments conducted with single individuals or single species. Trees of intermediate sensitivity may experience impacts either much greater or much lower than expected, due to changes in their relative competitive potential. Further, these results offer a different view of air pollution effects extrapolated over large regions than the traditional view constructed from single species response data.

2.2.6 Evaluating Impact of Regional Scale Problems Across Landscapes

The above examples display the effects on single stands. Using this type of model, we can also explicitly simulate the response of entire landscapes to environmental perturbations. First, the landscape can be subdivided into a matrix of internally homogeneous subunits (stands) based on differences in key environmental conditions. A model run can then be made using each set of environmental conditions as initializing states. Figure 9 shows an example of such an approach applied to a watershed in the Australian montane forest in the Brindabella Range (Shugart and Noble 1980). The watershed is broken into a two-dimensional matrix of environmental cells based on the two predominant environmental gradients of the area, fire frequency (along the abcissa) and changes in temperature with altitude (along the ordinant).

The gap model BRIND (Shugart and Noble 1981) for this region was run 10 times under appropriate conditions of climate and fire frequency. For each cell in the figure, the number above the box indicates the percentage of time of those 10 runs in which the indicated species contributed 90% or more to stand biomass, between 200 and 1,000 years of simulated time. The values on the lines between boxes are the percentages of instances in which the two named species contribute 90% or more of a stand's biomass. For example, under the initial conditions at an elevation of 1,300 m with an average of 1,600 annual degree days and an annual fire probability of 0.02, *Eucalyptus delegatensis*, dominated the simulated stands 16% of the time, and the dominance was shared by *E. delegatensis* and *E. pauciflora* 23% of the time. An annual fire probability of 0.02 corresponds to an average expected time lapse between fires of 50 years.

Results of the simulation for high elevation locations (1,500 m) show that as the frequency of wildfire decreases, the forest stands become dominated by a single species, *E. pauciflora*. Similar community dynamics have been reported from corresponding field sites (Anonymous 1973; Costin 1954). At lower elevations (1,200 m) a reduction in wildfire frequency in the simulation leads to a decrease in the dominance of *E. pauciflora* in the face of superior growth and dominance by *E. delegatensis*. At lowest elevation (below 1,050 m) simulated forests continue the pattern of increased dominance with decreased fire frequency, although here the species showing positive responses are *E. robertsonii* and *E. fastigata*.

In general, the absence or near absence of fire tends to increase the level of single-species-dominated stands, and more frequent disturbances (in the form of wildfire) tend to encourage more diverse, mixed communities. This trend is particularly evident at the higher altitudes (Fig. 9). Several species show fire responses that change with altitude. *E. dalrympleana*, for example, is most abundant in the presence of frequent fires (probability = 0.02) at 1,300 m, but it is more prevalent at lower altitudes (1,050 m) in sites that are burned less frequently. The indication is that the detailed, quantitative nature of the vegetative pattern in the Brindabella Range may be a consequence of fire events occurring over time periods longer than current historical records could be expected to reveal.

With the development of this matrix it is not only possible to comprehend better the causes for heterogeneity in this landscape, but also to understand the poten-

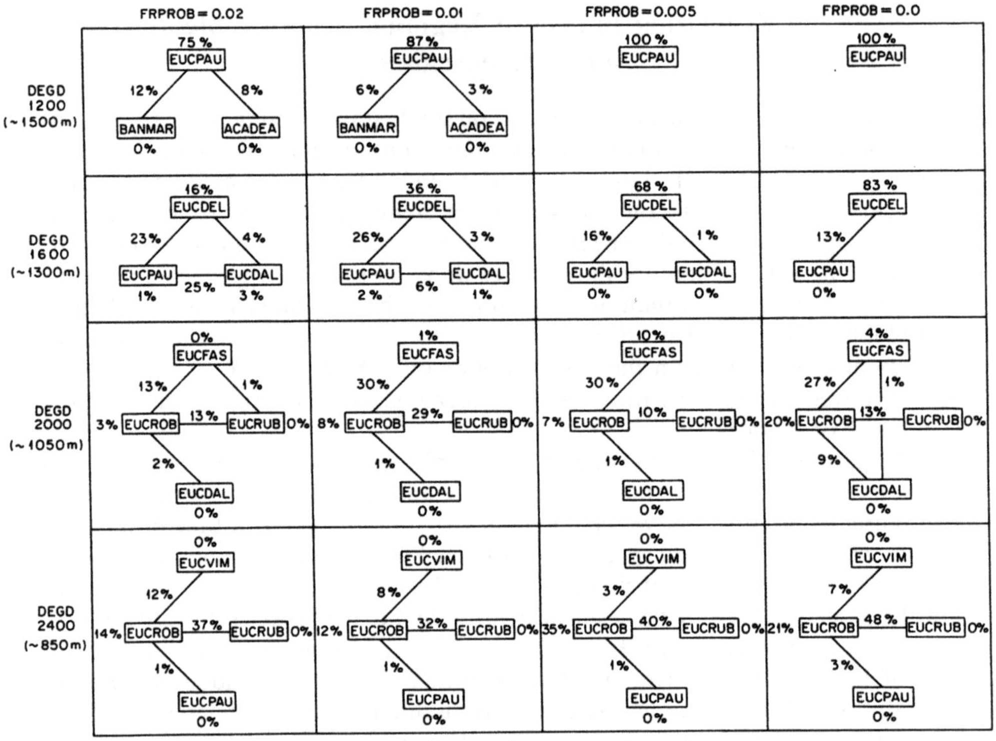

Fig. 9. Summary forest-type constellations for one- and two-species-dominated forest types simulated by the BRIND (Shugart and Noble 1980) model at four different altitudes (expressed by the heat sum, DEGD) by four different wildfire probabilities (FRPROB). The value on a box is the percentage of instances in which the indicated species contributed 90% or more to the stand's total biomass. The values on the lines between boxes are the percentages of instances in which two named species contributed 90% or more of a stand's total biomass. Species mnemonics are taken from the first three letters of the scientific bionomial: ACADEA = *Acacia dealbata;* BANMAR = *Banksia marginata;* EUCDAL = *Eucalyptus dalrympleana;* EUCDEL = *E. delegatensis;* EUCFAS = *E. fastigata;* EUCPAU = *E. pauciflora;* EUCROB = *E. robertsonii;* EUCRUB = *E. rubida;* EUCVIM = *E. viminalis*

tial shifts in vegetational dynamics in a heterogeneous landscape as a result of a management decision, such as a fire prevention policy applied to the region uniformly. An evaluation of such a policy using this method would show that the policy would have drastically different effects on the presence of communities at 1,500 m elevation than at 1,200 m.

2.2.7 Future Directions in Simulation Analysis of Landscapes

A similar segmenting of landscape into environmental cells for simulation purposes is being conducted in a mountain forest in the Great Smoky Mountains National Park. In this instance the managers are attempting to maintain the existing vegetation communities in perpetuity. These managers have chosen the logical course toward this end, that is, to avoid disturbance to existing communities by restricting the spread of fires, which have such an obvious immediate impact on these communities. Yet there are indications that the absence of fire is having as drastic an effect on these communities as the fire was feared to inflict.

The frequency of fire in this region previous to enforcement ranged from one fire burning each plot of land every 500 years to one fire every 20 years. The former frequency was experienced on low elevation stands and the latter in the high elevation dry ridge forests. In the years since the beginning of the application of fire prevention policy, the frequency of fires has been reduced to one every 1,000 years. Associated with this drop have been signs of a gradual disappearance of virginia pine (*Pinus virginiana*) and pitch pine (*Pinus rigida*), vegetation types commonly found occupying the dry ridge environments. Old pine stands are being decimated by pine bark beetle outbreaks, and although the area records indicate that these outbreaks are not historically uncommon, the scarcity of young pine stands does seem to be a relatively new situation. Sites with exposed soil and abundantly available light are much more infrequent in the absence of forest fires. These environmental changes enhance the establishment of light-tolerant oaks over intolerant pines. The competitive dynamics of the species on these localities is complex. However, it can be evaluated by the modeling approach. The region can be divided into cells, isolating the important environmental types, such as dry ridges in one cell and wet coves in another. The management policies that will be applied over the entire region can then be evaluated for each cell. Initial investigations have suggested that in both high and low elevational forests the dominance relationships are greatly affected by the fire frequency, while at mid-elevational cells the dominance remains unaffected. The method outlined above will enhance the comprehension of the problem greatly and allow managers to deal with a matrix of responses to a single management policy.

However, more is needed than merely an analysis of vegetational pattern shifts on a landscape. The landscape may have properties that emerge due to the pattern and heterogeneity of its composite communities and that cause interrelationships between these communities. In accurately evaluating the responses of different landscape elements, it becomes necessary to evaluate the exchanges of information, energy and materials between elements. Such exchanges can exert considerable influence over those responses. For example, when the long-term management of vegetational processes within one element raises the probability of intense fire (through the build up of dead organic matter, for example), adjacent cells become more susceptible to fire invasion. In addition, fire repression in upper slope forests can potentially affect the flow of nutrients to lower slope forests. In turn, such restrictions can indirectly affect lower slope fire probabilities (with vegetation weakened by nutrient limitation for example). Because these interrelationships affect the ability of a community to recover from disturbance, interrelated communities must be distinguished from those that exist in isolation from the rest of the landscape. Such properties in the Great Smoky Mountains system will be investigated through the application of information on mineral cycles to the shifting mosaic of vegetation. Simulation of simultaneously interacting variables will permit an evaluation of the effect of landscape pattern on the ability of the landscape to prevent the loss of essential substances, such as limiting nutrients, or to regulate perturbation regimes internally.

2.2.8 Conclusions

This paper has described the present applicability of simulation models to solving problems of regional impact. These modeling approaches offer a capability of breaking down a heterogeneous landscape into comprehensible units without ignoring the variable nature of the responses of forest organisms that is so much an essential part of that system. Several methods with complimentary dichotomies of underlying assumptions have been developed. These methods form a diverse array of tools for solving management problems if one can reasonably consider the landscape to be a mosaic of changing vegetation. We have attempted to provide examples of these models from our own research. Further, these methods can provide an avenue toward a basic understanding of the properties of these regional units. The comprehension of these properties is vital to our ultimate evaluation of human impact on natural systems.

Acknowledgements. Research supported by the National Science Foundation's Ecosystem Studies Program under Interagency Agreement No. DEB-77-25781 with the U.S. Department of Energy, under contract W-7405-eng-26 with Union Carbide Corporation. Publication No. 2218, Environmental Sciences Division, Oak Ridge National Laboratory.

Résumé

Pour modéliser la cinématique d'un paysage, les auteurs utilisent des équations aux différences quand les éléments de la mosaïque qui compose le paysage sont nombreux, et une simulation sur ordinateur digital quand sont peu nombreux.

En effet, dans le premier cas, les variations des recouvrements de chacune des espèces peuvent être combinées en un système d'équation différentielles, quand l'échantillonnage est continu, et dans un système d'équations aux différences quand il est discontinu, et c'est le cas le plus fréquent. Le défaut de ce modèle est d'utiliser des coefficients constants, alors que, dans la nature, les coefficients varient.

Dans le deuxième cas, trois types de modèles ont été utilisés: des modèles de croissance des arbres, des modèles de cinématique des populations et des modèles de régénérations dans les trouées. Ce dernier type de modèle est l'un des plus satisfaisants; il convient aussi bien pour simuler l'influence des ouragans dans une forêt tropicale que pour étudier les effets de la pollution atmosphérique sur des forêts tempérées.

Ce dernier modèle a été appliqué à des paysages très hétérogènes d'Australie, en tenant compte des probabilités d'incendie; il est en cours d'application dans un parc national des Etats-Unis. Son usage pourrait être sans doute généralisé, et il sera utile d'y ajouter les interactions entre éléments du paysage.

References

Anonymous (1973) A resource and management survey of the Cotter River catchment. Report prepared for the Forest Branch, Department of Capital Territory by the Consultancy Group. Dep For, Aust Nat Univ, Canberra

Cattelino PJ, Noble IR, Slatyer RO, Kessell SR (1979) Predicting the multiple pathways of plant succession. J Environ Manage 3:41–50

Costin AB (1954) A study of the ecosystems of the Monaro Region of the New South Wales. Government Printer, Sydney

Curtis JT (1959) The vegetation of Wisconsin: An ordination of plant communities. Univ Wisconsin Press, Madison, pp 657

Davis DD, Wilhour RG (1976) Susceptibility of woody plants to sulfur dioxide and photochemical oxidants. EPA-600/3-76-102. US Environ Protect Agency, Corvallis Oreg, pp 71

Doyle TW (1981) A simulation model of forest succession in the Puerto Rican tropical montane rain forest. In: West DC, Shugart HH, Botkin D (eds) Forest succession: Patterns and applications. Springer, Berlin Heidelberg New York

Heck WW, Brandt CS (1977) Air pollution effects on vegetation: Native crops and forests. In: Stern A (ed) Air pollution, 3rd edn. Academic Press, London New York, pp 157–229

Hool JN (1966) A dynamic programming-Markov chain approach to forest production control. For Sci Monogr 12:1–26

Horn HS (1976) Succession. In: May RM (ed) Theoretical ecology: Principles and applications. Blackwell Scientific Publ, Oxford, pp 187–204

Kessell SR (1976) Gradient modeling: A new approach to fire modeling and wilderness resource management. J Environ Manage 1:39–48

Kessell SR (1977) Gradient modeling: A new approach to fire modeling and resource management. In: Hall CAS, Day JW Jr (eds) Ecosystem modeling in theory and practice: An introduction with case histories. Wiley, New York, pp 575–605

McLaughlin SB, Lee NT (1974) Botanical studies in the vicinity of the Widows Creek Steam Plant. Review of air pollution effects studies, 1952–1972, and results of 1973 surveys. TVA I-EB-74-I. Tennessee Valley Authority, Chattanooga Tenn, pp 62

Mudd JB, Kozlowski TT (1975) Responses of plants to air pollution. Academic Press, London New York

Okubo A (1980) Diffusion and ecological problems: Mathematical models. Springer, Berlin Heidelberg New York

Shugart HH, Noble IR (1981) A computer model of succession and fire response of the high altitude eucalyptus forests of the Brindabella Range, Australian Capital Territory. Aust J Ecol 6:149–164

Shugart HH, West DC (1977) Development and application of an Appalachian deciduous forest succession model and its application to assessment of the impact of the chestnut blight. J Environ Manage 5:161–179

Shugart HH, West DC (1980) Forest succession models. Bio Science 30:308–313

Shugart HH, Crow TR, Hett JM (1973) Forest succession models: A rationale and methodology for modeling forest succession over large regions. For Sci 19:203–212

Smith OL (1980) The influence of environmental gradients on ecosystem stability. Am Nat 116:1–24

Waggoner PE, Stephens GR (1970) Transition probabilities for a forest. Nature (London) 225:1160–1161

West DC, McLaughlin SB, Shugart HH (1980) Simulated forest response to chronic air pollution stress. J Environ Qual 9:43–49

Whittaker RH (ed) (1978a) The classification of plant communities. Junk, The Hague, pp 408

Whittaker RH (ed) (1978b) The ordination of plant communities. Junk, The Hague, pp 388

Wilkins CW (1977) A stochastic analysis of the effect of fire on remote vegetation. Unpubl PhD thesis, Univ Adelaide, S Aust

Ziegler I (1973) The effect of air polluting gases on plant metabolism. Environ Qual Saf 2:182–208

Section 3 Ecosystem Functions

3.1 Research on the Characteristics of Energy Flows within Natural and Man-Altered Ecosystems

M. LAMOTTE

3.1.1 Introduction

The world contains an extraordinary variety of ecosystems, ranging from equatorial forests to deserts and including a multitude of different vegetation types. Diversity of biomass and structure is matched by a large variation in productivity. We have some knowledge regarding the productive capacity of these ecosystems, but this knowledge is incomplete and contains a large degree of error.

Ecosystem diversity is very dependent on environmental conditions, particularly climate, but also on conditions related to parent rock, soil and topography. Man's intervention – voluntary or involuntary – adds to this natural diversity, and this intervention has become increasingly important.

The plant community is the most evident, and in quantitative terms, the most important part of the ecosystem. It is also the central ecosystem feature for it determines the type and number of animals present, and thus plays an important overall role in ecosystem functioning. However, the interactions between the plant and the animal community must not be underestimated, and today, if man is considered as part of the animal community, it can be said that the animal community is playing an increasingly important role.

A natural ecosystem can be replaced by a completely artificial one, for example by an agro-ecosystem or grazing ecosystem. In the case of forest management, this transformation process may be less visible but still effective, by encouraging or eliminating certain tree species or by regular cutting. More limited types of ecosystem modification include mowing of grasslands, grazing by cattle or temporary destruction of grassland by fire. Man adds to this action in the animal world through managing his livestock, hunting and fishing. It is important to analyze how the environment can become exhausted through excessive sustained exploitation.

The gradual reconstitution of a climax formation which has been destroyed by fire or cutting can be considered as being initiated by man's action, but then progressing more or less independently. These transitory seres have been studied intensively but still little is known of the energy relations during succession. In some cases, the destruction of the climax community has brought about practically irreversible environmental changes, especially of the soil, in such a way that the plant communities which develop will not lead to the original plant formation but to another, generally poorer, type. Numerous examples can be found in the Mediterranean region.

All the ecosystem changes mentioned above are directly or indirectly related to intentional human actions. Other ecosystem modifications result from the secondary and tertiary effects of human actions–effects which are often unpredictable. Ex-

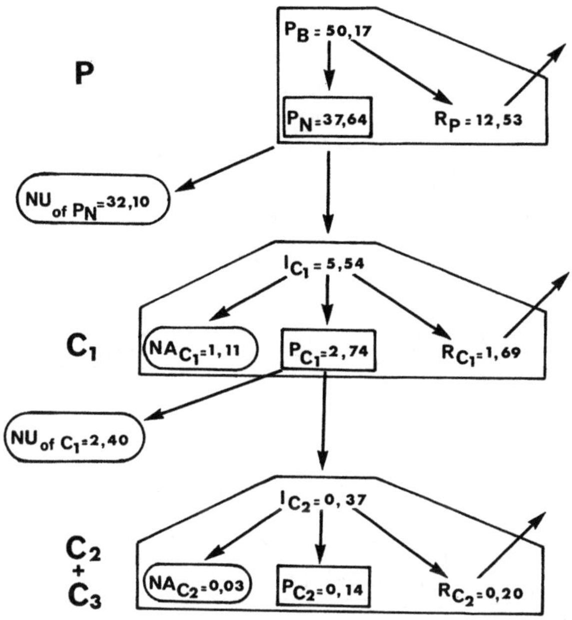

Fig. 1. Energy flow in the Lake Mendota ecosystem in the United States. (After Juday 1940 and Lindeman 1942) *P* Producers; C_1 *primary consumers;* C_2, C_3 Secondary, tertiary consumers, P_B gross primary production; R_p respiration of producers; I_{C1} ingested energy primary consumers; P_{C1} production by primary consumers; P_N net primary production; *NU* non-used energy; NA_{C1} energy not assimilated by primary consumers; R_{C1} respiration by primary consumers.
(Units are expressed in 10^6 kcal ha^{-1} $year^{-1}$)

amples are those brought about by pollution, accidental fire or epidemics. These changes affect both the qualitative and quantitative aspects of the structure and functioning of ecosystems.

3.1.2 Examples of Energy Flows in Selected Ecosystems

Despite the substantial progress made within the International Biological Programme (IBP), there are still few data which elucidate the energy functioning of ecosystems. A complete energy budget has been calculated for no more than 10 different ecosystems. These results are generally imprecise even though some of them are considered to be classical.

This lack of reliable examples makes comparisons among ecosystems difficult. The problem can be illustrated by examining the energy flows of some of the most intensely studied ecosystems, data from which can be expressed in summary form. These ecosystems are Lake Mendota, Wisconsin (Fig. 1), Lamto savanna in the Ivory Coast (Fig. 2), a Serengeti savanna in East Africa (Fig. 3), and dry, short-grass prairie at Pawnee, U.S.A. (Fig. 4).

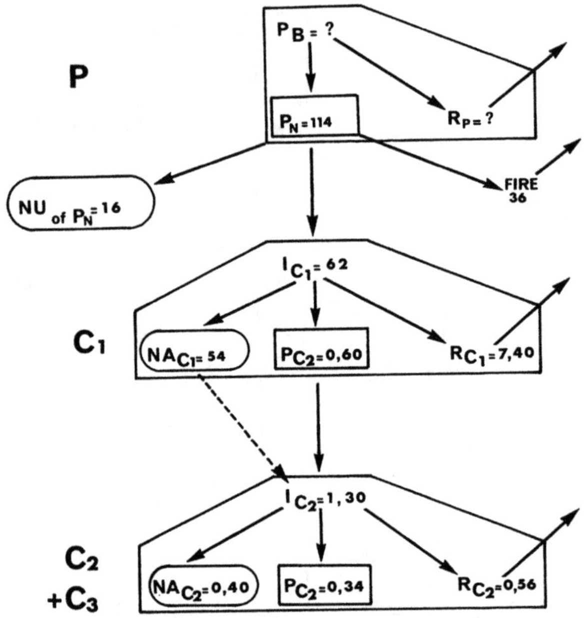

Fig. 2. Energy flow in the Guinean savanna ecosystem at Lamto in the Ivory Coast. (After Lamotte 1977) (for abbreviations, see Fig. 1; the units are expressed in 10^6 kcal ha^{-1} $year^{-1}$)

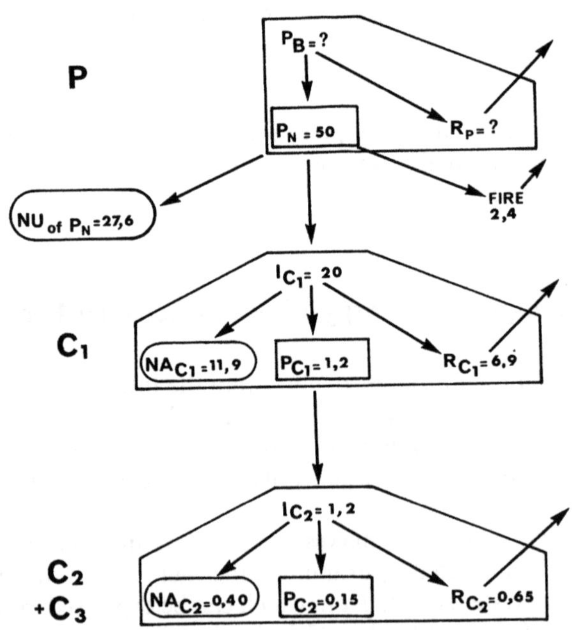

Fig. 3. Energy flow in the short-grass Serengeti savanna in eastern Africa. (After Sinclair 1975) (for abbreviations, see Fig. 1; the units are expressed in 10^6 kcal ha^{-1} $year^{-1}$)

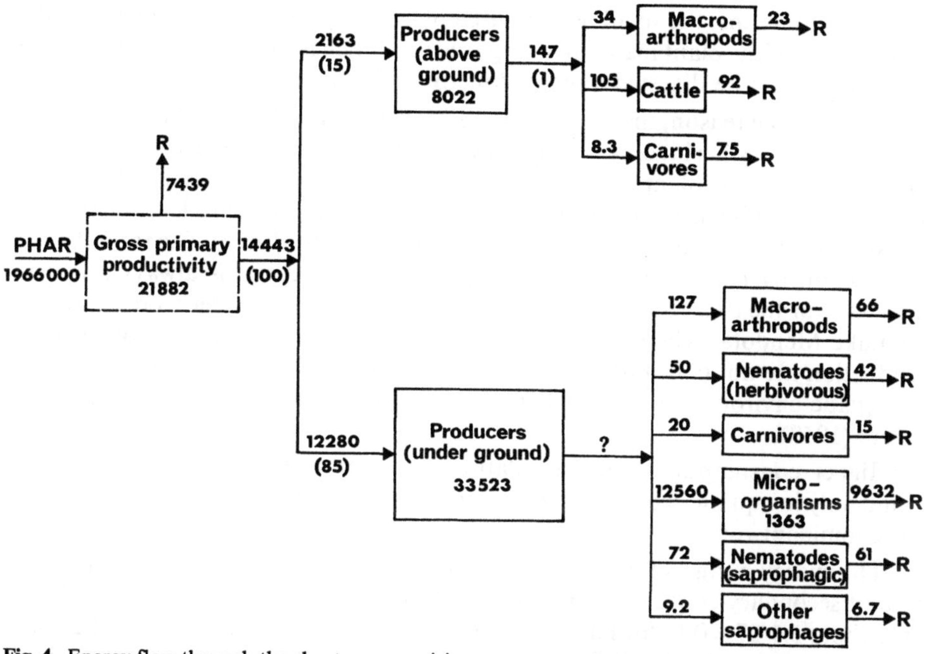

Fig. 4. Energy flow through the short-grass prairie ecosystem at the Pawnee site. The values *above the lines on the arrows* are annual rates of energy flow in kJm^{-2}), while those *below the lines* represent the percentage of net primary production that follow various routes. The *values in the boxes* represent the energy content in kJm^{-2}) of the mean standing crop. *PHAR* photosynthetically active radiation. (After Coupland and Van Dyne 1979)

These results are clearly not comparable in that they relate to different environments and different climates. In addition, the units employed in these studies are varied and are even imprecise. It would be particularly useful to compare different ecosystems occurring in the same region and within the same climatic zone. Other than the differences between grasslands and forests, or between aquatic and terrestrial ecosystems, special consideration should be given to ecosystems that have been more or less disturbed or modified by man, and particularly to a comparison of the functioning of agricultural systems with natural ecosystems.

The few known studies of energy flows, although corresponding to very different habitat types, do reveal certain energy characteristics of ecosystems that should be kept in mind, for these are probably generally applicable characteristics.

The first characteristic concerns the large portion of primary production which apparently is not used by herbivores and macroscopic detritus consumers, but which is rather used by microorganic decomposers. This fraction reaches more than 90% in the American Pawnee prairie (Coupland and Van Dyne 1979). For the Lamto savanna (Ivory Coast) where earthworms are very important, the proportion for microorganisms also attains about 90% of the unburnt primary production of 78×10^6 kcal ha^{-1} $year^{-1}$ (Lamotte 1977, 1978). In the Serengeti savannas, despite great numbers of large ungulates, microorganisms still consume 80% to 87% of the primary production dependent on site (Sinclair 1975).

In general terms, the small portion used by primary consumers is related to the composition of the plant material, a large part of which is made up of cellulose and lignin which can be digested almost alone by microorganisms.

For the same reason, the Assimilation Efficiency (A/I, A, assimilation, I, ingestion) of plants by herbivores is often low, and even lower for detritus consumers which feed off dead plant matter. This obviously means that there is a low value for ecological efficiency, i.e. production at the primary consumer level – *microorganisms excepted* – is low compared to that of primary producers. The ratio P_{C_1}/P_N (production by primary consumers/net primary production) is always low, less than 0.01 for the Pawnee prairie, 0.01 for Lamto, 0.02 for Serengeti; it reaches 0.07 for Lake Mendota, where 32.10 of the 37.64×10^6 kcal ha^{-1} year^{-1} of primary production are not consumed by macroscopic organisms. This is a high ratio for an aquatic environment, and higher than that for the Silver Springs studied by Odum (1957).

Higher consumption and assimilation rates for primary consumers can be found only in aquatic ecosystems where unicellular phytoplankton organisms play a predominant role.

The energy budgets for the next trophic levels, starting with P_{C_2}/P_{C_1} (production of secondary consumers/production of primary consumers), are quite different: about 3.1/27 = 0.11 for Pawnee, 0.12 for Serengeti, 0.26 for Lamto (?), but only 0.05 (?) for Lake Mendota. This means that energy in the form of organic animal matter is used with a relatively high efficiency. It will be seen later on that the ecological efficiency of certain cold-blooded carnivores (spiders, praying mantis, etc.) can reach and even exceed 0.25 or even 0.30. Only the warm-blooded carnivores have very low efficiencies, probably less than 0.02 and thus lie in same range as those of warm-blooded herbivores. However, warm-blooded carnivores represent only a small fraction of the overall animal population.

It should be noted that the energy budgets generally presented do not take account of microorganisms, which are rarely studied and require particular techniques. The classical energy pyramids therefore constitute only a partial representation of energy flows within ecosystems. This is true of the diagrams in Figs. 1, 2, and 3. In reality, the energy left unused by the macroscopic organisms, whose production and population dynamics have been studied, either accumulates (as it is the case for peat bogs) or is used by microorganisms, bacteria, and fungi.

If the ecosystem is in equilibrium, with no accumulation of organic matter, the energy consumed by microorganisms is the difference between the energy produced and energy consumed by macroorganisms. It has been shown, for example, through soil respiration studies of the Lamto savanna, that most of this energy is used by soil microorganisms (Lamotte 1977).

Given the paucity of available data, it is evidently difficult to advance a theory on the effects of man's activities on the energy balance of ecosystems. In the absence of more precise and comprehensive data, we are therefore obliged to consider the results of studies, fortunately more numerous, that exist for certain major ecosystem compartments. After discussion of some aspects of primary production, an examination is made of problems related to energy balance of populations of animal consumers at various trophic levels.

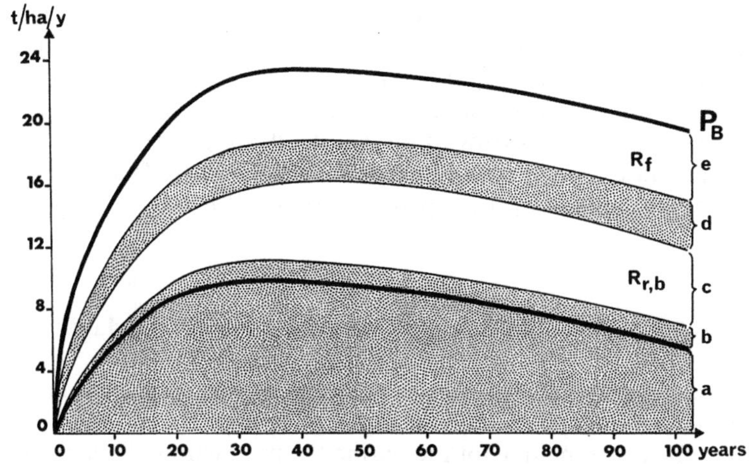

Fig. 5. Temporal change in dry matter production (t^{-1} $year^{-1}$) of a Danish beech forest. (After Möller et al. 1954) P_B Gross production; *a* Biomass increment (roots, trunks, and branches); *b* Loss of roots and branches; *c* Loss by trunks, branches, and roots respiration ($R_{r,b}$); *d* Loss of leaves; *e* Loss by leaves respiration (R_f)

3.1.3 Primary Production

The energy flow in an ecosystem depends mainly on its primary production, the quantitative and qualitative characteristics of which vary considerably, and are very easily modified by man.

Man can intervene in energy flow pathways by trying to increase the quality or the quantity of production of a given ecosystem, for example, a natural forest or grassland. Man can also lower primary production, but this is usually involuntary and is a consequence of over-exploitation, pollution or soil degradation. Man can also radically change the ecosystem and replace it – intentionally or otherwise – by other plant communities. This occurs after clear-cutting of a forest, burning or clearing in order to plant crops, or abandoning cropland. There is then a problem in comparing the production of completely different ecosystems, such as forests compared with grasslands or agrosystems.

Botanists, foresters and agronomists have studied vegetation transformations in some detail but have tended to emphasize the qualitative rather than the quantitative aspects of production. The complexity and diversity of the problem can be illustrated by a few examples.

3.1.3.1 Effect of Rejuvenating Forest Stands

It is known that the production of an even-aged stand decreases after the trees reach a certain age. Figure 5, taken from the classical work by Möller et al. (1954) on a beech forest, illustrates this effect and shows how cutting trees at 30–40 years prevents a decline in production.

It must be pointed out, however, that in certain cases, such as areas with nutrient-poor soil, excessive exploitation can rapidly lead not only to a decrease in production but even to an impoverishment, where the forest ecosystem is replaced by a sometimes completely different type of system. This is the case for clear cutting of certain Amazonian forests growing on sandy soil, and, as described subsequently in this paper, in the Mediterranean zone of France where Green oak (*Quercus ilex*) forest can be replaced by depauperate garrigues of Kermes oak (*Quercus coccifera*).

3.1.3.2 Effect of Cutting on the Production of Grasslands

There are unfortunately few precise experiments comparing production, even of above-ground matter, under different cutting regimes. This is due to the absence of a simple, efficient technique for measuring the production of perennial grasses, especially in respect to root production.

It seems however that, as in the case of forests, grass production can be increased by regular cutting, as is shown below, but on condition that only a reasonable amount of plant matter is harvested. This is generally true when there are sufficient reserves of water and nutrients in the soil, as with montane grasslands in temperate zones or for secondary – and artificial – grasslands of the humid tropics.

Conversely, cutting can cause a decrease in production when the soil is unable to provide a regular supply of water and necessary nutrients. Thus, the experiments carried out by César (1978) on Ivory Coast savannas growing under different climatic conditions showed that monthly cutting resulted in a lower total production compared with untreated savanna (Fig. 6). The drier the climate, the greater this difference, which means that the absence of soil water appears to be the limiting factor. It should be noted, however, that the grass grown after cutting has a much higher protein content than old grass. Only the grass regrowth can be eaten by cattle, which is the reason this technique of cutting is practised.

César's experiments also showed how a plot that is cut repeatedly can become exhausted; growth and production during the following year are significantly less than in a control plot (Fig. 7).

3.1.3.3 Effect of Grazing

Grass consumption by cattle has a fairly similar effect to cutting, but nutrients are at least partially restored through feces and urine. Kelly and Walker (1976) have obtained interesting results for pastures in Zimbabwe (Fig. 8). Light grazing favors production compared with total absence of grazing or excessive grazing. The response obtained differs, however, according to the seasonal distribution of rainfall.

In experiments undertaken in the Rwenzori National Park in Uganda, Strugnell and Pigott (1978) showed that maximum biomass was greatest in fenced-off plots (exclosures), but this experiment is not significant in respect to the level of production itself.

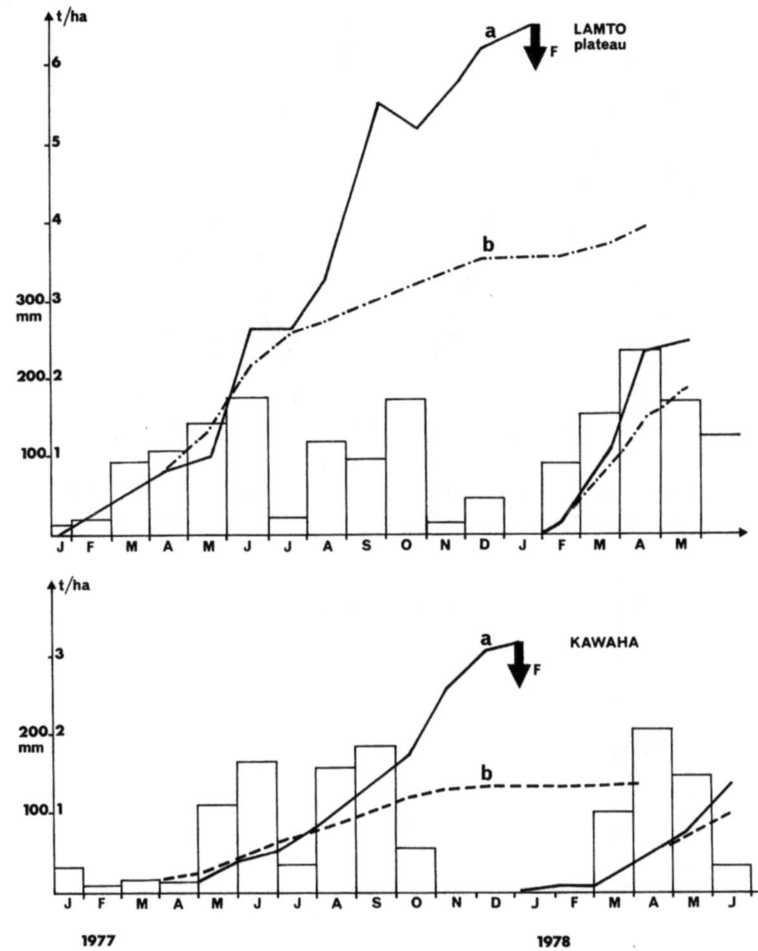

Fig. 6. Changes in the standing crop biomass (*a*) and the cumulative biomass of resprouts after cutting (*b*) in the savannas of Lamto and Kawaha in the Ivory Coast. *F* burning The histograms represent monthly rainfall. (After César 1978)

In Mali, Breman and Cissé (1977) showed that grazing stimulated short-cycle legumes, a non-palatable plant (*Elionurus elegans*) and a low-growing grass (*Microchloa indica*). On the contrary, there was a sharp decrease in the most palatable perennial grass *Andropogon gayanus,* which in the Sahel area does not withstand grazing.

The results published by Singh and Ambasht (1975) on the savannas of the Varanasi region in India showed that above-ground production was higher for protected sites than for grazed sites, but production in this case was measured by differences in biomass. One important fact is clear, however: there was a very significant increase in below-ground production (200 to 600 g m^{-2} in grazed areas).

For a savanna in Nigeria with 1,175 mm annual rainfall, Ohiagu and Wood (1979) reported a production of 2,741 kg ha^{-1} in ungrazed plots against 3,157 kg

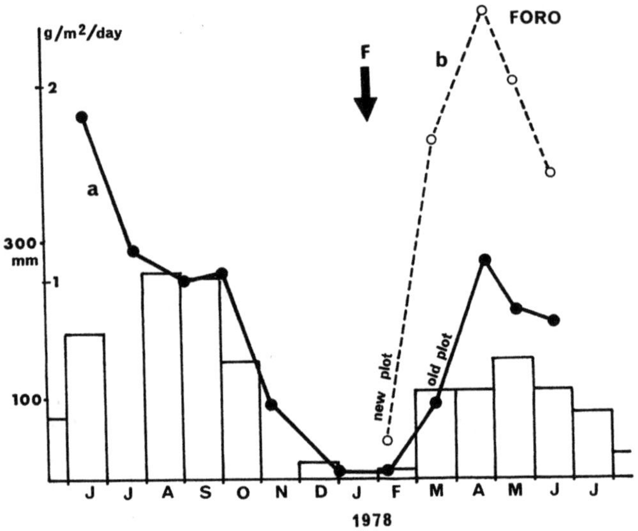

Fig. 7. Cycle of above-ground resprouts in the Foro savanna in the Ivory Coast. (After César 1978). Comparison of a new plot (*b*) and an old plot (*a*). *F* burning

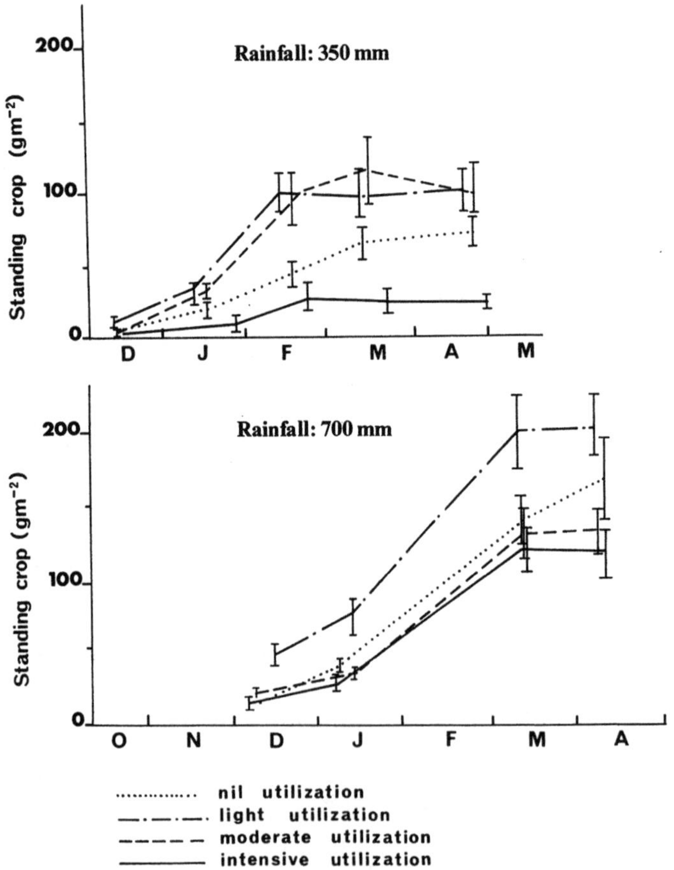

Fig. 8. The standing crop of above-ground herbaceous vegetation in Zimbabwe-Rhodesia. (After Kelly and Walker 1976)

ha^{-1} in grazed plots, where 1,405 kg ha^{-1} was consumed by cattle and 790 kg ha^{-1} by termites, particularly in the dry season. Grazing therefore clearly increases production in this relatively wet area.

In the American mixed-grass prairie, primary production is about 8% higher in grazed than in ungrazed plots and there is a higher transfer of photosynthates to underground parts (Sims and Singh 1978).

In sum, it appears difficult to draw clear general conclusions on how grazing affects production. All that is certain is that there are different responses according to climate, soil and intensity and timing of grazing. Research in this field should be continued in an organized manner and efficient techniques should be elaborated rapidly.

3.1.3.4 Effects of Fire on Grasslands

Fire is another widespread form of human intervention in grassland areas, but as yet there are few accurate quantitative data on its effects. An intense fire can entirely destroy the vegetation and initiate the development of a new set of plant formations. This is particularly true for most forests. In other cases the vegetation is adapted to fire and regenerates afterwards; it is rather the absence of fire that constitutes the abnormal situation. In Strugnell and Pigott's (1978) experiment in Uganda, maximum biomass and therefore very probably above-ground production, was less for burnt sites than for unburnt sites (459 g m^{-2} $year^{-1}$ compared to 536–553 g m^{-2} $year^{-1}$).

In the sub-Sahelian zone in Mali, Breman et al. (1978) showed that fire lowers production from 1.6 to 0.3 t ha^{-1} (the loss in total biomass is replaced in one year). On the other hand, fire increases the quantity of effectively useable forage from 0.1 to 0.3 t ha^{-1}; at the same time, fire induces an increase in protein content and digestibility, and therefore improves the overall food value per hectare.

In the humid savannas, it was always believed that production was invariably higher in burnt areas, but precise data are lacking on this very fundamental point. Pandey (1974) nevertheless noted that for *Dichanthium annulatum* savanna, production increased from 13.3 t ha^{-1} on an unburnt plot to 21.7 t ha^{-1} on a plot burnt in January and to 24.8 t ha^{-1} on a savanna burnt in January and in May; this represents an increase of 87%. In Venezuela, San José and Medina (1976) found that the production of *Trachypogon* savannas increased by 30% to 60% after fire (Fig. 9). In the same way, the maximum above-ground biomass on burnt *Andropogon* savanna in Nigeria was 30% greater, and was produced months earlier, than on unburnt savanna (Egunjobi 1973).

When fire occurs several times on a stand of savanna which had been previously protected, it often leads to a decrease in the tree stratum. This is particularly the case for Guinean savannas. Furthermore, this effect of fire decreases as the climate becomes drier and the herbaceous biomass – the fuel – decreases. This is the case for many herbaceous plant formations in Mali where many trees are still found. As in the case of cutting and grazing the only clear conclusion on the effect of fire on production is that various types of rigorous experiment under different climates are still needed.

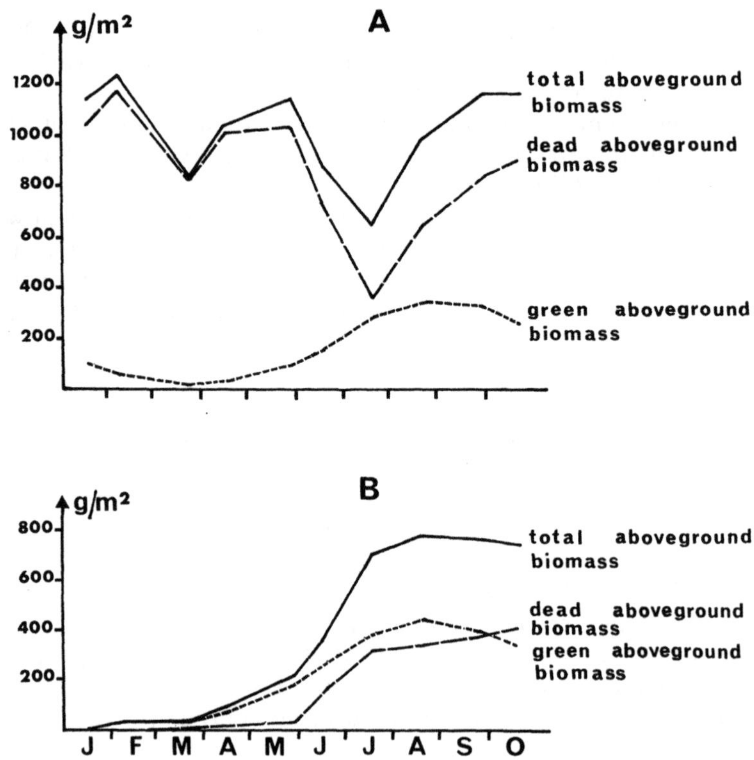

Fig. 9 A, B. Net above-ground production in protected (**A**) and burned (**B**) plots in Venezuelan savannas. (After San José and Medina 1976)

3.1.3.5 Destruction and Transformation of Wooded Areas

The vegetation that grows back once a wooded area has been destroyed can be very different from the original type. Green oak (*Quercus ilex*) forest in the Mediterranean region can thus gradually be replaced, following soil degradation, by a Kermes oak (*Quercus coccifera*) garrigue. The garrigue has a much lower biomass and primary production than the Green oak, with most of the biomass occurring below ground (Fig. 10). On the other hand, production can be higher when the soil has not been altered, but this is often only a temporary situation, although few reliable data are, as yet, available. In humid areas in Panama, Breymeyer's (1978) value for prairie production from altered forest is 72 t ha^{-1} $year^{-1}$, whereas the primary production of the equatorial forest probably does not exceed 30 t ha^{-1} $year^{-1}$. A similar situation is found in certain forest areas of Africa, where *Pennisetum* savanna, which often grows in deforested areas, produces 80 t ha^{-1} $year^{-1}$ which is much higher than for the pre-existing forest. Similarly, in India, near Varanasi, Singh (1975) found that the herbaceous vegetation, which grows after the destruction of a *Terminalia* and *Shorea* forest, produced 24 t ha^{-1} $year^{-1}$ in comparison with only 20.3 t ha^{-1} $year^{-1}$ for the forest itself. In the temperate

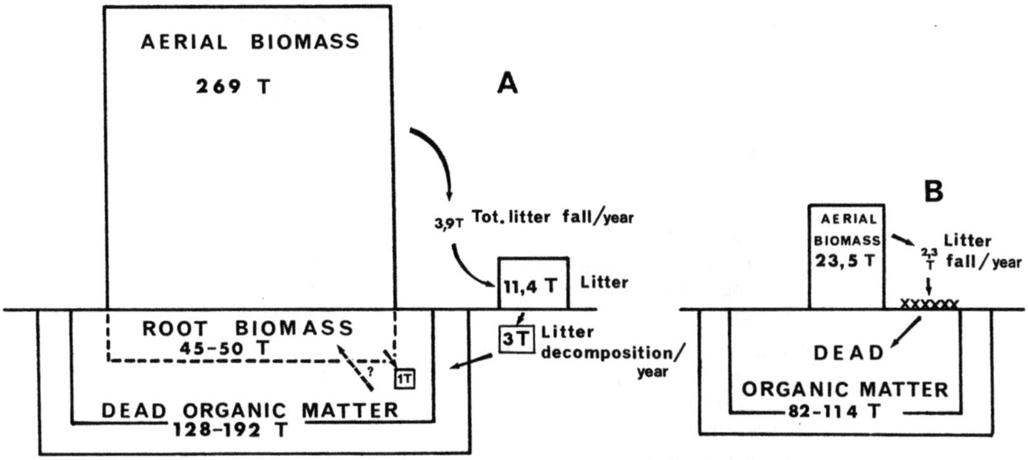

Fig. 10. Distribution of organic matter (t ha^{-1}) and litter fall and decomposition (t ha^{-1} $year^{-1}$) in two different oak ecosystems (*Quercus ilex* and *Q. coccifera*). (After Lossaint 1973)

zone, a Normandy grassland can produce almost 15 t ha^{-1} $year^{-1}$ (Ricou 1978), while Lemée et al. (1978) found that beech forest in the Paris region only produces some 10 t ha^{-1} $year^{-1}$.

3.1.4 Diversity of Energy Flows in Animal Populations

The energy budget for an individual organism is classically represented by the formula:

$$I \swarrow NA \quad I \searrow A \rightarrow P + R^{\nearrow}$$

where I = energy ingested in the form of food,
NA = non-assimilated fraction,
A = assimilated fraction,
P = production (i.e., essentially weight increase),
$R^{\nearrow}$ = energy dissipated through heat loss and respiration.

Three main parameters are needed to quantity such a budget, the third parameter being in fact the product of the first two:

A/I = Assimilation Efficiency,
P/A = Tissue-Growth Efficiency,
P/I = Ecological Efficiency.

As for the energy budget formula, these ratios can be applied at both the individual and population levels.

Table 1. Assimilation Efficiency (A/I) for selected organisms. (After various authors)

Taxa	A/I value
Internal parasites	
Entomophagous Hymenoptera *Ichneumon* sp.	0.90
Carnivores	
Amphibian (*Nectophrynoides occidentalis*)	0.83
Lizard (*Mabuya buettneri*)	0.80
Praying mantis	0.80
Spiders	0.80 to 0.90
Warm- and cold-blooded herbivores	
Deer (*Odocoileus* sp.)	0.80
Vole (*Microtus* sp.)	0.70
Foraging termite (*Trinervitermes* sp.)	0.70
Impala antelope	0.60
Domestic cattle	0.44
Elephant (*Loxodonta*)	0.30
Pulmonate mollusc (*Cepaea* sp.)	0.33
Tropical cricket (*Orthochtha brachycnemis*)	0.20
Detritus eaters	
Termite (*Macrotermes* sp.)	0.30
Wood louse (*Philoscia muscorum*)	0.19
Soil-eating organisms	
Tropical earthworm (*Millsonia anomala*)	0.07

3.1.4.1 A/I Assimilation Efficiency

The value of the Assimilation Efficiency ratio varies greatly according to species. Table 1 presents some values, in decreasing order, for various categories or organisms. The data clearly indicate that Assimilation Efficiency depends on the food regime and not on the taxonomic group.

3.1.4.2 P/A Tissue-Growth Efficiency

The Tissue-Growth Efficiency, which can also be written as 1-R/A, is also quite variable. Table 2 presents some values, in decreasing order of magnitude, for various categories of organisms. High P/A values correspond to low R/A values.

The P/A ratio, which thus varies from about 0.65 to 0.01, appears essentially to depend on the amount of energy the organism spends on maintenance, i.e., energy used for muscular activity and, for warm-blooded organisms, on maintaining a constant temperature. Warm-blooded organisms therefore form a separate group.

Amongst these, particularly low P/A values are found for long-lived animals which spend most of their life cycle in the adult stage (e.g., elephant and deer).

Table 2. Tissue Growth Efficiency (P/A) for selected organisms. (After various authors)

Taxa	P/A value
Immobile, cold-blooded internal parasites	
Ichneumon sp.	0.65
Cold-blooded, herbivorous and detritus-eating organisms	
Tropical cricket (*Orthochtha brachycnemis*)	0.42
Other crickets	0.16
Pulmonate mollusc (*Cepaea* sp.)	0.35
Termite (*Macrotermes* sp.)	0.30 (?)
Termite (*Trinervitermes* sp.)	0.20
Wood louse (*Philoscia muscorum*)	0.16
Cold-blooded, carnivorous vertebrates and invertebrates	
Amphibian (*Nectophrynoides occidentalis*)	0.21
Lizard (*Mabuya buettneri*)	0.14
Spiders	0.40
Warm-blooded birds and mammals	
Domestic cattle	0.057
Impala antelope	0.039
Vole (*Microtus* sp.)	0.028
Elephant (*Loxodonta*)	0.015
Deer (*Odocoileus* sp.)	0.014
Savanna sparrow (*Passerculus* sp.)	0.011
Shrews	Even lower values

However, P/A values also decrease for small-sized organisms, for highly active organisms, and for those that live in cold environments. There appears to be no general rule, but lowest values are found for small insectivores or carnivores.

3.1.4.3 P/I Ecological Efficiency

Values for Ecological Efficiency can be calculated from the examples cited above by the product $A/I \times P/A$. These are grouped in Table 3 along with some other values.

These Ecological Efficiency values clearly show that warm-blooded mammals and birds form a separate group with values of 0.01 to 0.025, with particularly low values for long-lived animals with a long adult period (e.g., elephant). However, as with the Tissue Growth Efficiency ratio, there does not seem to be a general rule. Fairly similar values are found for reptiles and amphibians, taking into account their shorter life span. The other non-vertebrate, cold-blooded organisms have very variable P/I values, ranging from 0.58 for the internal parasite *Ichneumon* to 0.005 for the earthworm *Millsonia anomala.*

Figure 11 shows the general characteristics of the energy budgets for natural populations of a certain number of organisms.

Table 3. Ecological Efficiency (P/I) for selected organisms

Taxa	P/I value
Herbivorous mammals	
Domestic cattle	0.026 (0.44 × 0.057)
Impala antelope	0.022 (0.59 × 0.039)
Vole (*Microtus* sp.)	0.020 (0.70 × 0.285)
Deer (*Odocoileus* sp.)	0.012 (0.80 × 0.014)
Elephant (*Loxodonta*)	0.005 (0.30 × 0.015)
Birds	
Savanna sparrow (*Passerculus* sp.)	0.010 (0.90 × 0.011)
Herbivorous invertebrates	
Termite (*Trinervitermes* sp.)	0.140 (0.70 × 0.20)
Tropical cricket (*Orthochtha brachycnemis*)	0.085 (0.20 × 0.42)
Other crickets (New Zealand taxa)	0.050 (0.31 × 0.16)
Pulmonate mollusc (*Cepaea* sp.)	0.130 (0.33 × 0.30)
Detritus-eating and soil-eating invertebrates	
Termite (*Macrotermes* sp.)	0.090 (0.30 × 0.30)
Wood louse (*Philoscia muscorum*)	0.030 (0.19 × 0.16)
Tropical earthworm (*Millsonia anomala*)	0.005 (0.076 × 0.06)
Carnivorous vertebrates	
Lizard (*Mabuya* sp.)	0.100 (0.80 × 0.14)
Amphibian (*Nectophrynoides occidentalis*)	0.180 (0.83 × 0.21)
Carnivorous invertebrate	
Spiders	0.350 (0.85 × 0.42)
Internal parasites	
Ichneumon sp.	0.580 (0.90 × 0.65)

3.1.4.4 Possibility of Changing the Energy-use Efficiency at the Individual Level

Man can intervene and change the energy-use efficiencies which occur within an ecosystem. Obviously one way to do this would be to replace a species with low productivity with a more productive species which has the same energy requirements. For example, warm-blooded species could be replaced by cold-blooded species. In a management context, however, the higher productivity replacements must be of use to man; even though caterpillars use energy more efficiently than sheep or rabbits, they would not make a useful replacement unless we change our eating habits. It is true however, that replacing a grazing-land ecosystem with an aquatic ecosystem where herbivorous fish like *Tilapia* are raised, can sometimes result in a considerable increase in the efficiency of energy use, assuming that permanent water is available.

The energy-use efficiency of a species already present in an ecosystem can be increased by altering the two components of the Ecological Efficiency (P/I), i.e., by acting on A/I, the Assimilation Efficiency, and on P/A, the Tissue Growth Efficiency. The Assimilation Efficiency value for a given species depends essentially on the type and digestibility of its food. This fact is well known for pasture land;

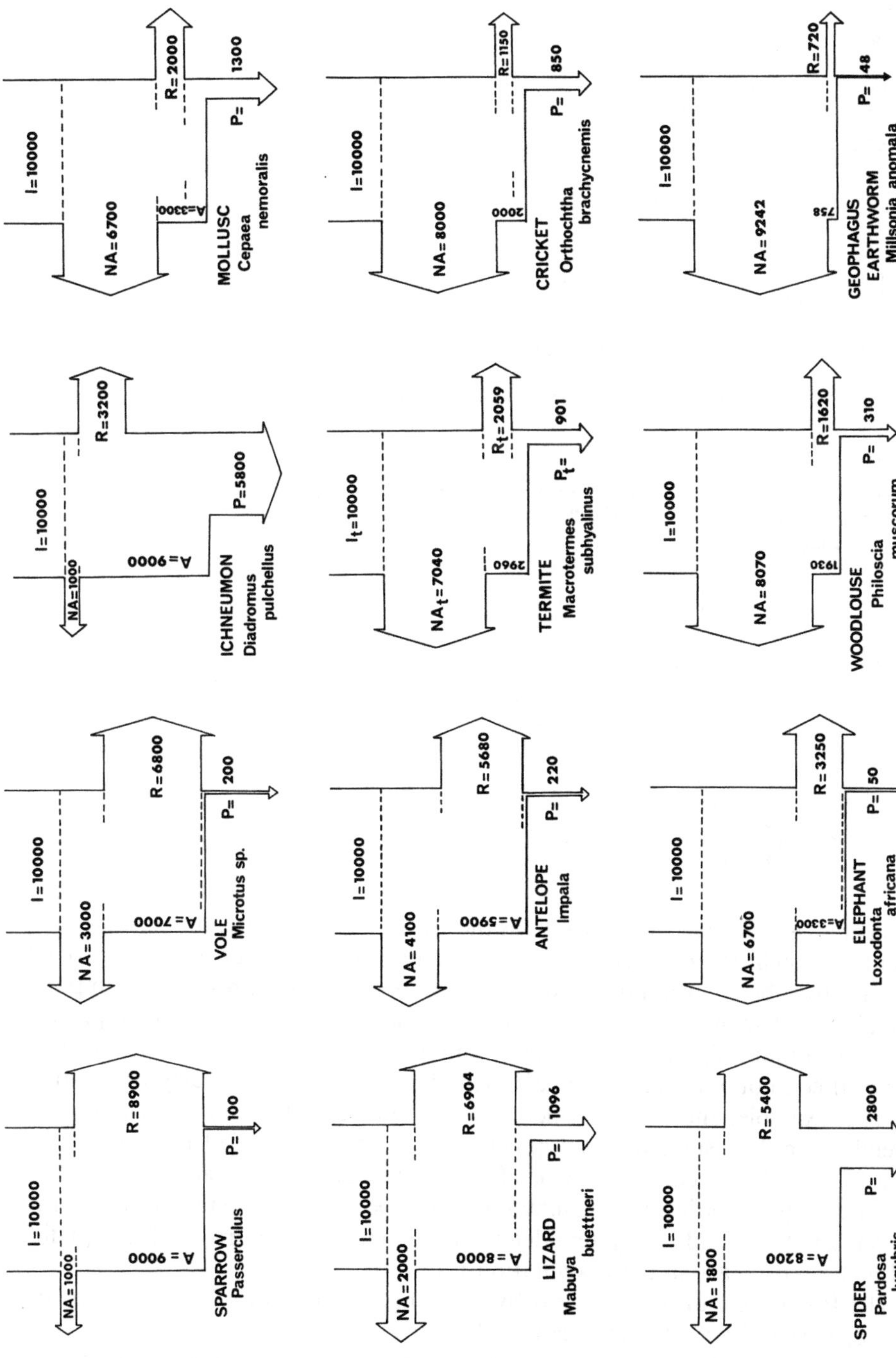

Fig. 11. Energy balances in several animal populations. (After different authors) (For abbreviations, see Fig. 12; units are arbitrary: a reference value for all species is ingestion I = 10,000)

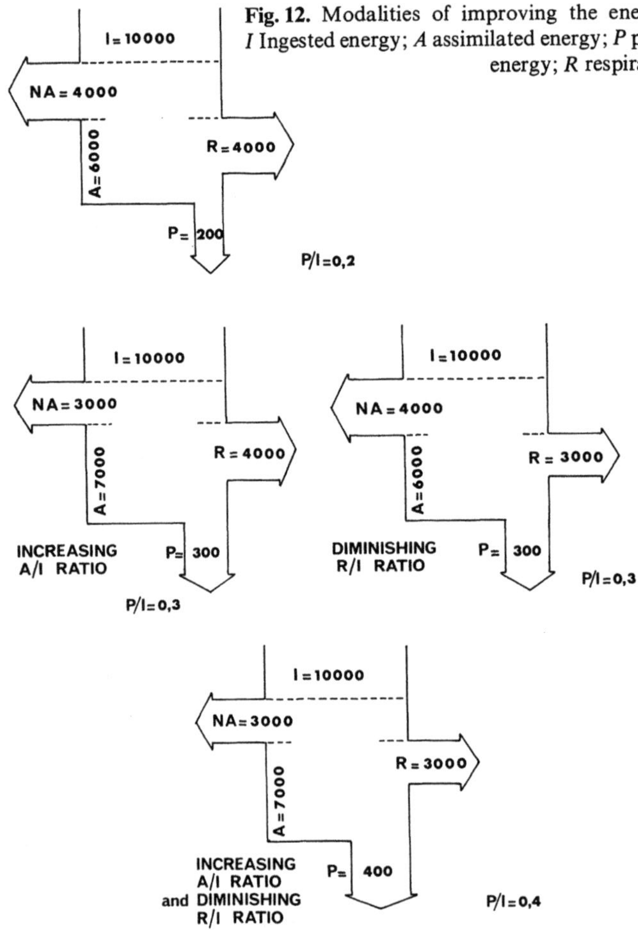

Fig. 12. Modalities of improving the energy balance of an individual: *I* Ingested energy; *A* assimilated energy; *P* production; *NA* non-assimilated energy; *R* respiration

young grass is preferred by cattle but, as the grass matures, it dries and loses a good part of its protein content. It becomes practically useless for cattle when the protein content drops below a minimum of 5%, a proportion which no longer supports normal animal growth. In most African savannas, for example, adequate protein content is found only in grass younger than 4 to 6 weeks. This influence of food quality affects not only higher animals such as mammals; it also clearly affects molluscs, for example, and crickets where the A/I value can change from 0.40 to 0.15 depending on their stage of development and on the age of the leaves consumed (data from Gillon 1968, 1972). This effect is probably widespread.

The energy-use efficiency of animals can thus be increased up to a certain point. But on the other hand, ecosystem deterioration may cause a drop in food quality and hence a decrease in the assimilation rate.

The P/A ratio can be improved by lowering the amount of energy spent for maintenance and activity. This can be done in three ways:

1. by decreasing the energy used for movement; this is the principle behind keeping animals in small cages and avoiding all types of stress; conversely, animals

grazing on sparsely vegetated land utilize more energy seeking food than can be obtained from the food ingested;

2. by treatments which lower rates of metabolism, such as castration;

3. by maintaining warm-blooded animals in an environment having a temperature near their own.

These various treatments are obviously possible only in artificial ecosystems. One must also add the possibility of genetic improvement, which has been highly developed for some species, as well as genetic adjustments to new environmental conditions.

Figure 12 shows how production can be increased at the individual level by improving A/I and increasing P/A through a decrease in energy expenditure, i.e., by respiration R.

3.1.4.5 Possibility of Changing the Energy-use Efficiency of Populations

The energy-use efficiency of many species changes during the organism's lifetime. This process has been described for trees (cf. Fig. 5) and a similar situation exists in many other plants, ageing being accompanied by a decrease in productivity. Animals undergo an identical process, sometimes to a much greater extent. This is the case every time the slope of the growth curve decreases with time. The change with age in the energy budget of the small toad *Nectophrynoides occidentalis* provides a good example (Fig. 13).

The extreme case is for mammals and birds which practically stop growing at the adult stage, production then becoming limited to reproduction. In these conditions, the P/I value is almost zero.

This slowing down of growth with age means that the Ecological Efficiency P/I of an entire population is closely related to its age composition. The value of the P/I ratio decreases as the proportion of old adults increases. The value of rejuvenating the population by eliminating some of the adults is therefore obvious. Man uses this principle to manage game and fish populations efficiently. It is also used for efficient management of domestic livestock. This principle is illustrated in Fig. 14.

3.1.5 Conclusions

Although the data presented are still rather subjective and incomplete, it is nevertheless possible to draw some conclusions on how human intervention – either voluntary or unintentional – can change bio-energetic budgets.

It can be said that moderate harvesting of biomass generally has a stimulating effect on primary production. This effect does not occur, and can even be reversed, when resources of water and of nutrients are exhausted. In the same way, destruction of climax ecosystems and their subsequent replacement by transitory seres often results in an increase in productivity depending on availability of water and nutrients, favorable soil structure, richness of soil microflora, etc. The crucial problem is therefore to maintain the integrity of environmental conditions, particularly

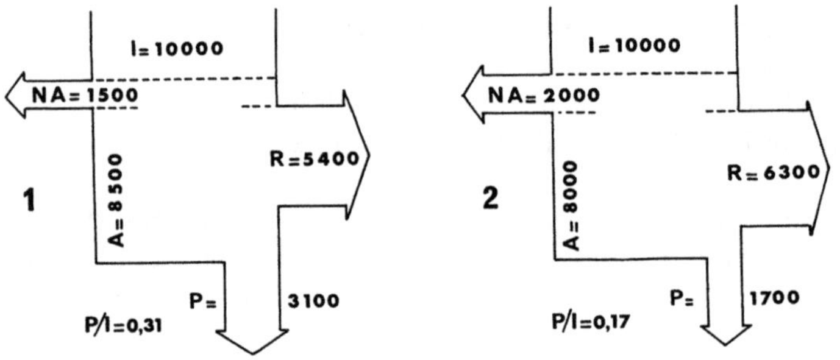

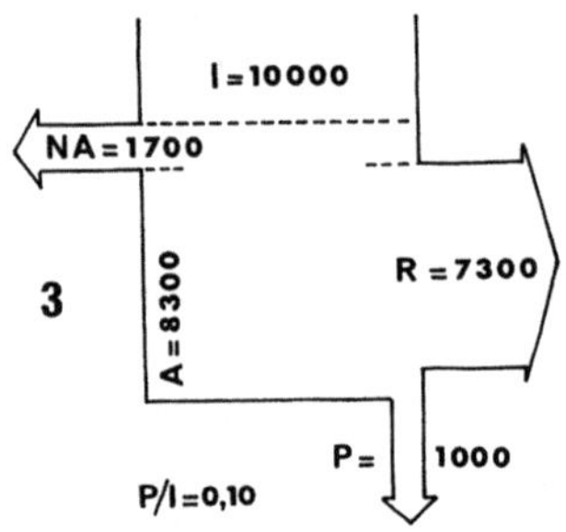

Fig. 13. Change in energy balance during the first three years of life of the Anuran amphibian *Nectophrynoides occidentalis.* (After Lamotte 1972) (for abbreviations and units, see Figs. 11 and 12)

Fig. 14. Effect of removal of old adults (*A*) on the energy balance of a population (*B*); with W = individual weight, *T* time (for other abbreviations and units, see Figs. 11 and 12)

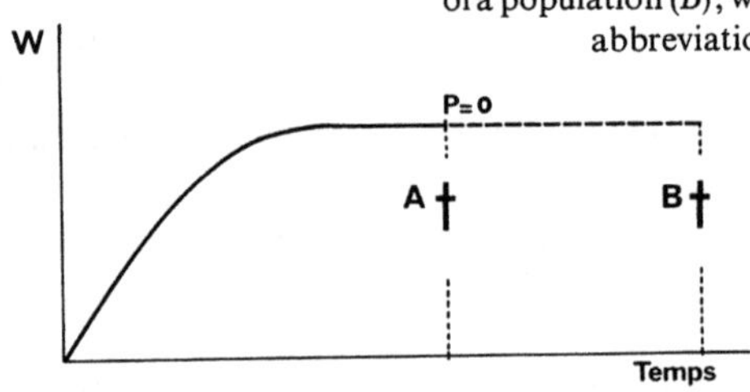

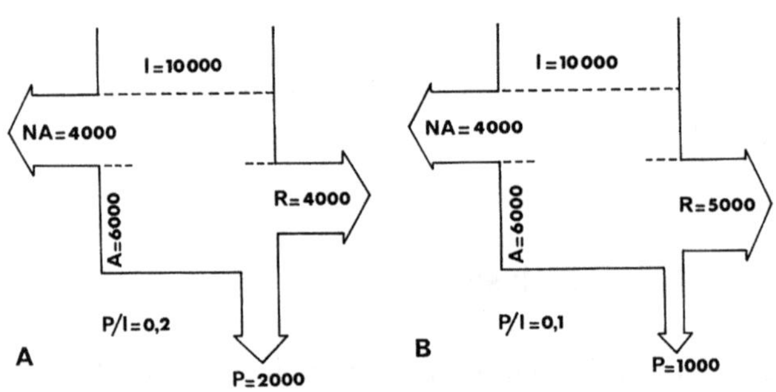

Fig. 14

soil conditions, but also climate. Unfortunately, this is often impossible, as shown by erosion and irreversible soil alteration in mediterranean and semi-desert areas, and sometimes even in the humid tropics. Very often, it is these soil changes that prevent the increased production which would have otherwise resulted from rejuvenating the plant cover.

Long-term pollution generally reduces and can sometimes even stop production. However some "pollutants" act as organic nutrients and can actually increase production. In this sense, eutrophication of aquatic environments can be considered as positive as long as it is not excessive. In most cases, pollution greatly modifies the composition of the community, usually resulting in a significant impoverishment.

Conversely, man can increase primary production in several ways. First, environmental conditions can be improved by irrigation, drainage, fertilization or ploughing. Next, the dynamics of the constituent populations can be managed so as to maintain them at the juvenile stage. Finally, low-productivity species can be replaced by species with higher rates of productivity. Increasing production in these ways always involves supplying more energy, whether through fertilizers or management practices. In grasslands, fire can also be used as a useful tool for increasing the part of production useful to man, but from the ecosystem viewpoint, this represents a considerable loss of energy.

Animal consumers present far more complex problems. Man can act directly upon animal populations, or indirectly through vegetation. Voluntary or involuntary modification of plant communities, for example, transformation into pasture or crops, induces the fauna to adapt to the new conditions. However, it is difficult if not impossible to forecast how the fauna will adapt and there is always the risk that human intervention will lead to a sudden increase in the numbers of a harmful species.

The fauna adapts to disturbance favoring the species with an "r" strategy at the expense of those species with a "K" strategy, i.e., those which were well adapted to the stable environment which has been destroyed. Production can therefore sometimes be increased, but this increase is related to qualitative changes, the species are no longer the same and are less numerous. Generally speaking, the larger animals are the first to disappear with disturbance, such that there is a significant decrease in biomass.

Man acts directly on the fauna in the same manner; the large, long-lived species are the most vulnerable to excessive hunting or fishing. Their disappearance can favor smaller, faster-reproducing species, which often counterbalance any loss of production. The quality of production is greatly altered, however; small rodents or grasshoppers, for example, are not appreciated as game by humans. If the fauna is particularly heavily exploited and even the smaller animals are removed, the energy from primary production is then rerouted to the food chain of the detritus-eaters and decomposers. The ecosystem continues to produce, but at a much smaller scale in respect to animals.

Only heavy pollution can break down the decomposer's food chain resulting in the accumulation of organic matter. Such "eutrophication" is far more frequent in aquatic environments when primary production can increase greatly due to surplus nutrients and neither consumers nor decomposers can keep up with such an increase.

Man can also act to increase the production of those animal species of use to him. He can first of all introduce and protect species with high rates of productivity, which is what happens with domestic animals, selected carefully for many hundreds of years. He can also provide domestic animals with special food, which increases production, or prevent them from wasting energy by limiting their movements. Animal populations can be managed to encourage maximum production by harvesting older individuals. Finally, man can select high performance strains, while recognizing that higher production levels also generally require more and better food and that these breeds are more susceptible to disease, requiring sophisticated medication and care.

The above discussion mainly concerns the overall energy aspects of production. In practice, there is generally a considerable difference of magnitude between total production and that part of production of use to man. Apart from using biomass for fuel, man usually utilizes only a small fraction of plant production, for example, fruits, seeds, buds, and shoots. On the other hand, a significant fraction of animal production is used, e.g. almost all the muscles of many domestic ruminants and even the whole animal of some molluscs or insects. However, there are a great many organisms which man does not use directly; many plants, bacteria, whole families of animals such as earthworms, spiders, sea-slugs, most orders of insects.

Together with increasing the production of plant matter and traditional foods, it would perhaps be rational to make more use of the large proportion of primary- or even secondary-production of ecosystems which is currently unexploited. We should consider the huge quantities of unconsumed plant matter, the large number of worms in certain soils, the microorganisms that are found in all natural systems and even the large amounts of soil organic matter. Even though such a perspective will not please the gastronomes, our continuously progressing technology may make such organisms available as food and, with time, it may become a necessity to turn to these foodstuffs.

Résumé

La comparaison des productions primaires nettes observées dans quatre types d'écosystèmes (un lac du Wisconsin, une savane de Côte d'Ivoire, une savane d'Afrique orientale et une „prairie" américaine) montre que celles-ci sont très différentes les unes des autres, mais que la fraction de cette production qui est consommée par les microorganismes y est toujours très importante (de 80% à plus de 90%). Quelle que soit l'efficacité de l'assimilation des producteurs primaires, le rapport PC_1/PN de la production des consommateurs primaires à la production primaire nette est toujours faible, de 0,01 ou 0,02 dans les formations herbacées, à 0,07 dans le lac considéré. Au contraire, le rapport PC_2/PC_1, où PC_2 est la production des consommateurs secondaires, est comprise entre 0,10 et 0,20 dans les milieux terrestres alors qu'il est voisin de 0,05 dans le lac.

L'Homme peut modifier la production primaire naturelle en utilisant des méthodes très variées, par exemple en coupant les arbres, en fauchant les plantes herbacées, en faisant paître du bétail, ou allumant des incendies. En règle générale, une intervention faible „rajeunit" le matériel végétal et peut augmenter la production primaire nette, en augmentant souvent aussi la „palatibilité" du fourrage, mais une intervention forte fait décroître la production et modifie la composition floristique dans un sens défavorable, en remplaçant des espèces productives et appréciées par des espèces moins productives et refusées par le bétail.

Les effets du pâturage semblent suivre cette règle. Dans l'un des exemples cités (en Inde) la production souterraine a été stimulée par le pâturage et elle est passée de 200 g m^{-2} à 600 g m^{-2}. Au Nigéria, la production des mises en défens était de 2,7 t ha^{-1} et celle des parcelles pâturées de 3,16 t ha^{-1}, dont 1,4 t ha^{-1} étaient consommées par le bétail et 0,79 t ha^{-1} par les Termites (surtout pendant la saison sèche).

L'Homme agit souvent par l'intermédiaire de l'incendie, qui entraîne généralement une diminution de la production (de 1,6 à 0,3 t ha^{-1} au Mali) dans les zones sèches, mais au contraire, dans les zones humides, une augmentation qui atteint 30% à 60% au Vénézuéla et au Nigéria. Les experiences de feux fréquemment répétés ont fait apparaître en revanche une diminution de la production, surtout dans les zones humides.

Le remplacement de la forêt par des savanes se traduit par une augmentation plus ou moins nette de la production: de 30 t ha^{-1} à 72 t ha^{-1} à Panama; de 20,3 t ha^{-1} à 24 t ha^{-1} en Inde; de 10 t ha^{-1} à 15 t ha^{-1} dans le Bassin Parisien.

Le budget énergétique d'une population animale peut être caractérisé par l'énergie consommée (I), l'énergie assimilée (A) et l'augmentation de poids (P). On en déduit trois rapports traduisant l'efficacité de l'assimilation (A/I), l'efficacité de la production (P/A), et l'efficacité écologique (P/I), produit des deux rapports précédents. Les tableaux 1, 2 et 3 donnent des exemples de ces rendements pour plusieurs types d'organismes (herbivores, carnivores, détritivores, parasites internes, vers de terre, oiseaux, mammifères, etc.). Les mammifères et les oiseaux, qui sont des animaux à sang chaud, forment un groupe à part, de faible efficacité écologique (0,01 à 0,025) alors que l'efficacité écologique des autres groupes est souvent située entre 0,10 et 0,30 et peut même attendre 0,35 pour des araignées et 0,58 pour un parasite interne (Ichneumon).

Les espèces animales utilisées par l'Homme ont souvent un mauvais rendement énergétique: c'est le cas notamment des mammifères, car ils sont homéothermes. Leur remplacement par de meilleurs transformateurs pourrait être étudié, mais les usages alimentaires devraient aussi être transformés. L'efficacité de l'assimilation des herbivores est améliorée quand l'herbe est jeune, et par ailleurs l'efficacité de la production du bétail peut être augmentée en limitant les mouvements des animaux, ou en diminuant leur métabolisme général (castration).

Comme le rendement écologique diminue au cours de la vie d'un individu, l'efficacité d'une population dépend de sa pyramide d'âges: elle s'améliore donc lorsque la population est maintenue jeune.

On voit qu'au total les actions de l'Homme peuvent augmenter la production d'un système écologique; celles-ci ne doivent cependant pas dépasser une valeur critique à partir de laquelle les potentialités du milieu diminuent. Ceci est vrai aussi bien pour les actions qui portent sur les facteurs physiques (irrigation, fertilisation, etc.) que pour celles qui modifient les facteurs biotiques (introduction d'espèces, rajeunissement des populations, sélection, etc.). Il faut toujours craindre en effet la déstabilisation des facteurs biotiques, car les réactions du système aux modifications biotiques sont souvent imprévues et néfastes. Or l'Homme néglige trop souvent les espèces qui ne lui sont pas directement utiles ou nuisibles, concentrant son attention sur une fraction minime de l'ensemble du système.

References

Bernhardt-Reversat F, Huttel C, Lemee G (1978) La forêt sempervirente de Basse Côte d'Ivoire. In: Lamotte M, Bourlière F (eds) Problèmes d'ecologie: Ecosystèmes terrestres. Masson, Paris, pp 313–345

Breman H, Cisse AM (1977) Dynamics of Sahelian pastures in relation to drought and grazing. Oecologia 28:301–315

Breman H, Diallo A, Traoré G, Djiteye MM (1978) The ecology of annual migrations of cattle in Sahel. Proc 1st Int Rangel Congr, pp 592–595

Breymeyer A (1978) Analysis of the trophic structure of some grassland ecosystems. Pol Ecol Stud 4:55–128

César J (1978) Cycles de la biomasse herbacée et des repousses après fauche dans quelques savanes de Côte d'Ivoire. Cent Rech Zootech, Bouake, n. 16 PAT, pp 16

Coupland RT, Van Dyne GM (1979) Systems synthesis. In: Coupland RT (ed) Grassland ecosystems of the world: Analysis of grasslands and their uses. Cambridge Univ Press, Cambridge, pp 97–106

Egunjobi JK (1973) Studies on the primary productivity of a regularly burnt tropical savanna. Ann Univ Abidjan Ser E 6:157–169

Gillon Y (1968) Caractéristiques quantitatives du développement et de l'alimentation de *Rhabdoplea klaptoczi* (Karny, 1915) (Orthoptera, Acridinae). Ann Univ Abidjan Ser E 1:101–112

Gillon Y (1972) Caractéristiques quantitatives du développement de l'alimentation d'*Anablepia granulata* (Ramme, 1929) (Orthoptera, Gomphocerinae). Ann Univ Abidjan Ser E 5:373–393

Juday C (1940) The annual energy budget of an inland lake. Ecology 21:438:450

Kelly RD, Walker BH (1976) The effects of different forms of land use on the ecology of a semi-arid region in South-Eastern Rhodesia. J Ecol 64:553–576

Lamotte M (1972) Bilan énergétique de la croissance du mâle de *Nectophrynoîdes occidentalis* Angel, Amphibien Anoure. CR Acad Sci Ser D 274:2074–2076

Lamotte M (1977) Observations préliminaires sur les flux d'énergie dans un écosystème herbacé tropical, la savane de Lamto (Côte d'Ivoire). Geo-Eco-Trop 1:45–63

Lamotte M (1978) La savane préforestière de Lamto (Côte d'Ivoire). In: Lamotte M, Bourlière F (eds) Problèmes d'ecologie: Ecosystèmes terrestres. Masson, Paris, pp 231–311

Lemeé G (1978) La hêtraie naturelle de Fontainebleau. In: Lamotte M, Bourlière F (eds) Problèmes d'ecologie: Ecosystèmes terrestres. Masson, Paris, pp 75–128

Lindeman RL (1942) The trophic-dynamic aspect of ecology. Ecology 23(4):399–418

Lossaint P (1973) Soil-vegetation relationships in Mediterranean ecosystems of southern France. In: Di Castri F, Mooney HA (eds) Mediterranean type ecosystems, Ecol Stud, vol VII. Springer, Berlin Heidelberg New York, pp 129–185

Möller CM, Müller, D, Nielsen J (1954) Graphic presentation of dry matter production in European Beech. Forstl Forsoegsvaes Dan 21:327–335

Odum HT (1957) Trophic structure and productivity of Silver Springs. Florida. Ecol Monogr 27:55–112

Ohiagu CE, Wood TG (1979) Grass production and decomposition in Southern Guinea savanna, Nigeria. Oecologia 40:155–165

Pandey AN (1974) Short term effect burning on the above-ground production of *Dichanthium annulatum* grassland stands at Varanasi. Trop Ecol 15:152–153

Ricou G (1978) La prairie permanente du nord-ouest français. In: Lamotte M, Bourlière F (eds) Problèmes d'ecologie: Ecosystèmes terrestres. Masson, Paris, pp 17–74

San José JJ, Medina E (1976) Organic matter production in the *Trachypogon* savanna at Calabozo, Venezuela. Trop Ecol 17:113–124

Sims PL, Singh JS (1978) The structure and function of ten western North American grasslands. IV. Compartmental transfers and energy flow within the ecosystem. J Ecol 66:983–1009

Sinclair ARE (1975) The resource limitation of trophic levels in tropical grassland ecosystems. J Anim Ecol 44:499–522

Singh RP (1975) Biomass, nutrient and productivity structure of a stand of dry deciduous forest of Varanasi. Trop Ecol 16:104–109

Singh UN, Ambasht RS (1975) Biotic stress and variability in structure and organic (net primary) production of grassland communities at Varanasi, India. Trop Ecol 16:86–95

Strugnell RG, Pigott CD (1978) Biomass, shoot production and grazing of two grasslands in the Rwenzori National Park, Uganda. J Ecol 66:73–96

3.2 "Natural" Mixed Forests and "Artificial" Monospecific Forests

D. Auclair

3.2.1 Introduction

Although some forests, particularly in Europe, have been managed for several centuries, they remain relatively natural ecosystems compared with agricultural crops. Ever since Man established himself in or near woodlands, he has tended to "domesticate" the forest, favoring certain species over others. For instance, the famous european oak-groves originated in the Middle Ages when the peasants favored the oak not for its wood but for its fruit. They used to let their swine loose in the forest to feed on the acorns. The oak was thus selected to the detriment of other species like beech and fir. Later oak was managed as coppice for the production of fire-wood and still later as high forests or coppice with standards for high quality timber.

This artificial modification of the forest has developed through the ages with more and more precise techniques which are well adapted to the environment, to the species cultivated, and to management requirements. But although forest managers have always attempted to simplify the forest ecosystem for a given objective (fruit, timber, fuel, protection, recreation) they have also tended to keep as near as possible to the natural characteristics of the environment in order to maintain the renewable resources.

Table 1 gives some figures concerning the fractions of forest land in different parts of the world which are unstocked or unproductive, and the fractions of forested area in which some form of management control is applied. In "unforested" areas reforestation or afforestation is intended "in the foreseeable future", evidently with "artificial" forests. "Unproductive" forests are certainly very nearly natural. They range from any forest that is now not in use (but may be used in future), to forest that is never likely to be in use in foreseeable circumstances. "Management control" concerns working plans or legal or contractual provisions limiting exploitation. This may be applied even in unproductive forests. Some data are lacking concerning management (North American data refer only to publicly owned and industry-owned forest in U.S.A.), but about 1,500 million ha of the world's forest (almost 50% of that on which information is available) have some form of management control.

One interesting point to note is the fact that while 40% of the world's forest is unproductive, in the less industrialized regions this proportion reaches 50% and is only 20% in the more industrialized areas.

Several examples of simplified forest ecosystems can be taken from the French forests:

Table 1. Land categories and management status of forests. From World Forest Inventory (F.A.O. 1963). Forest land includes unstocked areas from which forest has been clear-cut or burnt, or in which reforestation or afforestation is planned. For many countries data are lacking concerning the management status of forests (especially Asia and North America). Coverage is defined as the percentage of forest for which data are reported

Region	Forest land 10^6 ha	Unforested %	Unproductive as % of forested area	Forested area as percent of land area	Percentage of forested area with management control	
					%	Coverage
North America	750	5	41	38	100	12
Central America	76	7	28	26	18	71
South America	890	7	61	47	21	67
Africa	710	1	57	24	23	61
Europe	144	4	11	29	98	97
U.S.S.R.	910	19	4	34	100	100
Asia	550	9	32	19	67	35
Pacific Area	96	4	44	11	57	99
World total	4,100	7	40	30	60	61

a) Ecosystems in which the forester favored one species; the beech forests in Normandy, oak forests in central France, fir forests in the Vosges mountains.

b) Ecosystems in which the forester imposed a single treatment; the widespread coppices, the "natural regeneration" with shelterwood, seed-tree, or other silvicultural systems on relatively wide areas.

c) Artificial ecosystems which have been maintained for long periods; maritime Pine in the southwest of France (Landes), Cedars in the Ventoux and Lubéron mountains (southeast), Douglas fir in the Massif Central.

d) Several more recent artificial systems which have been planted; intensively managed *Populus* species. Several short-rotation coppices of these and other species have been tried.

Some ecological problems now appear in several of these forests; *Matsucoccus* in artificial pine forests, *Cryptococcus* in beech forests in Normandy, and the leaching of nutrients (potassium) under artificial spruce forests.

An important but controversial question in silviculture concerns the advantages of mixed versus pure forests. Pure forests, generally created artifically, either by selecting one species in the thinning operations or by planting, are more easily tended than forests of mixed species composition. The latter, resembling more natural stands also have many advantages, which are presented below.

This question has been discussed in silvicultural literature for over two hundred years, but very few conclusive experiments have been undertaken. Here several of the differences, advantages or drawbacks of these two types of forest management are reviewed. Most of the arguments detailed here are from Nisbet (1893), Bühler (1918), Köstler (1956), Nelson (1964), Assmann (1970), and must be considered more as the reflections of a forest manager than those of a basic ecologist.

3.2.2 Advantages of Mixed Forests

A greater density of crop is attainable in mixed forests. Most forests of any considerable extent exhibit almost constantly varying qualities of soil and microsite in regard to exposure, depth of soil, quantity of moisture, etc. A complete utilization of the soil can only be attained when each portion is stocked with the species best suited for growth there. Mixtures of species growing over patches of soil best adapted to their requirements do not thin themselves as early as they otherwise would on land less suited to their normal development. Further, they maintain a closed canopy for the longest possible period, thereby fully utilizing the light resources as well as the soil.

The different natural growth rhythms of the species involved in a mixture introduce new complications into management concepts. These natural rhythms undergo certain variations with age, the method of formation, the kind of mixture, and the methods of thinning practiced. The different growth rhythms can be utilized with suitable thinning methods in order to enable the maximum volume and dry matter production to be achieved. There will be however considerable variation in performance depending on the timing, degree and type of thinning, and the cultural methods employed.

The different light requirements of the species participating in a mixture may produce an increased assimilation efficiency of the total stand in comparison with pure stands. This is to be expected, especially if the upper story consists of light-demanding species, and the intermediate and lower story of shade- and semi-shade-tolerant species. These latter are capable of utilizing the light which has been transmitted through the crowns of light-demanding species in the upper canopy and thus producing an additive increment.

If the mixed species occupy different root horizons the sites can be utilized more fully. The species with strong roots are capable of opening soil layers in which rooting is difficult, enabling the whole stand to benefit from this and also from an improved nutrient supply by way of the litter from such mixed species. Generally it is to be assumed that the deepening and expansion of the rooting zone in the soil will increase productivity, because it results in a larger soil space to be penetrated by air, increases in water storage, and greater accessibility of soil fauna.

The character of the litter from mixed species, especially litter which is rich in nitrogen and easily decomposed, can improve productive capacity by stimulating the soil fauna. The amount of rainfall which penetrates into the soil, as well as the temperature of the surface soil layer depends on the overstory species. The addition of a broad-leaved species whose branches are bare in winter into a coniferous crop can improve the soil biology.

Relatively recently nitrogen-fixing legumes and nonlegumes have been planted as an admixture to more classical tree crops. The results appear most encouraging since such mixtures often give higher yields than in pure stands (Gordon and Dawson 1979). *Robinia pseudoacacia,* a pioneer legume species which colonizes poor soils, has been used for reforestation on abandoned mine sites. Its mixture with other species has been advocated and tried in several instances with promising results (Carpenter and Hensley 1979). *Lupinus arboreus* has been shown by Gadgil

(1971) to have a beneficial effect on the growth of *Pinus radiata,* due to the high nitrogen content of the litter. *Alnus* species have been studied intensively for the past 10 years and two symposia have been held in Washington State (U.S.A.) on its use: one on *Biology of Alder* (Trappe et al. 1968) and another on *Utilization and Management of Alder* (Briggs et al. 1978). A third symposium dealt with "Symbiotic nitrogen fixation in actinomycete-nodulated plants" (Torrey and Tjepkema 1979).

It has been shown in several experiments that the admixture of alder increases the total yield; Miller and Murray (1978) found a higher yield in douglas fir when mixed with red alder and De Bell and Radwan (1979) found that a coppice of red alder and black cottonwood had a higher yield when in mixture than for each species separately.

Plants can have favorable as well as unfavorable effects on one another. Chemical interactions between plants can occur, named allelopathy or allelophily, where root secretions and the effect of litter-casting play a part. It is not clear as yet what is the influence of these on the increment of forest trees. Leibundgut (1976) has shown that some chemical substances can have a very strong effect on germination but not on the subsequent growth of seedlings. Others have noticed not only some interaction with growth, but also on photosynthesis (Kolesnichenko 1964). The phenomenon called "alternation" which occurs in some forests should be noted in this context. Often the seedlings of one species do not grow under older trees of the same species but may grow quite well under a different species. This brings a change in the species composition during forest succession.

Natural reproduction and regeneration of mixed woods is, on the whole, easier than for pure stands. Mixed stands have characteristics which are more favorable for the evolution of abundant fruit production than those usually encountered in the pure stands. With the need to secure for each species the soil and situation best suited for it, healthier growth, larger production of starchy reserves and seed are logical results. Natural reproduction in small groups and patches is considerably easier and more likely to be successful than when one single species predominates over the whole area, notwithstanding the usually frequent changes in depth of soil and quantity of soil-moisture.

It is often stated (Assmann 1970) that the structure of mixed stands is more stable than that of pure stands. Mixed forests are less exposed than pure forests to external disruptions, whether of organic or inorganic origin. Shallow-rooting species, when mixed with deeper-rooting kinds of trees, are much less exposed to damage from storms than when grown alone. When grown along with deciduous broadleaved species, conifers of all kinds suffer much less from storms, fire, damage from snow, ice, or frozen sleet, and fungal diseases than when grown in pure crops, or even in mixed crops of conifers only. Associated along with broad-leaved trees conifers generally attain a better development than under other circumstances, so that they are less liable to attacks by insects. The latter are also better controlled by the birds that feed on them which are more numerous among deciduous trees than conifers. Even should attacks of insects occur, they seldom spread widely over all species in mixed stands and the total clearance of the entire crop, as often happens in the case of pure woods is generally not necessary. Species that are somewhat sensitive to late frosts during the early period of their growth are apt to suffer less damage when under the protection of hardier species with faster development. In mature mixed woods of conifers and broad-leaved species the technical value of the boles is less likely to be diminished by frost-shakes than when these latter are grown by themselves.

This stability of mixed forest ecosystems is related to the fact that mixed stands are nearer to natural ecosystems and have a higher biological diversity. But whether increased ecosystem complexity leads to greater ecosystem stability or not is a controversial problem which will not be discussed here (see Lawlor 1978).

3.2.3 The Yield of Mixtures

The above arguments tend to indicate that the mixed stands, presenting a more complete utilization of the site, with favorable interactions between species and with a more stable structure, should be more productive than the pure stands. The literature concerning the growth of mixed stands stems primarily from European observations near the turn of the century. Little has been written on the problem in the last 50 years.

In his book *Gemischte Wald* (1886) Gayer declared himself strongly in favor of mixed stands, and in this he found general support (Nisbet 1893; Bühler 1918). In a later treatise, Burger (1928), drawing upon the early work as well as on some 20 th century findings in relation to growth of pure vs. mixed stands, commented that, "however valuable from other points of view, one unfortunately misses as a rule the pure check plots, necessary for the problem under discussion ..." and, "the investigations of pure and mixed stands known hitherto have thus far proved only that, expressed in cubic meters of wood, sometimes one, sometimes others yield more, depending upon the local conditions of the site."

Burger's work probably is responsible for the shift in thinking to a statement in Hawley and Smith (1954) that it is almost impossible to generalize about the question of whether mixtures of species are more productive than pure stands.

The agronomists have considerably more data on the growth of mixtures than the foresters. With reference to fodder species, Donald (1963) concludes that, "it seems that when two fodder species are grown together they give no advantage in terms of the yield of dry matter over the higher-yielding pure culture." And with forage species in general, Donald further concludes that the yield of a mixture will usually be less than that of the higher-yielding pure culture and greater than that of the lower-yielding pure culture on the basis of yield data from 70 mixtures. Approximately 85% of the mixtures yielded less than the higher-yielding pure culture and 93% of the mixtures yielded more than the lower-yielding pure culture. It is generally accepted that when two species are cultivated on a soil where certain resources are limiting factors, the yield tends to reach that of the higher-yielding species. However much work has been done on the Graminaceae-legume association (Jacquard 1977) showing the higher yield of the mixture.

Nelson (1964) proposed the following generalization for forest stand production. In most cases, on a given site and with comparable stocking, the growth of an even-aged stand of mixed species composition will not exceed the growth of its

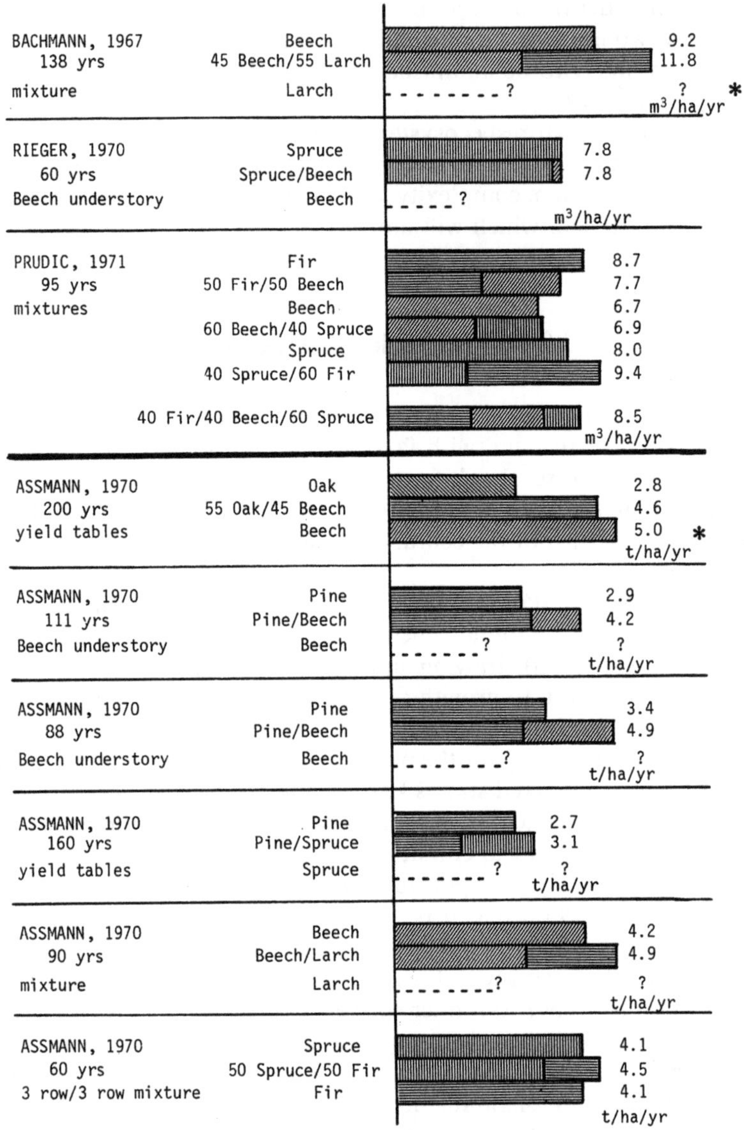

Fig. 1. Examples from the literature giving the production of mixtures in various situations; in existing mixtures, as understory or under-planting, or in yield tables. The comparison with pure stands of either species is not always rigorous. For instance yield tables are compared to others in "similar" situations, and experiments often lack controls

$1\ m^3\ ha^{-1}\ yr^{-1} = 10^{-4}\ m^3\ m^{-2}\ yr^{-1} = 14\ ft^{-3}\ acre^{-1}\ yr^{-1}$

$1\ t\ ha^{-1}\ yr^{-1} = 100\ g\ m^{-2}\ yr^{-1} = 0.44$ American tons $acre^{-1}\ yr^{-1}$

fastest growing species in a pure stand and will not be less than the growth of its slowest growing species in a pure stand.

Assmann (1970) notes that in order to make a quantitative assessment of the increment processes of mixed stands, and to compare them with those of pure stands, it is necessary to provide the following conditions which have received little attention in the past:

a) Knowledge about the performance in pure stands of species participating in mixture. This can be reliably achieved on a given site by simultaneous observation of such pure stands along with the mixed stand.

b) Knowledge about the true ratios of the species participating in the mixture. These are by no means identical with the basal area or volume ratios. An acceptable approximation is offered in the ground-cover area proportions found by vertical measurements of the crown projections. Even the ground-cover area ratios can be used only within each particular story. In a heavily-layered structure consisting of three or more stories, the comparison of efficiencies becomes difficult, because we do not yet know enough about the extent to which the efficiency of the intermediate and lower stories is reduced by the upper story.

A comparison of the efficiencies of the different species based on volume production is technically not free from objection because of the often considerable differences between dry wood weights. The volume production of Norway spruce with its low specific gravity (0.390) is not comparable to that of beech with a high specific gravity (0.560). The true specific gravities fluctuate from site to site and between individual trees as well as between stands. Because of this, local measurements of the specific gravity are indispensable. Another objection concerning the volume basis is the fact that volume is often calculated for wood larger than a given diameter. With complete tree biomass measurements more accurate values can be expected.

Assmann's conclusion is that "if it is not easy to discover the laws of growth in pure stands, the difficulties of assessing this for mixed stands seem almost insurmountable." He however compiled several observations of some typical mixed stands. His conclusions come very near to those of Burger; in some cases the yield may be increased by more than 50%, in other cases there is a slight loss in volume. Several examples of production of mixed stands shown in Fig. 1 show the difficulties of drawing firm conclusions. The use of nitrogen-fixing plants seems to be more reliably advantageous in mixed stands. For other species the result is much more irregular.

3.2.4 Economic Considerations

Although volume production is not necessarily increased by mixing species, there may be several other financial advantages. A silvicultural treatment qualified by Assmann as more "liberal" is frequently possible. But on the other hand, we must count on higher costs of silvicultural treatment, during the early years at least. A "liberal" treatment means that one is free to choose the best adapted species for each spot, taking into account the micro-environmental conditions which occur everywhere. But the necessity for directing the different growth patterns and the different values of the participating species complicates the techniques of management and demands a high intensity of intervention. The results actually achieved vary accordingly. It is necessary to pay attention simultaneously to the differing degrees of vigor, varying growth-rates and light requirements, and the changing crown overhang patterns, as well as the contrasts between the technical and biological value in operating such a management scheme.

Demands of varying nature for timber can more easily be met with mixed culture. Where several species of trees are grown together on the same area the classes of material periodically yielded during the necessary operations of tending, thinning, and removal of diseased individual stem will offer a much greater variety to the different classes of local consumers than is possible in the case of pure forests. When the mature crop comes to be harvested, the same holds true. Furthermore, while changes in the local demands can be more easily satisfied, the wider and more general requirements of the timber markets at various distant centers can be better met with the products of mixed than with those of pure forests. In mixed-stand culture there is also much less chance of over-production of any species or assortment to an extent which might possibly glut the market, even temporarily, and thus cause a decline in market prices. In mixed stands it is easier to modify or transform the crop at any time, so as to meet the present or the probable future requirements of the market.

On small plots of mixed stands, the operations of tending, harvesting, etc., can be conducted economically. On the other hand, this usually precludes mechanical harvesting as there may be large differences in the value and range of timber of the same dimensions.

Several foresters claim that many species in a mixed crop have a higher quality, not only in respect to their external structure, especially stem form, but also with regard to the internal quality of their timber. Natural pruning may occur more readily when a shade-tolerant species grows in the understory (Gayer 1886). But Burger claims that, owing to the marked phototropism of broad-leaved trees, the stem form in mixed stands is considerably poorer than may be expected on the same site in pure, broad-leaved stands.

One reason for the increase in quality of the products from a mixed forest is the fact that the more valuable timber trees (ash, maple, walnut, cherry) are secondary species, with specific habitat requirements. We generally find these high quality trees only in mixed stands.

The recreational value of mixed stands must also be considered. It is commonly accepted that diversity adds a high aesthetic dimension which should be taken into account in planning recreation areas or in urban forests.

In conclusion, although the higher cost of silvicultural treatment may discourage forest managers from using mixtures, a higher quality product and a lower risk from natural disasters should compensate this.

3.2.5 Conclusions – Research on Mixed Forests

Considering the many advantages of mixed stands why do not foresters always manage their forests in this way? There are several reasons for this:

a) The first is the difficulty of tending such stands. The stability of a mixed forest is maintained by the special care and attention of the forester who has artificially to maintain an equilibrium between the different species. This intensive management is more costly than in pure stands where silviculture is much simpler and can often be more easily mechanized.

b) The second is lack of verification of the benefits. Many of the advantages listed above have not been verified experimentally. The question of the economic output of mixed stands is still not fully resolved.

c) The yield, in volume or in total weight, is not known with certainty. The higher quality of timber from mixed stands is not yet proven.

These considerations lead to a number of questions about the growth of forest trees, alone or in stands, and their interactions:

a) Interactions between trees. The degree of competition for various resources such as light, carbon dioxide, water, and nutrients between trees of either the same or different species, has not yet been quantified. Chemical interactions like allelopathy, acting in the soil or in the atmosphere, are still to be described fully. The significance of root grafts and their influence on the growth of trees is not yet fully known.

b) Interactions with the physical environment. Now that the likelihood of whole-tree harvesting is being discussed it becomes important to know the amount and the effect of exportation of mineral nutrients from forest sites. The changes in soil composition and structure under certain crops need elucidation. We need to determine the effects of introducing various species into pure crops. The impact of fertilizers, pesticides, and herbicides utilized with intensive cultivation of pure stands needs to be assessed. The problem of forest fires needs to be addressed particularly in industrial countries.

c) Interactions with the biological environment. Little is known about the effects of mixing different forest species on the swarming of insects or the spread of diseases.

Traditional forest managers tend to keep their forests as natural as possible, attempting to maintain a good balance between human action and the effects of nature. With an increasing demand on renewable resources, however, there is a tendency to manage forests in the same manner as agricultural crops. Intensive mechanical cultivation and the use of chemicals are becoming common practice and monospecific crops are commonly planted.

There is however a big difference between forest trees and agricultural crops. If an annual crop is poorly managed and a mistake is made, or if a natural disaster happens, one year may be lost. If this happens in a forest, it may sometimes mean the loss of over a hundred years' work. This is why it is important to consider carefully the use of mixed forests and to study in detail the various points outlined above. Forests, even those managed for several centuries, are still relatively close to undisturbed ecosystems.

Résumé

La forêt, même cultivée depuis des siècles, est encore un écosystème relativement naturel comparé aux cultures agricoles. Le sylviculteur a de tout temps cherché à simplifier la forêt en fonction du but recherché (fruits, bois de feu, bois d'oeuvre), mais il est resté „proche de la nature".

Il est plus facile pour le sylviculteur de gérer une forêt en peuplements purs, monospécifiques, qu'en peuplements mélangés. Ces derniers semblent beaucoup plus attrayants selon de nombreux points de vue, mais posent des difficultés techniques de gestion.

Nous nous efforçons ici de faire ressortir les différences, avantages et inconvénients, de ces grands types de gestion, vus plutôt du point de vue du forestier que de celui de l'écologue.

Cette liste, qui n'est pas exhaustive, résume les principaux arguments qui ont pu être émis en faveur des peuplements mélangés:

La densité du peuplement peut être plus forte: ayant des structures différentes et des exigences différentes, les espèces variées pourront se répartir de manière plus complète dans l'espace, chaque espèce se trouvant à l'emplacement lui convenant le mieux. Dans le cas d'espèces à rythmes de croissance différents, des méthodes sylvicoles appropriées peuvent favoriser une production optimale en volume ou en matière sèche.

Les différentes exigences en lumière des espèces présentes peuvent être satisfaites pleinement, entraînant une bonne efficacité de la masse assimilatrice. Un exemple classique est celui d'un peuplement dominant, constitué d'une espèce de lumière avec un sous-étage formé d'une espèce d'ombre.

Le sol peut être prospecté de manière optimale si les racines des diverses espèces prospectent des horizons différents. Les espèces à enracinement puissant peuvent ainsi ouvrir des couches où le sol est difficilement pénétrable, au profit des autres espèces.

Des interactions entre espèces, du type allélopathie, sont parfois citées, mais sont parfois favorables, parfois défavorables. Certaines espèces (aulne) sont particulièrement améliorantes pour le sol et peuvent avoir une action immédiate sur la croissance des autres.

La litière provenant d'un mélange d'espèces, riche en protéines et facilement décomposée, peut augmenter la productivité par un recyclage des éléments minéraux, une augmentation de la porosité et de l'activité de la pédofaune. Un mélange d'espèces à feuilles persistantes et à feuilles caduques peut être intéressant pour la répartition des précipitations.

Les possibilités de régénération naturelle semblent meilleures en peuplement mélangé qu'en peuplement pur. Ceci peut être dû simplement au fait qu'un peuplement mélangé nécessite une sylviculture très suivie, ce qui facilite la régénération. Mais il faut aussi remarquer le phénomène dit „d'alternance", qui peut être observé dans certaines forêts. Dans les forêts „vierges" cette alternance peut suivre un cycle de plusieurs siècles. Dans des forêts de hêtre-sapin-épicéa, on constate l'installation de nombreux semis d'une espèce sous les arbres d'une autre espèce, alors qu'ils se régénèrent difficilement sous eux-mêmes.

La structure des peuplements mélangés est en général très stable. Ils présentent une bonne résistance contre les dangers, qu'ils soient climatiques, ou biologiques: le vent, la neige, les gelées, les incendies, mais aussi les insectes, les maladies, les champignons, voire les vertébrés petits ou grands. Dans la plupart des cas, même si l'une ou plusieurs des espèces est décimée, le couvert n'est pas complètement détruit.

Cette stabilité est liée au fait qu'une forêt mélangée est en général relativement proche du milieu naturel et qu'elle présente donc une diversité biologique assez forte: on y rencontre par exemple un plus grand nombre d'oiseaux qu'en forêt pure, que ce soit en nombre d'espèces, ou en nombre d'individus.

Un traitement sylvicole qualifié de plus „libéral" par Assmann (1970) peut être appliqué dans les forêts mélangées. Ceci est lié en particulier au choix des espèces: dans une même parcelle, on observe souvent des variations microstationnelles qui peuvent être utilisées au mieux par une diversification des espèces, chacune étant dans son site optimum. Le choix des espèces est partiellement lié à l'utilisation et un peuplement mélangé est le plus apte à faire face aux fluctuations de la demande au cours des années. Dans des forêts de faible surface, une concentration de produits variés sur une même parcelle peut sérieusement réduire le coût d'exploitation.

Une amélioration de la qualité du bois, lorsque plusieurs espèces sont en mélange, est affirmée par certains: structure externe améliorée, en particulier la forme du tronc, mais aussi la qualité interne du bois.

Bien que le coût du traitement sylvicole soit assez élevé dans les premières années, le rendement financier d'une forêt mélangée est plus élevé, d'après bon nombre d'auteurs, que dans une forêt pure. Ceci vient du fait que l'on rencontre souvent des essences précieuses dans le mélange, mais aussi que le mélange est plus stable et a beaucoup moins de risque de catastrophes naturelles.

Enfin, un argument relativement peu invoqué est celui de l'amélioration esthétique qu'apporte la diversité dans un peuplement mélangé. Cette proposition somme toute très subjective n'est pas à rejeter dans tous les cas, en particulier dans les forêts dites „de loisir" ou périurbaines.

Après avoir évoqué ces multiples avantages d'un peuplement mélangé, on peut se demander pourquoi toutes les forêts ne sont pas traitées ainsi. On peut invoquer deux raisons à celà:

– La première est la difficulté de la conduite d'un tel peuplement: la stabilité évoquée plus haut n'est en général maintenue que par une sylviculture attentive permettant d'éviter l'envahissement d'une espèce sur les autres. Ces difficultés peuvent être comparées à celles rencontrées dans les peuplements jardinés. Le coût d'une telle sylviculture semble donc plus élevé que dans des cas plus simples de peuplements purs.

– La deuxième raison invoquée est le manque de connaissance du résultat. Bien qu'un des arguments en faveur du mélange soit celui du haut rendement financier, de nombreux auteurs ont calculé l'inverse et, en l'absence de résultats probants, il semble difficile de généraliser dans un sens ou dans l'autre.

Mais il faut noter que les arguments pour ou contre les peuplements mélangés manquent de fondement scientifique. Il existe très peu d'études expérimentales permettant d'appuyer ces affirmations.

La simplification des écosystèmes forestiers pose un certain nombre de problèmes, au niveau du fonctionnement des arbres et des peuplements, ou de leurs interactions:

- Interaction entre arbres:
  compétition pour la lumière, le gaz carbonique, l'eau, les éléments minéraux; phénomènes d'allélopathie, au niveau racinaire ou dans l'atmosphère; anastomose des systèmes radiculaires, ...

- Interaction avec le milieu physique:
  exportation d'éléments minéraux par l'exploitation complète de toute la biomasse;
  évolution du sol, par exemple sous peuplements résineux artificiels, ou en présence de plantes „améliorantes";
  pollution par apport non contrôlé d'engrais, de pesticides, ...
  problèmes d'érosion, de protection contre les éléments physiques (incendies);
  impact sur les paysages et perception par le public.
- Interaction avec le milieu vivant:
  pullulation d'insectes dans les monocultures;
  maladies, champignons.

References

Assmann E (1970) The principles of forest yield study. Pergamon Press, Oxford New York

Bachmann P (1967) Baumartenwahl und Ertragsfähigkeit. Schweiz Z Forstwes 118:306–317

Bell De DS, Radwan MA (1979) Growth and nitrogen relations in black cottonwood and red alder in pure and mixed plantings. Bot Gaz (Suppl) 140:97–101

Briggs DG, Bell DE DS, Atkinson WA (1978) Utilization and management of alder. USDA For Serv Gen Tech Rep PNW-70. Pac Northwest For Range Exp Stn, For Serv. US Dep Agric, Portland Oreg

Bühler A (1918) Der Waldbau. Ulmer, Stuttgart

Burger H (1928) Reine und gemischte Bestände. Z Forst- Jagdwes 60:100–108

Carpenter PL, Hensley DL (1979) Utilizing N2-fixing woody plant species for distressed soils and the effect of lime on survival. Bot Gaz (Suppl) 140:76–81

Donald CM (1963) Competition among crop and pasture plants. Adv Agron 15:1–118

Gadgil RL (1971) The nutritional role of Lupinus arboreus in coastal sand dune forestry. I. The potential influence of undamaged lupin plants on nitrogen uptake by *Pinus radiata*. Plant Soil 34:357–367

Gayer K (1886) Der gemischte Wald. Parey, Berlin

Gordon JC, Dawson JO (1979) Potential uses of nitrogen-fixing trees and shrubs in commercial forestry. Bot Gaz (Suppl) 140:102–107

Hawley RC, Smith DM (1954) The practice of silviculture, 6th edn. Wiley, New York

Jacquard P (1977) Relations entre espèces dans les associations graminée légumineuse. Sel Fr 24:3–28

Köstler J (1956) Silviculture. Oliver and Boyd, Edinburgh London

Kolesnichenko MV (1964) The biochemical conformity of woody species when combined in a forest stand (in Russian). Lesn Zuhr 7:3–6

Lawlor LR (1978) A comment on randomly constructed model ecosystems. Am Nat 112:445–447

Leibundgut H (1976) Beitrag zur Erscheinung der Allelopathie. Schweiz Z Forstwes 127:621–635

Miller RE, Murray MD (1978) The effects of red alder on growth of Douglas-fir. In: Briggs DG, Bell DE DS, Atkinson WA (eds) Utilization and management of alder. USDA For Serv Gen Tech Rep PNW-70. Pac Northwest For Range Exp Stn, For Serv. US Dep Agric, Portland Oreg, pp 283–306

Nelson TC (1964) Growth models for stands of mixed species composition. In: Proc Soc Am For, Denver Colo, pp 229–231

Nisbet J (1893) On mixed forests, and their advantages over pure forests. Eyre and Spottiswoode, London

Prudic Z (1971) Influence of the stand composition on the production of fir-beech woods and the deduction of the perspective operational objective (in Czechoslovak). Lesnictvi 17:271–286

Rieger G (1970) Fichten-Reinbestand und Fichten/Buchen-Bestand im Vergleich. Allg Forstz 25:258–260

Torrey JG, Tjepkema JD (1979) Symbiotic nitrogen fixation in actinomycete-nodulated plants. Bot Gaz (Suppl) 140:i–ii

Trappe JM, Franklin JF, Tarrant RF, Hansen GM (1968) Biology of alder. Pac Northwest For Range Exp Stn, For Serv. US Dep Agric, Portland Oreg

3.3 Disturbance and Basic Properties of Ecosystem Energetics

W. A. Reiners

3.3.1 Introduction

Ecologists have traditionally been inclined to envisage the natural environment as relatively benign and disturbance-free, nurturing a diversity of steady-state systems. Thus, it has been customary to view the human-dominated world as a harsh and strange place for the native biota because of the predominance of disturbance associated with human activities. In the last decade, however, the realization that disturbance was and is a natural and frequent component of unpeopled landscapes has taken firm root in our thinking. As a result, a different paradigm for natural systems is emerging: one that recognizes natural disturbance and concomitant recovery mechanisms as integrated aspects of normal ecosystem behavior (Loucks 1970; Levin and Paine 1974; Connell and Slatyer 1977; Grime 1977; Trudgill 1977; Cattelino et al. 1979; White 1979; Holling 1981; Shugart and West 1980; Vogl 1980).

The paradigm of nature as a mosaic landscape composed of stages or phases of change originating from a wide range of natural disturbances underlies the philosophy of this paper. Together with that conceptual base, there is an acceptance of all that it implies in terms of natural adaptations for recovery. The thrust of comparison in this paper, therefore, is not between undisturbed and disturbed ecosystems, but between natural and human-engendered disturbances and their consequent effects. What, if anything, is different about the nature of human-caused disturbances and what, if any, difference in ecosystem behavior can be expected?

Some human disturbances closely mimic natural disturbances in kind, intensity and frequency; for example, certain silvicultural and transhumance grazing practices may contrast only in detail with certain windfall patterns and ungulate migrational patterns. More generally, however, human-induced disturbances contrast sharply in kind, scale, intensity, and frequency; for example, plowing grasslands or distributing exotic biocides are essentially unique disturbances in kind compared with natural disturbances. Furthermore, the scale of conversion of grassland to cropland in North America and Eurasia, or the extent of global deposition of polychlorinated biphenyls (PCB's) are probably unparalleled by natural disturbances. Human disturbances may or may not be more intensive than natural disturbances. For example, flood control has reduced the frequency as well as amplitude of floods in some riverine systems. In contrast, fire suppression policy in the U.S.A. ironically has led to a higher frequency of crown fires in forests, totally different in their impact from ground surface fires. Urbanization, too, represents an intensity of disturbance without comparison in the natural world. Finally, many kinds of human disturbances have a higher frequency than natural disturbances.

This is most obvious in agricultural practices ranging from increasing the frequency of gap formation in tropical forests in shifting agriculture to annual disturbance by tillage agriculture.

The objective of this paper is to examine the ways fundamental aspects of ecosystem energetics change as a result of any disturbance, and to seek the most general predictions for change in these properties which can be expected with variation in the disturbance regime. Several definitions of disturbance are used in this book. The definition followed in this paper is the destruction of living biomass or accumulated detritus. The approach is to trace changes in ecosystem energetics in the simplest case and then examine modifications one might reasonably expect under increasingly realistic conditions. Whereas this paper examines the behavior of energy flow, potentially one of the most "holistic" topical areas of ecology, it will become clear that the author's approach is fundamentally mechanistic or "reductionist". Predictions at aggregated behavior for the ecosystem are derived from generalizations about the behavior of dominant populations, not from a "quasi-organismic" system. For a review of the holistic versus reductionist concept of succession, see McIntosh (1980).

3.3.2 Biomass and Energy Flow in Infrequently Disturbed Ecosystems

"Infrequent disturbance" is an arbitrary, relative term. Its use here will be confined to a disturbance frequency less than the time required for dominant species to complete their life cycles. It is a frequency for which we would expect recovery mechanisms to be present and recovery to be relatively rapid. In general, this kind of recovery can be regarded as secondary succession but the relationship with primary succession will be discussed later. This review largely follows well-established concepts promulgated by Margalef (1963), Woodwell and Whittaker (1968), Odum (1969), Whittaker (1975), Bormann and Likens (1979) and others.

3.3.2.1 Net Primary Production and Energy Flow Pathways

The conventional view of primary production following an infrequent disturbance is diagrammed in Fig. 1 A. The initial lag, if any, the exact shape of the curve, the maximum rate, and final "steady state" value are products of the local environment and particular ecosystem. Peet (1981) summarizes the evidence for the hump-shaped rather than asymptotic curve. The hump is exaggerated in Fig. 1 A for forest development, but probably underrepresents the extent of change in shrub-dominated ecosystems such as chaparral.

The material produced through net photosynthesis can follow very complex food webs but for purposes of this analysis, I will generalize three major pathways. The first of these is a long-term storage in "living" plant biomass which may meet any of three fates. It ultimately may be consumed as detritus, be consumed more or less episodically following export to other systems, or be "consumed" non-bio-

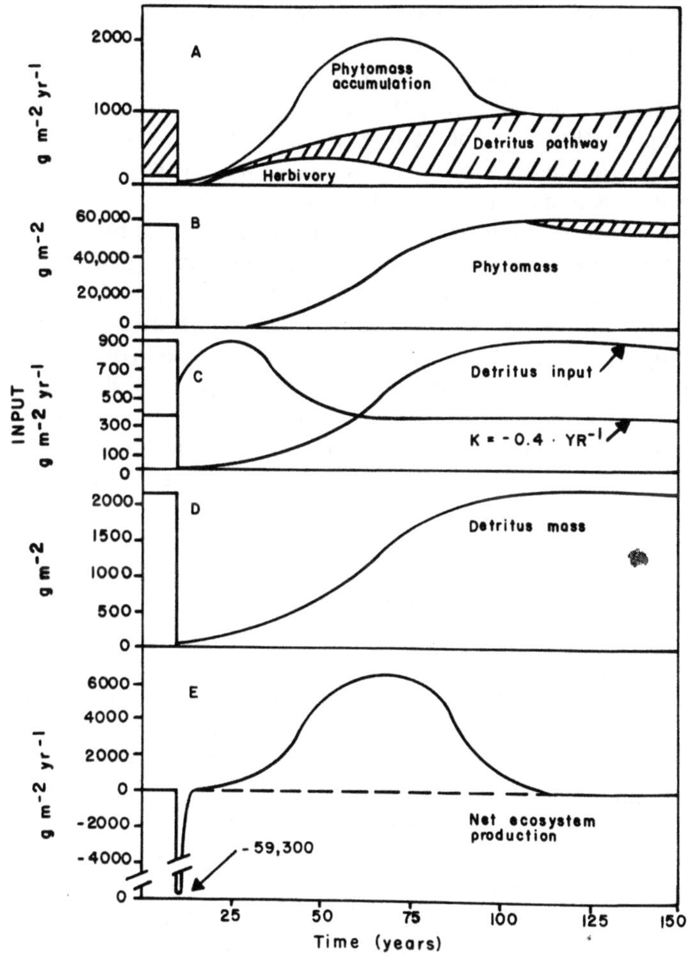

Fig. 1. *A* Change in annual net primary production and the apportionment of the energy represented by that mass following a single, non-unique disturbance. The scaling is for a forest but the allocation to herbivory is made unusually high and the period of positive accumulation is made unusually short for illustrative purposes. *B* The accumulation of phytomass based on the integration of the phytomass accumulation area in *A*. A shallow, hump-shaped maximum is suggested by the *shaded area* representing a possible shunting of accumulated biomass to detritus. *C* Detritus input calculated from the integral of the detritus area in *A* over the disturbance recovery time course. Also shown is a suggested change in the decay coefficient (k) with changing conditions following the disturbance. *D* Change in detritus mass calculated from input and decay variables in *C*. *E* Net ecosystem production calculated from the changes in phytomass and detritus mass above. The sudden consumption or export of all phytomass and detritus at the time of disturbance is considered to be net negative production. Animal biomass is considered too trivial to consider

logically by fire. The second broad pathway of energy flow is the classical "herbivory" pathway in which living plant tissues are consumed directly by heterotrophs, mostly animals, and energy is transferred along associated food chains. The third broad pathway is the biological consumption of dead plant tissue by saprovorous organisms and its transfer along their associated food chains.

These three broad pathways have been chosen because they are much more amenable to generalization than would be more detailed patterns, and because they relate well to human exploitation practices. A few more comments on the significance and characteristics of each pathway are appropriate. First, the long-term biomass storage pathway is, of course, associated with ecosystems containing woody plants. The sometimes massive amounts of woody tissue are protected morphologically and chemically from direct consumption by herbivores. Second, such accumulations of massive biomass are associated with freedom from disturbance, but when disturbances do occur, the amplitude of change is greater than in a non-woody system. The degree of synchrony in death or removal of woody tissues underlies this amplitude and is obviously critical to the status of the ecosystem. Combustion by wildfire or clearcut logging are extreme cases of complete and synchronous "harvest events." Finally, woody plants, especially shrubs, may also invade semi-arid grasslands under overexploitative regimes; this is a form of disturbance, because of differential selection of grazers for graminoid plants over shrubby plants, and because of diminution of competitive effects of the reduced grasses.

The herbivory energy flow pathway is associated mostly with herbaceous plants or plant parts incompletely protected from direct consumption by morphological or chemical means. Large animals are associated with herbivory in natural and man-managed systems, in part because of the relative ease and speed of digestion of such plant material compared with woody materials. This tendency is accentuated in managed grazing systems because large animals are more easily exploited than small ones such as insects. Whereas energy flow tends to be episodic in the woody biomass pathway, it tends to be more regular in grazing pathways. Outstanding examples to the contrary exist, however (Southwood 1968).

The third major pathway – the detritus pathway – is usually the largest single pathway in most terrestrial ecosystems (Coleman et al. 1976). Most of the energy is metabolized by microorganisms. These have very high metabolic rates and a rich suite of enzymatic capabilities but are limited by immobility within the substrate and sensitivity to environmental factors, especially temperature. Energy flow is highly stable from year to year and with the minor exception of some fleshy fungi, the members are not very susceptible to exploitation by man. It is significant that there is no predator–prey type of feedback between saprovores and plants as there is between herbivores and plants, yet there is a clear essentiality in the overall ecosystem function of saprovores that is not often clear for herbivores and associated food chain members.

A fundamental question relevant to disturbance is, "how does the apportionment of energy flow to these three pathways vary under different disturbance regimes?" A comparative analysis over a range of such regimes is needed to provide the simplest empirical answer to this question. A pattern of change in the low frequency regime of a woody plant-dominated system is suggested in Fig. 1 A. This pattern is simply based on the expected change in proportion of herbaceous tissue mass to total biomass (Assmann 1970), and the expected changes in chemical protection of this herbaceous tissue (Cates and Orians 1975). (A better-based estimate of this apportionment would almost certainly develop with a thorough review of available literature from a cross-section of kinds of low-frequency disturbance systems.) The effects of changing the disturbance regimes themselves will be discussed later.

3.3.2.2 Biomass and Detritus Accumulation

Following the same limitation to low frequency disturbances described at the beginning of this section, a postulated pattern of biomass change is graphed in Fig. 1 B. Peet (1981) discusses the alternative models of asymptotic, damped oscillation, and a mid-successional maximum behavior for biomass change in such a disturbance-succession regime. The asymptotic model is adopted here for simplicity but a midsuccessional maximum is suggested by the shaded area in Fig. 1. B where biomass is shunted to the detritus pathway, thus reducing biomass. For the purposes of this analysis, it makes no difference which model is followed. The time and mass scaling in Fig. 1 approximates that for forests but published data illustrate similar curves of lower amplitude and faster response in grasslands, shrublands, and tundra.

Energy flow to detritus discussed above leads to an accumulation of detritus that equilibrates at the point at which energy flow is balanced by decomposition. The simple, linear feedback theory of Olson (1963) adequately describes this process at this level of resolution. According to this theory, decomposition, or output, is the product of accumulated detritus multiplied by a decay coefficient.

$$\frac{\Delta M}{\Delta t} = L - kM,$$

where M equals detritus mass and L equals detritus input. The decay coefficient (k) is, in fact, a function of the physical environment and the chemistry of detritus tissue so that it is likely to be a variable through the changing conditions of succession. Thus, both inputs and outputs are probable variables in a disturbance-recovery sequence.

A possible scenario for changes in these variables is graphed in Fig. 1 C and for the resulting detritus accumulation in Fig. 1 D. The situation simulated in Fig. 1 is a simple case of the disturbance removing all of the detritus and biomass, as a hurricane might do to a mangrove swamp. Such a total harvest is now being considered in the U.S.A. for the conversion of agricultural crops to alcohol. A more common type of disturbance would be the partial removal of above-ground biomass with no direct effect on detritus. A case history for such a disturbance has been simulated for a real forest by Aber et al. (1979). As in their case, Fig. 1 C includes an increase in the decay coefficient following disturbance because of microclimatic enhancement accompanying removal of the canopy.

Two points about detritus deserve special emphasis. The first is that its mass is in the same range or even larger than plant biomass in virtually all terrestrial ecosystems, if humus is regarded as detritus (Ajtay et al. 1979). This is not true of the simulated system in Fig. 1 D because a very high decay coefficient was chosen to give a rapid response which would fit in the time span of the figure. Second, the detritus itself plays numerous essential roles in the ecosystem through passive physicochemical properties as well as through the actions of the enormous diversity of organisms residing in the detritus. The status of detritus following disturbance is therefore one of the most important indices of disturbance severity and of recovery potential.

3.3.2.3 Net Ecosystem Production

Woodwell and Whittaker (1968) defined net ecosystem production as the net change of organic matter in the ecosystem. It is the difference between gross photosynthesis and total ecosystem respiration plus organic export. Net ecosystem production is positive during periods of accumulation, negative during periods of declining mass, and zero when the system is in a steady state. It is a useful index of overall rate and direction of change of a system in terms of stored energy or mass (Fig. 1 E).

3.3.2.4 Variation in Infrequent Disturbance Events

Up to this point, only the conventional wisdom of ecosystem behavior following disturbance has been presented for the simplest case: a single, non-unique disturbance. The patterns portrayed in Fig. 1 represent a basic or "starting" model of behavior from which variations can be derived. This representation of Fig. 1 as even a basic "model" is somewhat presumptuous. More data from a broad range of ecosystems types are needed to confirm its generality and make corrections. Nevertheless, a discussion of variations in disturbance regimes and their significance follows. Much of the subsequent discussion is an application of Trudgill's organization of the subject (Trudgill 1977).

The first variable in disturbance is the "kind" or "uniqueness" of the disturbance. Species composing an ecosystem must have, to varying degrees, adaptations to normally occurring disturbances. The more frequent the disturbance, the stronger the representation of better adapted species and theoretically, the stronger the selection pressure for adaptive traits. Thus the impact of a unique disturbance is likely to be more profound than that of a "normal" disturbance. Whereas we can predict the behavior of a forest harvested in a manner simulating blowdowns, we cannot so easily predict the recovery of prairie from cultivation, deserts from ionizing radiation, or tundra from vehicular traffic. To a degree, uniqueness can be considered in terms of frequency of occurrence. The incredible diversity of exotic impacts that man can perpetrate on an ecosystem, however, makes such a system of scaling on the basis of frequency alone impractical. A taxonomic or scaling system for disturbances in terms of uniqueness or kind is needed.

The second variable is intensity of disturbance. This refers, for example, to the severity of a fire, the degree of chemical loading, or the completeness of mortality resulting from a defined disturbance type. Scaling intensity is relatively simple within a disturbance type and for a particular ecosystem, but we presently have no means of making comparisons of intensity between disturbance types, or of predicting responses of different ecosystems to the same intensity of a particular disturbance. Some of these problems are discussed in Lugo (1978) and Odum et al. (1979). Intensity-responses may be non-linear and threshold behavior is particularly likely. These thresholds may be critical points at which recovery rates are significantly slowed, or more importantly, at which recovery to the original state is impossible. A fuller discussion of such possibilities is given by Peterman (1980) in terms of "boundaries" and "multiple absorption states". Long-term fumigations at Copper Hill, Tennessee and Sudbury, Ontario are probably examples in which such final thresholds have been exceeded.

It is in this context that it should be noted that there really is no practical distinction between primary and secondary succession. Operationally, we distinguish between them on the basis of former occupancy of a site by an ecosystem. It can readily be observed, though, that disturbance may be so severe in secondary succesion as to produce conditions equivalent or worse than those of primary succession. Thus, prior occupancy is not the best criterion for predicting successional patterns. Rather, we should recognize a full range of disturbance, especially of soil, and a continuum of responses ranging from those characteristic of conventional primary succession to those of most rapid secondary succession see Gorham et al. (1979) and Reiners (1980) for characteristics of clear examples at each end of the continuum.

The third pertinent variable is areal scale. The impact of a disturbance must be in some way a function of areal extent. This, in turn, will be amplified by the degree of contiguity of the disturbed area and its shape. These relationships of impact to scale can be promulgated in at least two ways: environmental modification and recolonization potential. A larger disturbed area will, within limits, create more extreme microclimates for regenerative processes. It also creates a larger catchment for runoff leading to a higher potential for erosion on upland sites and deposition in lowland sites. A larger disturbed area, particularly as it takes a round and fully contiguous shape, creates a longer dispersal distance for recolonizers, including both the spores and seeds of plants and vectors such as specific and critical pollinators. The importance of this size function has been suggested for recolonization by tundra plants in scarified areas of Alaska (Chapin and Chapin 1980).

3.3.3 Biomass and Energy Flow in Multiple Disturbance Ecosystems

The foregoing describes changes in a few basic aspects of energy flow and biomass in systems subjected essentially to a single disturbance. In fact, most of our landscapes are covered by ecosystems responding to multiple disturbances that vary in frequency as well as kind, intensity and scale. As human disturbances are particularly marked by higher frequencies, this variable is especially relevant in the context of this paper. The following discussion of the relationship between frequency and basic properties of ecosystem energetics owes much to the ideas of Trudgill (1977) and Olson (1981).

3.3.3.1 Constant Species Composition and Site Quality

If we hold species composition and site quality constant, and increase only frequency of a single kind of disturbance, change in net primary productivity can be predicted as shown in Fig. 2. Continuous disturbance such as constant grazing can be considered a special case of high frequency.

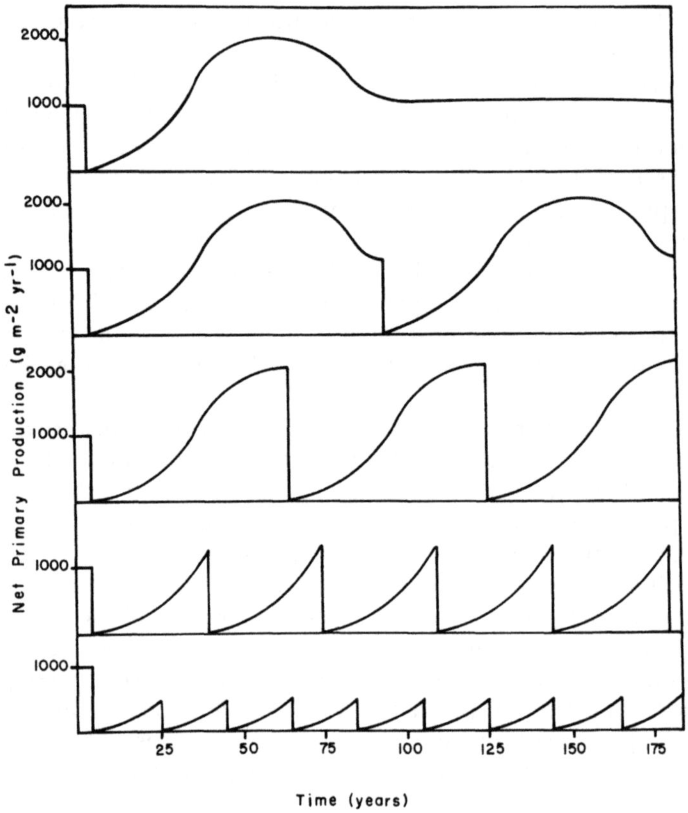

Fig. 2. Changes in net annual primary production with sequential increases in disturbance frequency assuming no change in recovery potential with disturbance. The curves all follow the time course of Fig. 1 A

Given the conditions and behavior graphed in Fig. 2, we can calculate the net primary production for the entire time period graphed and find the maximum point on this curve highly skewed, in this case, to the left (Fig. 3 A). Such an optimization is, of course, the basis of harvest management for many forests and grazing lands. The peak of this skewed curve is very possibly less than would occur in grassland systems, at least, because of the growth response of plants to clipping (McNaughton 1979).

It is obvious from Fig. 2 that average plant biomass must decrease as disturbance frequency increases. A graph of this relationship is shown in Fig. 3B assuming the same apportionment of energy flow to biomass accumulation illustrated in Fig. 1 A.

To the extent woody plants occur, there will be a larger proportion of herbaceous tissue (foliage, young twigs, and roots) in younger stands. Assuming that successional species have lesser chemical defenses against animal consumption, then the integrated total for energy flow to herbivory should increase to an optimum, then decrease as total net primary productivity declines with increasing

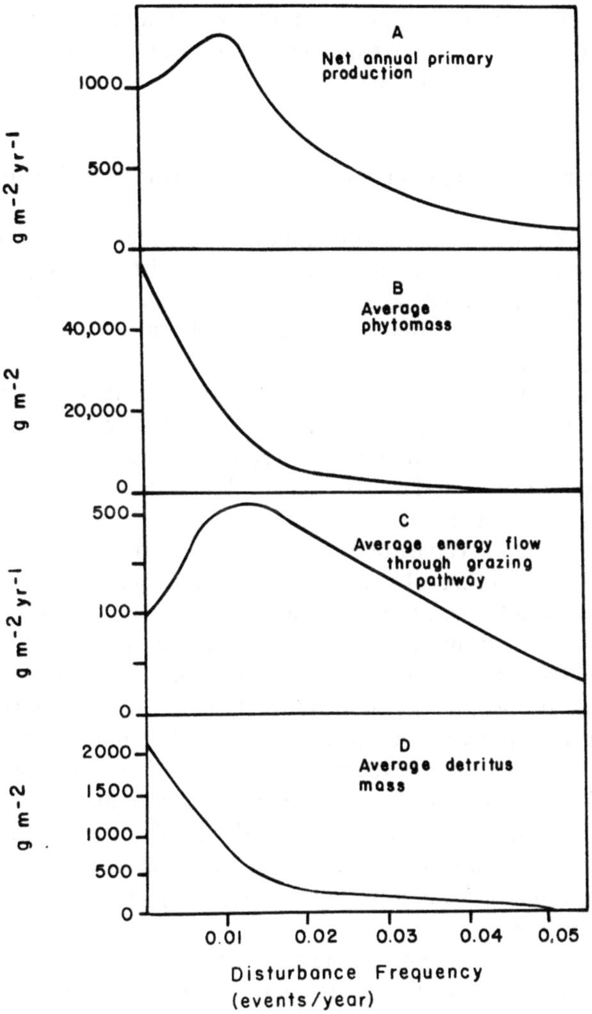

Fig. 3. *A* Change in integrated net primary production over an 180-year period as a function of disturbance frequency (see Fig. 2 for disturbance patterns). *B* Change in average phytomass over the 180-year period as a function of disturbance frequency. *C* Changes in energy flow through grazing or herbivory pathways over a 180-year period. *D* Change in detritus mass over an 180-year period as a function of disturbance frequency

frequency of disturbance (Fig. 3 C). This same pattern apparently occurs in non-woody plant systems as well (McNaughton 1979).

By following the same changes in decomposition coefficient specified in Fig. 1 C, productivity changes in Fig. 2, and apportionment to detritus as illustrated in Fig. 1 A, the change in detritus mass as a function of disturbance frequency is calculated as graphed in Fig. 3 D. Obviously there is a diminution in detritus and its generally beneficial attributes with increasing frequency of disturbances.

Net ecosystem production oscillates around zero through each disturbance-recovery episode. The amplitude of change in net ecosystem production will vary inversely with frequency of disturbance.

3.3.3.2 Changing Ecosystem Structure with Disturbance Frequency

The preceding section demonstrates changes in basic energy flow properties one would predict from first principles, and from the unrealistic assumption that species composition and site quality do not change with increasing disturbance frequency. This section briefly discusses some possible changes that could occur if this last assumption were altered.

Given change with disturbance, site quality could either deteriorate or improve with increasing disturbance frequency. The first alternative has many examples and underlies the fears of environmentalists regarding intensification of land use. Degeneration of semi-arid grasslands of North America under domestic grazing pressure (Hastings and Turner 1965), deterioration of soils under accelerated shifting cultivation of the tropics (Nye and Greenland 1960), and degeneration of some farmlands today in the U.S.A. (Brink et al. 1977) are possible examples of this trend. On the other hand, there are examples of good maintenance, if not actual improvement of site quality under some disturbance-management regimes. Some prime agricultural land has been maintained at a high level of productivity although not without important energy subsidies in the form of industrial fertilizers. Fire management of *Calluna* moors in Britain seems to enhance productivity, although long-term site quality is still open to question after centuries of such use (Gimingham et al. 1979). Improvement or maintenance of site quality seems definite in some well-managed humid pasture lands (Swaby 1966). While improvement may be possible without fossil fuel energy subsidies, it is still likely that improvement will maximize around a particular frequency rate above which deterioration is likely to follow.

The other aspect of ecosystem change deserving consideration is species composition. In order to persist within a particular disturbance regimen, species must be adapted to complete their life cycles within this regimen. This is one of the fundamental aspects of life history attributes that underlies the model of multiple pathway succession (Cattelino et al. 1979; Noble and Slatyer 1980). As disturbance frequency goes up, some species are eliminated. On the other hand, other species may enter the system which can reproduce between, or in spite of, disturbance events and are no longer eliminated through competition with the original species. Where disturbance frequency has been accelerated for some time, new ecotypes or even new species may evolve that are better adapted to the new disturbance regimen. The European origin of the majority of weeds in the U.S.A. probably derives from the much longer time for evolution of disturbance-adapted species in Europe. In any case, increased disturbance frequency can lead to a loss of native species which can be compensated for in time by the migration, introduction, or evolution of better adapted ecotypes.

One cannot predict as a general case whether increased disturbance frequency will change ecosystem structure in a way that will buffer the impact of disturbance

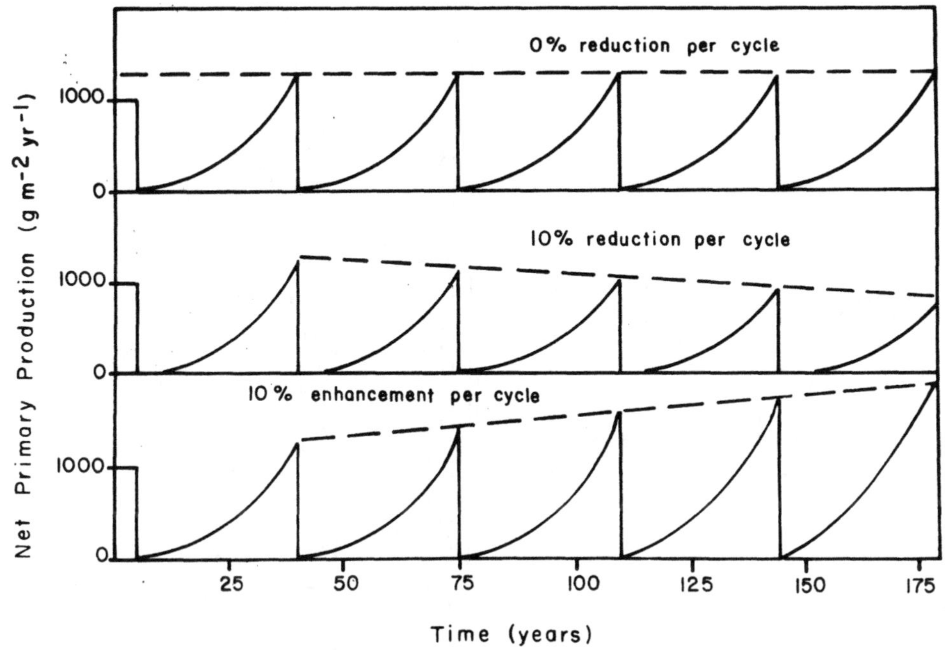

Fig. 4. Upper block-net primary production with 35-year disturbance intervals with no change in ecosystem properties. Middle block-change in net primary production with a 10% decrease in recovery with each disturbance event. Lower block-change with a 10% improvement in recovery with each disturbance event

or aggravate it. Both possibilities are graphed in terms of net primary production in Fig. 4. Not illustrated is the possibility that changes will not be gradual with increasing frequency of disturbance. Thresholds may be penetrated in the same way described earlier for increasing intensity, particularly for cases of deterioration.

Possible changes in structure as a result of changes in disturbance frequency have been described, but it is not clear that further generalization on such a complex problem is possible. Given this hesitation to pursue changes in frequency of a *single kind* of disturbance further, it is even less realistic to attempt generalizations about *multiple kinds* of disturbance. At this level of complexity, mass balance and thermodynamic considerations become of limited value and one must turn to more specific cases armed with information on the attributes of key populations of the system in question. For a population-level analysis of disturbance see the paper by Bazzaz in this volume.

3.3.4 Conclusions: Integration

A small number of ecosystem properties have been discussed in terms of disturbance effects. A number of disturbance variables have been outlined. The last section of this paper is devoted to the summarization of these relationships in a dia-

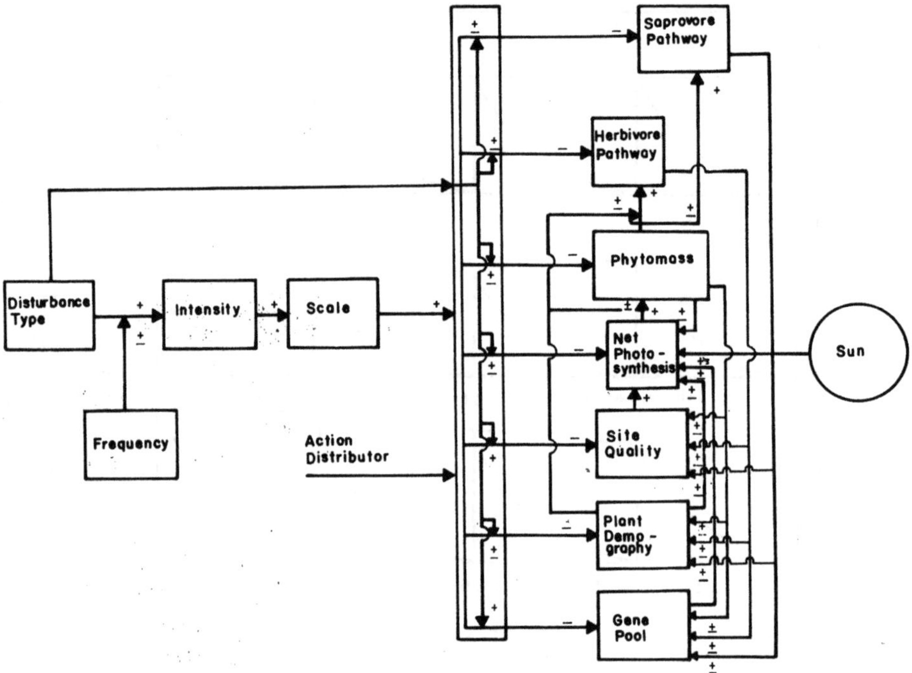

Fig. 5. A diagrammatic model illustrating the relationships between disturbance and energy flow with special attention to the variables of disturbance (*left side*) and the ecosystem attributes and feedbacks regulating the rate and form of response

grammatic form that may be useful for further consideration of this subject (Fig. 5). This diagrammatic model is a hybrid between a causal loop diagram and an energy flow scheme.

Focus is first directed to the disturbance module which integrates the several aspects of disturbance discussed. According to this model, "kind" of disturbance is expressed in two ways. It defines the nature of the impact and at the same time, the specific targets to be affected. The nature of the impact is modified by intensity and scale. These may be considered to be two different kinds of analogs of variable amplifiers. Frequency acts as a gate or valve controlling the release of the disturbance. In nature, disturbances are basically stochastic processes, whereas human-caused disturbances are controlled by management decisions, which in turn are determined by a plethora of human and social factors.

The separate action of kind of disturbance on targeting the impact may be visualized as through an action distributor. Kind of disturbance will actuate gates to specific components of the ecosystem. Thus a surface fire will mainly affect the detritus component of the ecosystem, whereas an SO_2 fumigation will mainly affect the photosynthetic transfer process.

The second module includes the basic attributes controlling net primary production (lower right corner) which is visualized in Fig. 5 as a process for the transfer of sunlight to biochemical energy. The basic attributes controlling net primary production are the kinds of plants (referred to here as the "gene pool"), the numbers and ages of plants (referred to here as "demography"). Each of these attributes influences net primary productivity in different ways and some play more important roles in different parts of disturbance-recovery cycles than others. For example, gene pool is principally important in the reestablishment phase, as a reservoir of species adapted to different disturbance regimes and stages of succession (see Mooney and Chapin chapters in this volume on the nature of such adaptations). Demography and site quality are always influential. Net primary production probably maximizes at a demographic state of intermediate age and density for most ecosystems, whereas it is directly proportional to site quality by definition. For a fuller discussion of the relationships between life history characteristics ("gene-pool") and demography with net ecosystem production, see the chapter by Bazzaz this Volume.

The third module (upper right corner) contains the energy flow circuits from net primary production to plant biomass and thence to the herbivory or "grazing" energy pathway and the saprovory or "detritus" energy pathway. The obvious energy flow connections between these two pathways are deleted in this diagram for the sake of simplicity, and because most of these interconnections occur at higher trophic levels, where the energy involved has been diminished by several orders of magnitude.

The numerous, complex feedbacks between phytomass and each of the two food chains on the one hand, and gene pool, demography, and site quality on the other, are symbolized in the model by single arrows. For example, an increase in herbivory could have positive feedbacks to the richness of the plant gene pool and possible site quality (Mattson and Addy 1975). These feedbacks are labeled with plus and minus signs because they may follow complex, non-monotonic functional relations. For example, an increase in energy flow through the detritus pathway might at first increase site quality and then decrease it if litter accumulated. Similarly, gene pool, demography and site quality could have influences on phytomass and the energy flow pathways aside from their effects on net primary production. Much of the most essential research challenge in ecology lies in the elucidation of these feedbacks.

Finally, the effect of disturbances can be connected from the action distributor to the ecosystem attributes and energetic units as shown in Fig. 5. The model makes clear how a particular event such as a fire can be scaled by intensity and extent, metered out in time by the frequency function, and then affect one or more ecosystem attributes. A ground fire (low intensity) may have two direct effects: acceleration of energy flow of detritus via combustion, and the killing of small plants. Oxidizing detritus might feed-back positively to site quality by temporarily increasing the ion concentration of the soil solution. Killing young plants could decrease species richness and shift demography toward older, fewer individuals.

As another example, SO_2 fumigation might affect only the photosynthetic process directly. If photosynthesis decreased, rates of energy flow would decrease, activating changes in the feedback loops between ecosystem components and producing commensurate adjustments. For both of these examples and any other disturbance, repeated effects controlled by the frequency function would continue change until new equilibria were reached, perhaps after the penetration of thresholds.

Even with a fair amount of parameterization, this model would have limited predictive ability for a particular ecosystem. The reason for suggesting such a model here is to promote consideration of how major components of ecosystems might be linked with respect to disturbance. As with many, if not most, models, its principal value may lie in the stimulation of thought on the major questions we need to address before planning critical experiments and collaborative research on the general problem of ecosystem response to disturbance.

Résumé

Les perturbations étudiées dans cette contribution sont les actions humaines qui entraînent une destruction de phytomasse vivante ou de détritus (litière + humus).

Après une perturbation exceptionnelle, l'évolution spontanée d'un système écologique moyen, sous climat non aride, est résumée dans la Fig. 1:

– la production annuelle de phytomasse croît jusqu' à 50 ou 80 ans, passe par un maximum voisin de 2 kg m^{-2} an^{-1}, et décroît ensuite jusqu' à 100 ans, pour atteindre un palier où une production annuelle d'environ 1 kg m^{-1} an^{-1} est exactement équilibrée par la décomposition des détritus;

– la phytomasse sur pied augmente en suivant une courbe logistique;

– la production de détritus et la masse de détritus suivent aussi une courbe logistique;

– la production nette de biomasse passe par un maximum pour un peuplement compris entre 50 et 80 ans.

Les flux d'énergie se répartissent entre trois „voies" principales: l'accumulation dans les végétaux ligneux, le transit par le tube digestif des herbivores, la décomposition par les saprovores.

Les perturbations peuvent différer par leur fréquence et leurs types, par leur intensité et par l'étendue qu'elles affectent. Il n'en existe pas encore de typologie générale; en conséquence, il est logique d'examiner en premier lieu les cas théoriques simples. Le premier est celui où la composition spécifique et la qualité de la station ne sont pas fortement modifiées. La courbe de la production annuelle de phytomasse est alors périodiquement interrompue, mais elle repart en gardant la même forme (Fig. 2). La production moyenne et la phytomasse disponible pour les herbivores sont alors maximales quand la perturbation se répète environ tous les 100 ans; la phytomasse sur pied moyenne et la masse des détritus sont des fonctions monotones décroissantes de la fréquence des perturbations (Fig. 3).

Si la qualité de la station et la composition floristique sont un peu affectées par les perturbations, les schémas précédents sont légèrement modifiés (Fig. 4).

Un schéma général de l'ensemble des interactions est présenté sur la Fig. 5; ce schéma n'a évidemment pas de valeur prédictive, mais il peut stimuler la réflexion.

References

Aber JD, Botkin DB, Melillo JM (1979) Predicting the effects of different harvesting regimes on productivity and yield in northern hardwoods. Can J For Res 9:10–14

Ajtay JL, Ketner P, Duvigneaud P (1979) Terrestrial primary production and phytomass. In: Bolin B, Degens S, Kempe S, Ketner P (eds) The global carbon cycle. Sci Comm Probl Environ. Wiley, Chichester, p 129
Assmann E (1970) The principles of forest yield study. Pergamon Press, Oxford New York
Bormann FH, Likens GE (1979) Pattern and process in a forested ecosystem. Springer, Berlin Heidelberg New York
Brink RA, Densmore JW, Hill GA (1977) Soil deterioration and the growing world demand for food. Science 197:625–630
Cates RG, Orians GH (1975) Successional status and the palatability of plants to generalized herbivores. Ecology 56:410–418
Cattelino PJ, Noble IR, Slatyer RO, Kessell SR (1979) Predicting the multiple pathways of plant succession. Environ Manage 3:41–50
Chapin FS III, Chapin MC (1980) Revegetation of an arctic disturbed site by native tundra species. J Appl Ecol 17:449–456
Coleman DC, Andrews R, Ellis JE, Singh JS (1976) Energy flow and partitioning in selected man-managed and natural ecosystems. Agro-Ecosystems 3:45–54
Connell JH, Slatyer RO (1977) Mechanisms of succession in natural communities and their role in community stability and organization. Am Nat 111:1119–1144
Gimingham CH, Chapman SB, Webb NR (1979) European heathlands. In: Specht RL (ed) Heathlands and related shrublands of the world, A. Descriptive studies. Elsevier, Amsterdam
Gorham E, Vitousek PM, Reiners WA (1979) The regulation of chemical budgets over the course of terrestrial ecosystem succession. Annu Rev Ecol Syst 10:53–88
Grime JP (1977) Evidence for the existence of three primary strategies in plants and its relevance to ecological and evolutionary theory. Am Nat 111:1169–1194
Hastings JR, Turner RM (1965) The changing mile. Univ Arizona Press, Tucson
Holling CS (1981) Forest insects, forest fires, and resilience. In: Mooney H, Bonnicksen TM, Christensen NL, Lotan JE, Reiners WA (eds) Fire regimes and ecosystem properties. USDA For Serv Gen Tech Rep WO-26. US Dep Agric For, Washington DC, pp 445–464
Levin SA, Paine RT (1974) Disturbance, patch formation, and community structure. Proc Natl Acad Sci USA 71:2744–2747
Loucks OL (1970) Evolution of diversity, efficiency, and community stability. Am Zool 10:17–25
Lugo AE (1978) Stress and ecosystems. In: Thorp JH, Gobbons JW (eds) Energy and environmental stress in aquatic systems. DOE Symp Ser. CONF-771114. Natl Tech Inf Serv, Springfield VA, p 62
Margalef R (1963) On certain unifying principles in ecology. Am Nat 97:357–374
Mattson WJ, Addy ND (1975) Phytophagous insects as regulators of forest primary production. Science 190:515–522
McIntosh RP (1980) The relationship between succession and the recovery process. In: Cairns J (ed) The recovery process in damaged ecosystems. Ann Arbor Sci Inc, Ann Arbor Mich, p 11
McNaughton SJ (1979) Grazing as an optimization process: grass-ungulate relationships in the Serengeti. Am Nat 113:691–703
Noble IR, Slatyer RO (1980) The use of vital attributes to predict successional changes in plant communities subject to recurrent disturbances. Vegetatio 43:5–21
Nye PH, Greenland DJ (1960) The soil under shifting cultivation. Tech Commun No 51. Commonwealth Bureau of Soils, Harpenden. Commonw Agric Bur, Farnham Royal, Bucks
Odum EP (1969) The strategy of ecosystem development. Science 164:262–270
Odum EP, Finn JT, Franz EH (1979) Perturbation theory and the subsidy-stress gradient. Bio Science 29:349–352
Olson JS (1963) Energy storage and the balance of producers and decomposers in ecological systems. Ecology 44:322–331
Olson JS (1981) Carbon balance in relation to fire regimes. In: Mooney H, Bonnicksen JM, Christensen NL, Lotan JE, Reiners WA (eds) Fire regimes and ecosystem properties. USDA For Serv, Gen Tech Rep WO-26 US Dep Agric For, Washington DC, pp 327–378
Peet RK (1981) Changes in biomass and production during secondary forest succession. In: West DC, Shugart HH, Botkin DB (eds) Forest succession: concepts and application. Springer, Berlin Heidelberg New York, pp 324–338
Peterman RM (1980) Influence of ecosystem structure and perturbation history on recovery processes. In: Cairns J (ed) The recovery process in damaged ecosystems. Ann Arbor Sci Inc, Mich, p 125

Reiners WA (1980) Nitrogen cycling in relation to ecosystem succession. In: Clark FE, Rosswall T (eds) Terrestrial nitrogen cycles. Processes, ecosystem strategies, and management impacts. Ecol Bull (Stockholm) 33:507–528
Shugart HH Jr, West DC (1980) Forest succession models. Bio Science 30:308–313
Southwood TRE (ed) (1968) Insect abundance. Symp R Entomol Soc No 4. Blackwell, Oxford
Swaby RJ (1966) Cultivation practices in relation to soil organic matter levels. In: Use of isotopes in soil organic matter studies. FAO/IAEA Tech Meet, Brunswick-Volkenrode, 1963. Spec Suppl Appl Radiat Isotopes. Pergamon Press, Oxford New York, p 21
Trudgill ST (1977) Soil and vegetation systems. Clarendon Press, Oxford
Vogl B (1980) The ecological factors that produce perturbation-dependent ecosystems. In: Cairns J (ed) The recovery process in damaged ecosystems. Ann Arbor Sci Inc, Ann Arbor Mich, p 63
White PS (1979) Pattern, process, and natural disturbance in vegetation. Bot Rev 45:229–299
Whittaker RH (1975) Communities and ecosystems, 2nd edn. MacMillan, New York
Woodwell GM, Whittaker RH (1968) Primary production in terrestrial ecosystems. Am Zool 8:19–30

3.4 Ecosystem Water Balance

R. LEE

3.4.1 Basic Concepts

Balancing water inflows and losses from an ecosystem involves the application of either mass or energy conservation principles. The quantitative methods, *hydrometric* (water budgeting) and *physical* (energy budgeting), may be used singly or in concert to describe water movements and phase changes that reflect interactions of physical and biological mechanisms. The relevant laws and formal relationships are conceptually adequate, and extremely useful in identifying and quantifying the hydrologic effects of human modifications of ecosystems, but observational problems impose some practical limitations.

3.4.1.1 Water Budgeting

Water budgeting is a simple accounting procedure that is based on mass conservation precepts; its is most useful in isolating and estimating quantities that cannot be directly assessed. The central idea is simply that the algebraic sum of all flow elements can be equated to zero when additions and losses to the ecosystem are given opposite signs. In general form,

$$\text{Inflow} + \text{Outflow} + \text{Storage change} = 0 \qquad (1)$$

where, by convention, the sum of inflow (+) and outflow (−) determines the sign of storage change. As illustrated in Fig. 1, storage increase (−) in a system is a loss to the immediate flow cycle, and storage decrease (+) is an addition.

Inflow to a terrestrial ecosystem may be predominately precipitation, but it also includes any other subsurface flows *into* the system. *Outflow* may consist entirely

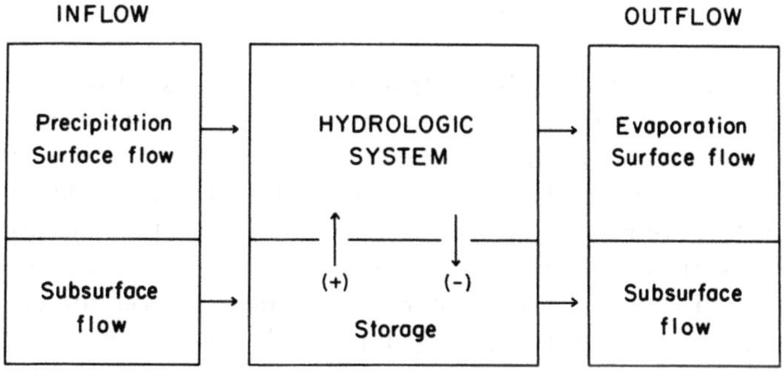

Fig. 1. Water budget components

Table 1. Mean annual water balance components for the earth. (After Baumgartner and Reichel 1975)

Item	Land	Ocean	Earth
Area (10^6 km^2)	148.9	361.1	510.0
Volume (10^3 km^3)			
Precipitation	111	385	496
Evaporation	− 71	− 425	−496
Discharge	− 40	40	0
Mean depth (mm)			
Precipitation	745	1066	973
Evaporation	−477	−1177	−973
Discharge	−269	111	0

of stream discharge and vapor flow (evaporation and transpiration), but it frequently includes other minor drainages *from* the system. Likewise, for particular purposes, it may be useful to identify the components of storage (soil moisture, groundwater, or snowpack storages).

Climatological (long-term average) water-budget equations for extensive hydrologic systems can be written most succinctly; over the total earth surface, for example, absolute values of precipitation inflow (P) and evaporation outflow (E) are equal, and

$$P+E=0 \tag{2}$$

since the storage change (S) is zero. The last column of Table 1 illustrates the numerical equality. In practice the sign convention is often tacitly ignored, positive and negative flow symbols are simply assembled on opposite sides of the equality sign, and P=E is understood as an equality of absolute values.

Climatological water-budget equations for entire land and water masses (Table 1) are only slightly more complex. With inflows to the left of the equality sign, average flow volumes are given by

$$P+Q=E \text{ (Oceans)} \tag{3}$$

and

$$P=Q+E \text{ (Land),} \tag{4}$$

where Q is the total liquid discharge from land to oceans (Q is also equal to the atmospheric vapor flow from oceans to land). Here it is also assumed that ocean and land storages are constant (i.e., S=O).

The long-term water budget for a small catchment or ecosystem is frequently written in the form of Eq. (4); the assumptions are that (1) precipitation is the only inflow (no subsurface leakage into the area), (2) stream discharge is an adequate measure of liquid water outflow (underflow and deep percolation are negligible), and (3) ecosystem storage is steady (subject only to random or seasonal fluctuations). None of these assumptions is universally true, and they are especially suspect as applied to small ecosystems. Yet, in carefully selected research areas, $E=P-Q$ [from Eq. (4)] is still the most accurate estimate of total evaporation loss.

The short-term water budget equation for a terrestrial ecosystem must include a storage term; in the form of Eq. (4),

$$P=Q+E+S, \tag{5}$$

where S is defined as a storage increase (representing a loss to the immediate flow cycle) and negative values are to the right of the equality sign. This means simply that if $P<(Q+E)$,

$$P+S=Q+E \tag{6}$$

in absolute terms, and water withdrawn from storage is an addition to the immediate flow cycle. Ordinarily the value of Q is taken as the observed stream discharge, and Eqs. (5) and (6) do not account for subsurface flows.

In controlled experiments where it is possible to measure subsurface flows, the water budget during periods of drying (storage decrease) is given as

$$P+Q_i+L_i+S=E+Q_o+L_o, \tag{7}$$

where Q_i and Q_o are surface inflow and outflow respectively, and L_i and L_o are the corresponding subsurface flows. Equation (7) is used most frequently to estimate total evaporation loss. During a drying period, with zero precipitation,

$$E=S+(Q_i-Q_o)+(L_i-L_o), \tag{8}$$

where Q_i-Q_o are net inflows; if, under these conditions, the soil surface is sealed to prevent direct evaporation, E is a measure of plant transpiration.

3.4.1.2 Energy Budgeting

Energy budgeting is a simple accounting procedure that is based on the energy conservation principle; in hydrology it is used to quantify water phase changes in terms of equivalent energy flows. As in water budgeting, the central idea is simply that the algebraic sum of all energy flows can be equated to zero when inputs and losses of energy from a surface element are given opposite signs. In general form the energy budget equation is analogous to Eq. (1), and the sum of energy inflow (+) and outflow (−) is equal to a heat storage change (±).

Energy transfer in an ecosystem occurs primarily as *net radiation* (R), *conduction* (B) in solids, *convection* (H) in moving fluids, and latent heat exchange (λE) associated with the vaporization of water. Metabolic energy (photosynthesis and respiration) which rarely exceeds one percent of R is ordinarily ignored in computations that are relevant to the water balance. Energy flow is usually expressed as a flux per unit area of surface (flux density), power/area, in langley/min (1 ly = 1 cal cm^{-2}) or $W\ m^{-2}$ (1 ly min^{-1} = 697 $W\ m^{-2}$).

The energy-budget equation for a terrestrial ecosystem (observing the sign convention) is

$$R+B+H+\lambda E=O \tag{9}$$

Table 2. Mean values of the components of water and energy budgets for land areas of the earth. (Adapted from Baumgartner and Reichel 1975)

Latitude (degrees)	Water budget ($mm\ yr^{-1}$)			Energy budget ($kly\ yr^{-1}$)		
	P	Q	E	λE	H	R
Polar (N)	176	−106	− 70	− 4	4	0
70–60	428	−227	− 201	−12	− 8	20
60–50	577	−259	− 318	−19	−11	30
50–40	535	−155	− 380	−22	−23	45
40–30	534	−122	− 412	−24	−36	60
30–20	611	−245	− 366	−21	−48	69
Tropics (N)	1,292	−436	− 856	−50	−22	72
Tropics (S)	1,576	−546	−1,030	−60	−13	73
20–30	564	− 88	− 476	−28	−42	70
30–40	660	−165	− 495	−29	−33	62
40–50	1,302	−914	− 388	−23	−18	41
50–60	993	−605	− 388	−23	− 8	31
60–70	429	−369	− 60	− 4	− 9	13
Polar (S)	148	−120	− 28	− 2	6	− 4

or, during periods of positive net radiation (daytime, or as an annual or seasonal average for other than polar regions) when, as a rule, B, H, and λE are negative,

$$R = B + H + \lambda E \tag{10}$$

is understood as an equality of absolute values. The conduction term (B) represents heat storage, analogous to the water storage term (S) in Eqs. (5)–(8), so that over an extended time period (with $B = O$),

$$R = H + \lambda E \tag{11}$$

or, in combination with the water balance of Eq. (4),

$$R = H + \lambda(P - Q), \tag{12}$$

where λ is the latent heat of vaporization (590 cal g^{-1} at mean earth temperature). The meridional distribution of the components of the water and energy budgets for land areas of the earth (mean values for latitude zones) is given in Table 2.

Net radiation (R), the dominant term in the energy budget, is the balance between incoming and outgoing fluxes of shortwave (solar) and longwave (atmospheric and terrestrial) radiation as illustrated in Fig. 2 for a forest ecosystem; symbolically

$$R = (1 - r)G + \varepsilon(C - \sigma T^4), \tag{13}$$

where G and C are incoming fluxes of shortwave and longwave radiation respectively, σ is a physical constant (for R in ly min^{-1}, $\sigma = 81.7.10^{-12}$), and r (reflectivity coefficient), ε (emissivity coefficient), and T (temperature) are ecosystem surface characteristics. Generally $\varepsilon \simeq 1$ for natural mineral and organic surfaces, but albedo

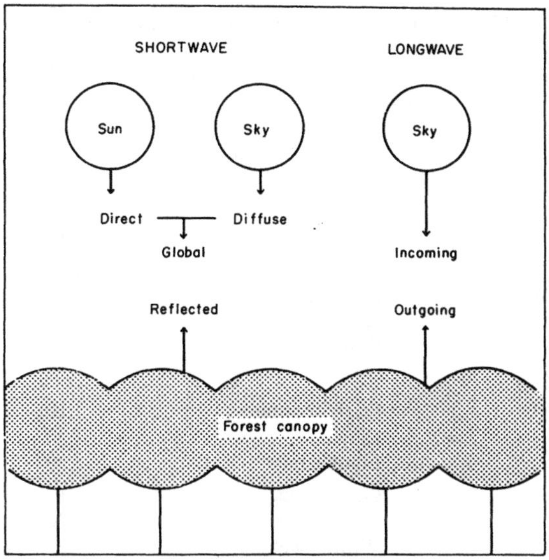

Fig. 2. Radiant energy fluxes over a forest canopy

(i.e., r expressed as a percentage) varies widely among ecosystem types, and net radiation generally increases as albedo decreases. Some typical albedo values for various surfaces are listed in Table 3.

3.4.1.3 Practical Limitations

The water- and energy-budgeting concepts have been applied experimentally to a wide variety of ecosystems in many parts of the world. The accumulating empirical data have been extremely useful in documenting ecosystem influences, as well as the gross effects of ecosystem modifications, and in promoting the continuing development of functional relationships and water balance theory. At the same time, several significant and practical limitations to precise understanding have become apparent.

Precipitation measurement accuracy is a matter of continuing concern; the raw data obtained with standard instruments should be interpreted realistically as indices rather than true quantitative measures of water input to ecosystems. The basic technique of rain gaging has not changed appreciably for hundreds of years; the standard gage is a cylindrical open-topped vessel that modifies the windfield in its immediate vicinity, causing a deficiency in catch that increases with windspeed. The estimated measurement errors, given in Table 4, are in general agreement with the empirical data.

Precipitation measurement errors are especially important in water budgeting; for example, when solving for evaporation (E) as a residual in Eq. (4), the relative error increases with the ratio P/E (if $P/E = 2$, a 10% error in P represents a 20% error in the estimate of E). The errors can be reduced to some extent by applying

Table 3. Typical albedoes for some natural surfaces

Surface	Description; conditions	Albedo
Water		
Liquid	Solar altitude: 60°	5
	30°	10
	20°	15
	10°	35
	5°	60
Solid	Fresh snow	75
	Old snow	50
	Glacier ice	30
Ground		
Soil	Dark organic	10
	Clay	20
	Light sandy	30
Sand	Gray: wet	10
	dry	20
	White: wet	25
	dry	35
Rock	Sandstone spoil, dry	20
	Black coal spoil, dry	5
Vegetation		
Grass	Typical field: green	20
	dry	30
	Dry steppe	25
	Tundra, heather	15
Crops	Cereals, tobacco	25
	Cotton, potato	20
	Sugar cane	15
Trees	Rain forest	15
	Eucalyptus	20
	Red pine forest	10
	Hardwoods, in leaf	18

Table 4. Estimated rainfall catch deficiencies as percentage of true rainfall. (After Lee 1980)

Rainfall intensity ($mm\ h^{-1}$)	Windspeed ($m\ s^{-1}$) 2	4	6	8	10
1	6	18	31	42	51
2	5	17	29	40	48
5	4	15	26	37	45
10	4	13	24	34	43
20	3	12	22	32	40
50	3	11	20	30	38
100	3	10	19	28	36
Mean	4	14	24	35	43

corrections based on the data in Table 4, or by the use of mechanical wind shields around gage orifices. But in addition to measurement errors per se, the extreme local and microscale variability of rain and snow at ground level introduces a profound degree of uncertainty with regard to areal interpretations; even in an intensive network of gages, the area sampled usually exceeds the total collecting area of the gages by a factor of from 10^6 to 10^9.

The measurement of total liquid water discharges from ecosystems is also problematical. Streamflow measurements are usually given an "excellent" or "good" rating if 95% of the discharges are expected to be within 5% or 10% of the true values (in carefully controlled experimental areas somewhat better accuracy may be attainable). In most instances there is no satisfactory method of measuring or estimating subsurface flows, and the uncertainty increases if surface and subsurface discharge areas differ.

Total vaporization losses from ecosystems cannot be measured directly, but must be estimated as the residual in either the water- or energy-budget equations, or obtained independently by aerodynamic methods; not uncommonly, as a last resort, loosely empirical or quasi-rational formulations are applied. Of these, the water-budgeting approach, with an average accuracy of perhaps 10% under ideal conditions, has undoubtedly provided the most accurate and abundant information to date; energy-budget, aerodynamic, and combination approaches, fraught with even greater sampling and logistical problems, have served primarily in a supportive role. For the present purpose, Budyko's (1974) "radiative index" will be useful in characterizing ecosystem influences:

$$E = P - P \exp(-R/\lambda P) = P(1 - e^{-I}), \tag{14}$$

where the index (I) is the ratio of net radiation (R) to the latent heat of precipitation (λP), both taken as annual averages, and e is the natural logarithm base.

3.4.2 Ecosystem Influences

"Ecosystem influences" is used in an attempt to identify the independent influences of what are, in fact, various combinations of interdependent phenomena; the goal is practically unattainable, and we can end with no more than a descriptive model that may satisfy a transigent intellectual need for a simple definitive explanation. The ecosystem is by definition an interaction, and where the interaction can be qualified (e.g., with the atmospheric processes that generate precipitation), it may be possible to set quantitative limits on its relative importance. Otherwise, ecosystem water balance components are a function of infinitely variable combinations of physical and biological entities that can be identified for practical purposes only in terms of dominant characteristics.

3.4.2.1 Gross Precipitations

The natural occurrence of various ecosystem types generally follows a recognizable sequence with regard to precipitation depth (e.g., P increasing with the se-

Table 5. Precipitation sources. (Adapted from Sellers 1965)

Region	Area (10^6 km²)	P (mm yr^{-1})	P_v (mm yr^{-1})	P_v/P (%)
Arizona	0.1	314	295	94
European Russia	4.9	487	433	89
U.S. and Canada	17.8	675	493	73

P_v is precipitation from advected vapor

quence: desert→grassland→forest); quantitatively, according to Petterssen (1969), the dividing line between geographic zones of steppe and forest climate is:

$$P_m = 20T_a = 140, \qquad (15)$$

where precipitation depth (P_m in mm yr^{-1}) is the minimum requirement for tree growth (assuming that it is not strongly seasonal), and T_a (in °C) is mean annual air temperature. Similar patterns have also been observed with regard to total water use (evaporation and transpiration) by ecosystem types, and there has been some confusion with regard to causal relationships. It is frequently alleged, for example, that the process is reversible, or that since forests use more water than other cover types, precipitation *must* be greater in order to satisfy the atmospheric water balance, P=E, from Eq. (2).

Taken on a global scale, the argument that P=E is irrefutable, yet is appears almost meaningless when the relative magnitude of the terrestrial effect is seen; proper perspective can be achieved by noting from Table 1, that evaporation from *all* terrestrial ecosystems amounts to only 14% of the total (and therefore contributes only 14% to total precipitation). Furthermore, the argument for global equality of precipitation and evaporation clearly does not apply to individual land segments, except perhaps in some tropical areas where internal circulations are dominant; water evaporated from a particular ecosystem is mixed by atmospheric turbulence and moved downwind, usually at the rate of several hundred kilometers per day. The data of Table 5 show that the contribution of evaporation to precipitation in the same general area is small, ranging from 6% to 27% for extensive land masses, and decreasing as the area of the region decreases; consequently any increases in local precipitation that are attributable to evaporation from smaller ecosystems must be immeasurably small.

Other arguments in support of an ecosystem influence on gross precipitation have been developed, and may have greater validity, especially for taller vegetation and in certain local areas. Forests, for example, tend to add to the effective elevation of upland areas, and it is conceivable that total precipitation increases with forest height as a direct result of the orographic effect; theoretical considerations suggest, however, that the maximum influence could not exceed 1%–2%. In coastal or mountainous areas where fog moves horizontally into a forest canopy, fog droplets are deposited by contact on the foliage, and the total accumulation of this "occult" precipitation at the leading edge may constitute a significant fraction of the annual input of water to the ecosystem.

Table 6. Precipitation in various ecosystem types, observed and adjusted for windspeed. (Lee 1980, based on data of Hursh 1948)

Ecosystem zone	Mean windspeed ($cm\ s^{-1}$)	Annual precipitation (mm)	
		Observed	Adjusted
Denuded	226	1,277	1,458
Grass	167	1,339	1,462
Forest	33	1,459	1,466

Numerous studies have been undertaken to find a definitive answer to the problem, but inherent difficulties and observational limitations have led to ambiguous interpretations. By way of example, Hursh (1948) selected an area where 2,800 ha of forest land had been denuded by smelter fumes, and measured rainfall and windspeed over a 4-year period at stations in the denuded area, in an intermediate grassed area, and in the surrounding forest; the data in Table 6 show that forest rainfall exceeded that in the denuded area by 14%, and in the grassed area by 9%. The data also show that rainfall depth was inversely related to windspeed (the latter being reduced by surrounding vegetation); when the rainfall data were adjusted for windspeed measurement errors (based on the method of Table 4), the residual ecosystem influence was found to be negligible.

3.4.2.2 Evaporation Losses

Ecosystems, identified in terms of dominant vegetation types, occur in geographic zones under relatively uniform regimes of water and energy exchange. The observed uniformity is, of course, a common basis of climatic classifications, and it has frequently been used as a point of departure in the development of more quantitative descriptions of surface hydrologic processes. As an example, Table 7, adapted from Budyko (1974), gives the evaporation coefficients (E/P):

$$E/P = 1 - \exp(-R/\lambda P) = 1 - e^{-I} \qquad (16)$$

from Eq. (14) for broad ecosystem types, where for constant normal precipitation (P) the ecosystem influence varies as a function of net radiation (R).

Numerous theoretical and empirical models have been developed and used to estimate evaporation losses from land segments. The more rational formulations suffer, in application, from a variety of practical and logistical problems that invalidate much of their apparent sophistication and precision; empirical models, generally based on relatively crude correlations with climatological data, are conceptually less impressive but perhaps no less serviceable. One of the most important unsolved problems in ecosystem hydrology is that of quantifying specific biological influences on latent heat exchanges, a problem that has frequently been obscured by resorting to a "potential evaporation" concept.

Lee (1980) developed arguments to the effect that "potential evaporation" has no definition that is useful, either theoretically or practically, beyond the idea that

Table 7. Evaporation coefficients (E/P) as a function of the radiative index (R/λP) for the range of broad ecosystem types. (After Budyko 1974)

Radiative index	E/P (%)	R<50	Net radiation (R, kly yr^{-1}) $50 \leqq R \leqq 75$	R>75
0.0–0.2	0– 18	Arctic desert	–	–
0.2–0.4	18– 33	Tundra (with sparse forest in south)	Subtropical forest, marshland	Equatorial forest swamps
0.4–0.6	33– 45	Northern and middle taiga	Pluvial	Heavily swamped equatorial forest
0.6–0.8	45– 55	Southern taiga and mixed forest	Subtropical forest	Swamped equatorial forest
0.8–1.0	55– 63	Deciduos forest and forest steppe	Subtropical forest and shrub	Equatorial forest, and forest savanna
1.0–2.0	63– 86	Steppe	Subtropical steppe	Dry savanna
2.0–3.0	86– 95	Semidesert of middle latitudes	Subtropical semidesert	Desert savanna, tropical semidesert
3.0–	95–100	Desert of middle latitudes	Subtropical desert	Tropical desert

when a surface is wet, evaporation will occur at a rate determined by local weather and the external characteristics of the surface; but this is true even if the surface is not wet, and the concept adds nothing to the solution of the problem. Certainly the formal relationships may be solved for a range of atmospheric and ecosystem conditions, alternately assuming that the biological effect is zero or some positive number, and the results may be instructive with regard to relative biological effects in various hypothetical situations. At the same time, however, it makes no sense to assume that the results of such an endeavor have contributed to an understanding of actual ecosystem water losses.

3.4.2.3 Discharge Losses

In a given climate, with fixed normal precipitation, liquid water discharges from ecosystems must satisfy Eq. (4); this means that any ecosystem difference in evaporation (ΔE) must be equal but opposite in sign to the discharge difference (ΔQ), or

$$\Delta Q = -\Delta E \tag{17}$$

and that the discharge coefficient (Q/P) can be given as

$$Q/P = 1 - (E/P) = e^{-1} \tag{18}$$

in terms of the evaporation coefficient [E/P, Eq. (16)]. Also, since E generally varies directly with the energy load or net radiation (R) at a place,

$$\Delta Q = -\Delta E = -f(\Delta R) \tag{19}$$

may be assumed as a first approximation. Specifically,

$$dQ/dI = -Pe^{-I} = P\exp(-R/\lambda P) \tag{20}$$

or

$$dQ = -Q\,dI \tag{21}$$

from Eqs. (14) and (18), where R expresses the ecosystem influence.

For any finite difference in R (given constant P) the corresponding difference in Q is given by

$$\Delta Q/Q = e^{-\Delta I} - 1 \tag{22}$$

for a given change in I ($=R/\lambda P$). In a given climate with normal global radiation (G), a conservative estimate of ΔR is

$$\Delta R = -\Delta r(G), \tag{23}$$

where r is the reflectivity coefficient, so that

$$\Delta I = r(G/\lambda P) \tag{24}$$

and

$$\Delta Q/Q = 1 - \exp[-\Delta r(G/\lambda P)] \tag{25}$$

by substitution in Eq. (22). In simpler form, letting $k = G/\lambda P$,

$$\Delta Q = Q(1 - e^{-k\Delta r}) \tag{26}$$

and

$$\Delta Q \simeq Qk\Delta r \simeq Q\Delta R/\lambda P \tag{27}$$

for small values of Δr and ΔR; in the eastern United States, for example, where $k \simeq 2$ for most regions, a difference in reflectivity of one percentage point ($\Delta r = 0.01$) denotes a 2% difference in discharge.

Application of these formal relationships yields, at least, a first approximation of the differences in normal annual discharge among ecosystem types; an example, using Eq. 25 and regional averages of G, P, Q, and average reflectivities for ecosystems in the eastern United States, is given in Table 8. It is important to recognize, of course, that ecosystem reflectivities vary with incidence angle (time of day, season, and latitude). Estimated average growing season albedoes for the ten major timber types of this region are given in Table 8 (grassland albedo was assumed to be 20%, but with greater seasonal and latitudinal variability).

With respect to forest types, considerable evidence has accumulated in support of the idea that coniferous forests yield less water to streamflow than do hardwood types in the same climate. The conservative data of Table 8 imply that the difference may be of the order of 50–100 mm yr^{-1}, but other estimates range from zero to more than 300 mm yr^{-1} (Anderson et al. 1976). In a most carefully controlled experiment, Swank and Douglass (1974) found that when two catchments were converted from mixed hardwoods to white pine, the pine cover discharge, after 15 years, was 200 mm yr^{-1} less than that of the original hardwood cover.

Table 8. Differences in normal annual water yield (grassland minus forest)

Forest type	Albedo (%)	Yield difference (mm yr^{-1})	
		Mean	Range
Coniferous			
Spruce-fir	10	167	86–196
White-red-jack pine	12	124	77–161
Loblolly-shortleaf pine	12	90	59–140
Longleaf-slash pine	12	74	59–112
Deciduous			
Oak-pine	15	59	37–101
Oak-gum cypress	15	52	37– 76
Elm-ash-cottonwood	17	49	24– 79
Oak-hickory	18	34	17– 62
Maple-beech-birch	19	33	17– 47
Aspen-birch	20	18	10– 33

Most of the above analysis, based on albedo differences among ecosystem types, excludes any appraisal of other physical and biological variables, for example, topographic exposure and ecosystem physiology and phenology; these and other influences introduce a complexity that is poorly defined in the context of contemporary theory. Additional understanding has been achieved in studies of individual hydrologic processes, and by development of computerized models that integrate the effects of numerous ecosystem components. Yet the natural integrator is the ecosystem itself, and water balance studies of these systems have produced the most reliable data, especially with regard to the effects induced by human activities.

3.4.3 Human Influences

Human influences on the water balance of terrestrial ecosystems are most apparent when major disturbances occur in taller climax vegetation types; as a result, disruptive effects on mature forest ecosystems have been most studied. It has been demonstrated repeatedly, by comparing discharges from adjacent forested and non-forested ecosystems, and by noting the effects of deforestation, reforestation, and afforestation, that the presence of forest cover is associated with reduced annual discharge, including annual volumes of both quickflow and delayed flow, and with characteristic seasonal flow regimes. The effects of other disruptions, for example grazing, mining, recreational activities, wildfire, and phreatophyte removal, have also been documented, if less comprehensively.

3.4.3.1 Major Disturbances

Discharge increases following forest removal have been obtained from numerous experiments under a wide variety of conditions; a selected list of results is given

Table 9. First-year water yield increases following forest cutting. (Anderson et al. 1976)

Location (Forest type)	Fraction Removed (F)	Q ($mm\ yr^{-1}$)	ΔQ_1 ($mm\ yr^{-1}$)	$\Delta Q_1/QF$ (%)
North Carolina				
(Mixed hardwoods)	1.00	792	370	47
	1.00	607	127	21
	0.50	1,275	198	31
	0.22	1,222	99	37
West Virginia				
(Mixed hardwoods)	0.85	584	130	26
	0.36	660	64	27
	0.22	762	36	21
	0.14	635	8	9
Colorado				
(Aspen, Conifer)	1.00	157	34	22
	0.40	283	86	76
Oregon				
(Douglas-fir)	1.00	1,448	462	32
	0.30	1,448	150	35

in Table 9. The increases tend to be greater from areas where normal discharge is greater, and from less intensively insulated topographic facets in the same general climate. Complete deforestation generally increases annual water yield by from 20% to 40% of normal (last column, Table 9), but maximum increases rarely exceed 400 $mm\ yr^{-1}$.

The confirmed general rules are that (1) forest removal increases total liquid discharge, (2) the increases are greater when a greater fraction of the forest is removed, (3) maximum increases occur the first year after removal, (4) effects decrease logarithmically with time, and (5) effects persist longer when the initial increase is greater. Douglass and Swank (1975) attempted to quantify these rules for hardwood forests of eastern United States; the first-year increase (ΔQ_1 in $mm\ yr^{-1}$) was estimated as

$$\Delta Q_1 = 44(F/G_e)^{1.45}, \tag{28}$$

where F is the fraction of forest removed, and G_e (in 10^6 ly yr^{-1}; ly = langley = gram calorie per square centimeter) is extraterrestrial solar radiation (Table 10). Solutions to Eq. (28) for the common ranges of F and G_e are given in Table 11.

As the forest begins to regenerate, the removal effect diminishes. The discharge increase (ΔQ_i) for any year (i) can be estimated as

$$\Delta Q_i = \Delta Q_1 - \Delta Q_1(\ln i/\ln n), \tag{29}$$

where n is the persistence period in years. The influence of forest removal persists until i approaches its limiting value (n) defined by

$$n = i(\text{maximum}) = \Delta Q_1/16, \tag{30}$$

where n is an integer [rounded from Eq. (30)].

Table 10. Extraterrestrial solar radiation on slopes (in 10^6 ly yr^{-1})

Slope inclination (%)	Slope aspect				
	N	NE-NW	E-W	SE-EW	S
Latitude 30° N					
10	0.267	0.271	0.282	0.291	0.295
20	0.249	0.259	0.280	0.298	0.305
30	0.230	0.246	0.278	0.303	0.310
40	0.211	0.233	0.275	0.306	0.317
50	0.192	0.220	0.271	0.307	0.319
Latitude 40° N					
10	0.235	0.240	0.253	0.266	0.271
20	0.214	0.226	0.252	0.276	0.285
30	0.193	0.212	0.251	0.284	0.296
40	0.172	0.198	0.250	0.290	0.305
50	0.153	0.186	0.248	0.294	0.310
Latitude 50° N					
10	0.198	0.205	0.220	0.234	0.239
20	0.176	0.190	0.220	0.247	0.257
30	0.156	0.176	0.220	0.258	0.272
40	0.138	0.163	0.220	0.267	0.284
50	0.123	0.152	0.220	0.274	0.293

Table 11. First-year water yield increases (ΔQ_1) in mm yr^{-1} based on Eq. (28)

G_e (10^6 ly yr^{-1})	Fraction of forest removed (F)				
	0.2	0.4	0.6	0.8	1.0
0.18	51	140	252	383	529
0.20	44	120	216	328	454
0.22	38	105	188	286	395
0.24	34	92	166	252	348
0.26	30	82	148	225	310
0.28	27	74	133	202	279
0.30	24	67	120	182	252
0.32	22	61	109	166	230
0.34	20	56	100	152	210

The total yield increase (ΔQ_t) integrated over the persistence period is frequently of special interest. The sum is given by

$$\Delta Q_t = \sum_i^n \Delta Q_i = \sum_i^n \Delta Q_1 - \Delta Q_1 (\ln i / \ln n) \tag{31}$$

but since, for consecutive integers,

$$\sum_i^n \ln i = \ln n\,! \tag{32}$$

Table 12. Persistence periods and total water yield increases based on first-year increases

Item	First-year increase (mm yr^{-1})				
	50	100	200	300	400
Persistence period (n, yrs)	3	6	12	18	25
Total increase ΔQ_t (mm)	80	235	270	1,605	2,840
Average increase $\Delta Q_t/n$ (mm yr^{-1})	27	39	64	89	114

(Stirling's formula), the summation is

$$\Delta Q_t = n\Delta Q_1 - \Delta Q_1(\ln n!/\ln n) \quad (33)$$

and this reduces to the approximation,

$$\Delta Q_t \simeq \Delta Q_1(0.25n + 0.85) \quad (34)$$

which is accurate to about 2% over the common range. Solutions to Eqs. (30) and (33) are given in Table 12 along with the average yield increases for persistence periods.

Contemporary theory and empirical knowledge are insufficient for the task of predicting exact quantitative effects of ecosystem disturbances on water yield. The formulations given in this section and in Section 3.4.2.3 can be used as first approximations when more precise local information is unavailable. In doing so, however, it is important to note that the overall annual effects may obscure more or less dramatic seasonal influences on the flow regime.

3.4.3.2 Flow Regimes

Seasonal patterns of streamflow are determined largely by regional climate; ecosystem influences are secondary, superimposed upon the general flow regime. The idea of streamflow "regulation" by ecosystems, in the sense of control of unfavorable extremes, has been overemphasized. The ecosystem influence is real nonetheless, and though its role may be less dramatic than sometimes supposed, objective knowledge of its workings can yet be put to beneficial use.

In general, the ecosystem influence on streamflow is greater following the season of greater evaporation; Douglass and Swank (1972) found that when a hardwood forest was cut, about 60% of the streamflow increase came during the low-flow period (the pattern illustrated in Fig. 3 is typical in the eastern United States where most precipitation occurs as rain). The general rule is that any ecosystem change that results in a water yield increase will have a greater effect during the low-flow periods; an important exception occurs where snowmelt discharge dominates the flow regime (Fig. 4). When ecosystem changes reduce flow, for example when hardwood forests are converted to pine, water yield reductions occur primarily during the dormant season, evidently as the result of greater interception losses by the coniferous type (Swank and Miner 1968).

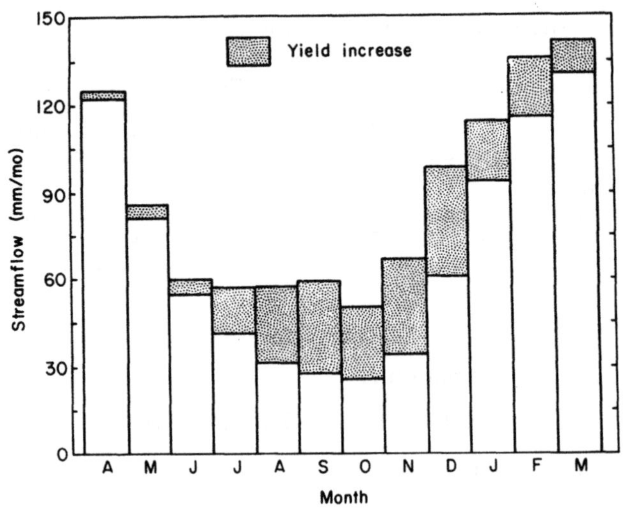

Fig. 3. Seasonal streamflow increases following forest cutting. (Adapted from Douglass and Swank 1972)

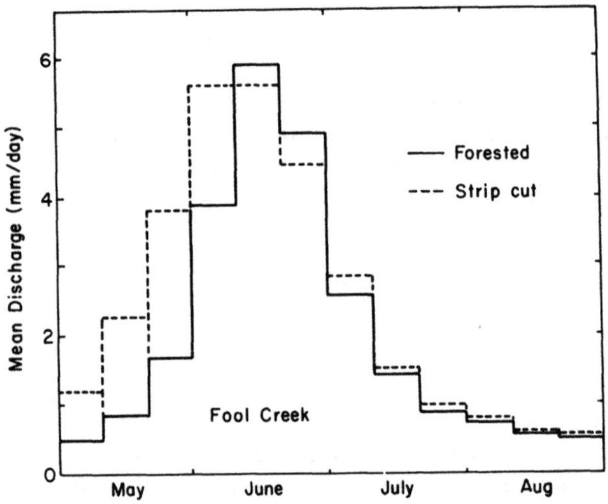

Fig. 4. Discharge rates during the snowmelt season before and after cutting Rocky Mountain conifers. (Adapted from Leaf 1975)

Clearcutting coniferous stands (versus hardwoods) usually produces greater discharge increases, and the increases occur more uniformly by seasons because the reductions in both summer transpiration and winter interception are reflected in the flow regime. None of the ecosystem influences on streamflow is independent of climate, however, and both precipitation type and its seasonal distribution must be considered. In western Oregon, for example, where most precipitation occurs during the winter half-year, forest removal may cause especially high flows during fall months when soils on cutover areas become saturated earlier than those in the forest (Haar 1976).

The popular notion that major ecosystems tend "to retard and lower flood crests and prolong increased flow in low-water periods" (Kittredge 1948) was still in vogue until the middle of the current century. It is true, to a degree, that forests lower flood crests, but the statement requires careful interpretation; some of the largest floods on record have occurred in forested drainages. The attractive notion that the existence of forest cover will prolong increased flow during low-flow periods is clearly false.

3.4.3.3 Miscellaneous Influences

Ecosystems are used by humans for a variety of purposes; timber, wildlife, forage, and minerals are harvested along with water, while simultaneously the land is used for recreation and aesthetic appreciation. Human use inevitably leads to ecosystem modifications that affect the water balance. With few exceptions, the initial influence of human disturbance of virgin cover is such that total evaporation is decreased, and liquid water discharge increased by the same amount; also, as a rule, peak discharges are greater, and the seasonal flow regime is modified.

The hydrologic effects of human modifications of ecosystems depend on both the character of the system, including its climate, soil, geology, topography, and vegetation, and on the nature and severity of the disturbance. Minor disturbances, for example moderate recreational use or the grazing of domestic livestock, may have negligible effects on the water balance as such, although accelerated erosion and organic wastes can affect the quality of discharge water. On the other hand, human-caused wildfire, for example in a forest ecosystem, can decrease infiltration capacities and increase overland flow, peak discharge, and water yield to an even greater extent than that which follows complete forest removal as a planned modification.

Ecosystem modifications are often deliberate, undertaken for the purpose of increasing water yields; timber harvesting on municipal watersheds is a case in point, as is the removal of phreatophytes from streambanks in arid and semiarid regions, where reduced evaporation and increased discharge are exceptionally great (as much as 800 mm yr^{-1}, according to Anderson et al. 1976). More commonly, however, hydrologic effects of human activities are fortuitous. Surface coal mining alone accounts for more than 500 km^2 yr^{-1} of forest land disturbance in the United States; in this case, water yield increases must be at least as great as those associated with forest removal, but changes in flow extremes and discharge quality may be much greater, and undoubtedly the effects are more persistent.

3.4.4 Conclusions

The water balance of an ecosystem is amenable to interpretation in terms of physical laws, functional relationships, and quantifiable characteristics of the biosphere. This mass balance is, moreover, an integral part of the energy balance, such that simultaneous solutions are mutally correctional, both with regard to theoretical considerations and the evaluation of problematic empirical data. Whereas, at present, quantitative exactitude is wanting, the nature, direction, and relative magnitudes of biological influences have been extensively documented.

Résumé

En règle générale, les bilans hydriques sont fondés sur le principe de la conservation de la matière; ils se traduisent par des équations exprimant que la variation de la réserve du sol en eau est égale à la somme algébrique des entrées et des sorties. L'hydrologie s'intéresse aussi aux bilans énergétiques, puisque l'évapotranspiration dépend directement du rayonnement solaire.

L'auteur analyse les formules qui expriment ces bilans, et souligne l'incertitude des mesures correspondantes. Il présente des tableaux qui montrent comment les éléments de ces bilans varient à l'échelle du globe, ou à l'échelle du paysage. En particulier, le Tableau 7 classe les principaux types de végétation en fonction du quotient E/P (évoporation réelle/précipitations) et de la radiation nette, R.

Le Tableau 8 montre que, sous le climat de l'Est des Etats-Unis, l'albedo de l'espèce dominante détermine très directement la quantité d'eau qui peut sortir de l'écosystème; ce tableau confirme que les résineux (et en particulier l'Epicea) à faible albedo retiennent moins d'eau que les feuillus.

Quand l'homme déforeste un bassin versant, la quantité d'eau qui en sort augmente sensiblement (Tableau 9); inversement, les reboisements retiennent l'eau.

References

Anderson HW, Hoover MD, Reinhart KG (1976) Forests and water. For Serv Tech Rep PSW-18. US Dep Agric, Berkeley Ca

Baumgartner A, Reichel E (1975) The world water balance (in German and English). Oldenbourg, Munich

Budyko MI (1974) Climate and life. Academic Press, London New York

Douglass JE, Swank WT (1972) Streamflow modification through management of eastern forests. For Serv Res Pap SE-94. US Dep Agric, Asheville NC

Douglas JE, Swank WT (1975) Effects of management practices on water quality and quantity. In: Proc Municipal Watershed Symp. For Serv Tech Rep NE-13. US Dep Agric, Upper Darby Pa

Haar RD (1976) Forest practices and streamflow in western Oregon. For Serv Tech Rep PNW-49. US Dep Agric, Portland Oreg

Hursh CR (1948) Local climate in the copper basin of Tennessee as modified by the removal of vegetation. US Dep Agric Circ 774, Washington DC

Kittredge J (1948) Forest influences. McGraw-Hill, New York

Leaf CF (1975) Watershed management in the Rocky Mountain subalpine zone: the status of our knowledge. For Serv Res Pap RM-137. US Dep Agric, Fort Collins Colo

Lee R (1980) Forest hydrology. Columbia Univ Press, New York

Petterssen S (1969) Introduction to meteorology. McGraw-Hill, New York

Sellers WD (1965) Physical climatology. Univ Chicago Press, Chicago

Swank WT, Douglass JE (1974) Streamflow greatly reduced by converting deciduous hardwood stands to pine. Science 185:857–859

Swank WT, Miner NH (1968) Conversion of hardwood covered watersheds to white pine reduces water yield. Water Resour Res 4:947–954

3.5 Some Problems of Disturbance on the Nutrient Cycling in Ecosystems

M. RAPP

3.5.1 Introduction

Among the numerous parameters, structural as well as functional, characterizing ecosystems, the distribution and the fluxes of nutrients are among the most central.

These parameters can be summarized by a series of compartments of immobilization and accumulation of nutrients, such as the biomass, the litter layer, or the nutrients included in the soil and available to the vegetation. These compartments are joined together by fluxes, such as the uptake of minerals by the plants, and their restitution to the soil by litter fall and subsequently litter decay. At the same time the plant-soil system is continually affected by the abiotic environment.

These pools and fluxes are generally not in a steady state, except in climax vegetation. Even in this case, the relationships between the different levels of organization of the nutrient cycle are very fragile, and the slightest abiotic, biotic or human disturbance can change them very easily.

The modifications introduced by humans into "natural ecosystems," such as forests, are very frequent and numerous. This is the case for many silvicultural practices, such as the reforestation by coniferous species in the place of deciduous ones, the use of fertilizers, and more intensive forest harvesting methods. Burning of vegetation, used to improve pasture in mediterranean and tropical countries, strongly modifies the nutrient budget of the ecosystem and produces new communities, most of them less rich in comparison with the original vegetation.

The turnover of nutrients is not only disturbed by these direct influences, but indirect effects of human activities can also cause changes, which are not always negative. For example, in industrially-developed countries, more and more minerals are given off to the lower atmosphere, and their effects are numerous and increasing.

The increase of nitrogen in the atmosphere could be a source of this element to ecosystems. Sulfur, an anion in "acid rain," could be an important factor modifying the mineral cycling processes. The recycling of nuclear fallout after nuclear testings or thermo-nuclear accidents, are also important indirect effects due to human activities.

Here we do not discuss these indirect disturbances to nutrient cycling in ecosystems. Rather we only report on studies of direct human intervention and disturbance on the mineral budget equilibrium and balance. Several problems will be examined including the role of fire in maintaining the structure of vegetation and the effect of forest management on the modification of the biogeochemical cycles after reforestation or fertilization.

3.5.2 Fire

Although fires interfere with nutrient cycling, if their effects are limited in intensity and frequency, the ecosystem tends to restore to the initial nutrient cycling condition. On the other hand, frequent, especially yearly, fires have large effects on nutrient cycling. This occurs especially in the mediterranean and tropical zones where managers attempt to maintain the best structure of the vegetation for pasture. With each fire, some elements are lost from an ecosystem, by run-off and wind erosion especially, and the available pool of nutrients decreases.

Villecourt et al. (1979) studied this problem in the Lamto savanna, in the Ivory Coast. Each year, that savanna is burned at the end of the dry season, during the second half of the month of January. The pool of N, P, and K, included in the aerial biomass at the beginning of January, amounted to 1.1, 0.2, and 1.21 g m^{-2} respectively. In February, this had fallen to 0.14, 0.02, and 0.16 g m^{-2}.

As the minerals in the roots did not decrease significantly, with respective amounts of N, P, and K of 2.47, 0.31, and 1.65 g m^{-2} in January and 2.84, 0.26, and 1.38 g m^{-2} in February, the losses of the aerial parts could be considered as losses for the whole ecosystem. Some authors, such as Leon (1968) and de Rham (1971), suggest that part of these nutrients were translocated before the beginning of the fire. This does not seem to be the case for the Lamto savanna. The decrease of nutrients in the above-ground biomass which is observed during September and November is not correlated with an increase of the same elements in the below-ground biomass.

Villecourt and Roose (1978) indicate that for the 1.19 g m^{-2} of nitrogen, 0.3 g of phosphorus and 1.21 g of potassium lost by the vegetation during fire, 0.11, 0.23, and 0.65 g respectively continue to exist in the ecosystem as ash on the top of the soil whereas 0.99, 0.07, and 0.55 g are lost from the system by subsequent erosion by wind or by volatilization of nitrogen.

At Lamto, the losses of some of these elements are compensated by the fixation of gaseous nitrogen and by input by rain. Nevertheless, there is a loss of potassium from that savanna which may be increased if leaching occurs in the soil. If this element is excluded, it is possible that yearly fires could continue for a long time without affecting the primary production and the mineral budget and the mineral availability of these ecosystems.

3.5.3 Reforestation by Conifers

During the last decades, plantation of coniferous species have been enlarged by replacing deciduous forests or natural shrublands.

Although this has been an economically favorable substitution, the ecological consequences, even though they are not yet fully known, seem to be unfavorable for the soil.

The vegetation replacement is accompanied by an acidification of the soil, a reduction in the cation exchange capacity and of the pool of exchangeable cations, and the transformation of a mull or moder humus to a moder or mor humus type. This development reduces the pool of mineral elements in the humus layer, Cole

Table 1. Nitrogen partitioning (g m^{-2}) within two forest ecosystems (Nihlgard 1972)

	Fagus sp.	*Picea abies*
Nitrogen in aerial biomass	107.1	72.0
Nitrogen in litter on the soil surface	8.6	24.5
Nitrogen in the soil	780.0	690.0
Nitrogen fixed annually in the biomass	12.9	6.7
Nitrogen returned annually to soil with litter fall	7.0	6.6

and Rapp (1980) indicate that this mostly affects elements such as nitrogen, potassium, phosphorus, calcium, and magnesium.

The longer life (several years) of the coniferous needles, opposed to the yearly fall of leaves of deciduous trees, slows the cycling of nutrients through the ecosystem. This is balanced by a lower consumption of nutrients, since coniferous species are more nutrient frugal then deciduous ones. Comparing a beech and a *Picea abies* forest, Nihlgard (1972) indicated a very small immobilization of nitrogen in the biomass of the coniferous forest (Table 1).

The evergreen morphology of the needles, and the structure of the canopy of the coniferous stands also results in the more efficient capture of the atmospheric aerosols (Ulrich et al. 1973), and this increases the nutrient input into the biological cycle.

The differences in nutrient cycling between coniferous and broad-leaved forests have not been well studied (Nihlgard 1972; Ulrich in Cole and Rapp 1980; Cole et al. 1978). Most of these scientists focused their attention on the influence of changing the forest type on pedogenesis and the availability of nutrients in the soil.

Bonneau et al. (1979) did not observe any major differences in a comparison of a mixed *Quercus*, *Fagus*, and *Carpinus* forest with a *Pinus sylvestris* plantation. However, a comparison between a *Quercus* and *Castanea* forest and a *Pinus laricio* var. *corsicana* stand indicated an increasing podzolisation in the latter type, without a change in the pool of exchangeable cations. The podzolisation produces nevertheless important losses of total K_2O and Na_2O. The soil losses were respectively 3.3 and 1.28 kg m^{-2} under the pine stand, against only 2.3 and 0.86 kg m^{-2} under the broadleaved trees.

Nyss (1977) compared a heath vegetation of *Calluna*, *Genista*, and *Deschampsia* to a neighboring 40-year-old *Picea abies* plantation on the same soil. He found annual losses of:

1.64 kg m^{-2}	of K_2O	or 4%	of the reserve found in the heath
0.01	CaO	0.4%	
0.06	MgO	1.9%	
0.02	MnO	9%	

Similar results were found for exchangeable cations:

	K	Ca	Mg	Mn g m^{-2}
Exchangeable cations in the heath soil	21.2	28.0	26.8	0.23
Exchangeable cations in the *Picea* soil	25.1	17.1	19.9	0.48
Difference: *Picea abies* – heath	+ 3.9	−10.9	− 6.9	+0.26

These differences resulted from leaching from the rooting zone and pedogenetic processes as well as from a reorganization of the nutrient pathways between the different compartments of the ecosystem. The *Picea abies* plantation included in its biomass 41.7, 43.8, 13.0, and 2.8 g m^{-2} more potassium, calcium, magnesium, and manganese respectively than in the biomass of the heath stand. The pool of mobile nutrients (nutrients included in biomass and exchangeable cations in the soil) was 45.6 (K), 32.9 (Ca), 6.1 (Mg), and 3.1 (Mn) g m^{-2} higher in the coniferous stand versus the heath vegetation.

At the same site, Bonneau et al. (1979) also observed that the coniferous plantation produced an alteration of the mineralogic skeleton of the bedrock. From an initial amount of around 170 kg m^{-2} of biotite, muscovite, and clay in the bedrock, the soil under the heath lost 27.5 kg m^{-2} and the soil under *Picea abies* 50.5 kg m^{-2}, or 23.0 kg m^{-2} more then the heath.

The development of coniferous stands, in place of other vegetation, induces a loss of mineral elements in the bedrock and a change of the pathways and the compartments of the biogeochemical cycle. The final result is a lower nutrient availability in the soil and more elements immobilized in the biomass or litter layer without important changes in the amount of nutrients included in the nutrient cycle.

3.5.4 Forest Fertilization

Foresters have attempted to find new techniques of forest management, often based on the use of fertilizers. Fertilizers are added to forest soils especially during their establishment. In addition to the use of mineral fertilizers, research on the effect of organic wastes, such as sludges or wastewaters, as forest fertilizing products has also been undertaken.

Ranger (1981) compared fertilized and non-fertilized stands at the ages of 15 and 18 years. The stem density was 4,500 stems per hectare. Half of the stand was fertilized at planting with additions of 9.3 g m^{-2} of nitrogen (27.5 g m^{-2} of urea), 16.3 g m^{-2} of phosphorus (38.0 g m^{-2} of P_2O_5) 57.1 g m^{-2} of calcium (80.0 g m^{-2} CaO), and 13.3 g m^{-2} of potassium (16.0 g m^{-2} of K_2O).

Fertilization largely increased the productivity of the stand (+122%) (Table 2). This increment was correlated with a higher uptake of nutrients. This effect changed with time. Between 15 and 18 years the biomass increased 39% in the unfertilized plot and only 30% in the fertilized one. At 18 years the biomass of the fertilized plot was only 107% greater than the control stand, whereas at 15 years it was 122% greater.

The same pattern was evident for nutrients. The fertilized stand, which produced 122% more biomass, used 153% more of the five nutrients studied than did the control plot. There was no direct relationship between biomass production and nutrient uptake. On the contrary, the increment of productivity was accompanied by an "over-consumption" of some, but not all, nutrients. Potassium was not "over-consumed," and more surprising, neither was nitrogen.

Table 3 indicates the amounts of elements included in a thousand kilograms of woody biomass and in a thousand kilograms of 1-year-old needles, at 15 and 18 years, and also the amounts of these same nutrients found in the same quantity of litter.

Table 2. Biomass and the included nutrients in a fertilized (F) and control (C) plot of *Pinus laricio* after 15 and 18 years

Age of the stand	Site	Biomass kg m^{-2}	Mineral elements in biomass g m^{-2}				
			N	P	K	Ca	Mg
15	F	11.14	17.02	2.21	14.57	19.07	4.06
	C	5.01	7.05	0.46	7.15	6.40	1.47
	F-C	6.13	9.97	1.75	7.42	12.67	2.59
18	F	14.46	21.47	2.70	20.00	23.94	5.15
	C	6.98	9.61	0.62	9.90	8.64	2.00
	F-C	7.48	11.86	2.08	10.10	15.30	3.15

Table 3. Nutrients required to build up 1,000 kg of biomass in fertilized (F) and control (C) plots and in the yearly litter fall of a *Pinus laricio* stand (kg of nutrients)

Age	Type of material	Site	N	P	K	Ca	Mg	Total
15	Woody biomass	F	1.53	0.20	1.31	1.71	0.36	5.11
		C	1.41	0.09	1.43	1.28	0.29	4.50
	1-year-old needles	F	13.16	1.20	6.78	4.01	1.06	26.21
		C	13.74	0.77	6.81	3.10	1.16	25.58
18	Woody biomass	F	1.48	0.19	1.38	1.66	0.36	5.07
		C	1.38	0.09	1.42	1.24	0.29	4.43
	1-year-old needles	F	13.22	1.20	6.85	4.06	1.08	26.41
		C	13.89	0.76	6.87	3.20	1.16	25.88
	Litter	F	7.21	0.49	1.90	6.59	0.95	17.14
		C	8.00	0.23	1.31	5.35	1.00	15.89

For building a thousand kilograms of perennial material, the fertilized plot used, at 15 and 18 years respectively, 122% and 111% more phosphorus, 33% and 34% more calcium, 24% and 24% more magnesium, and 8.5 and 7.2% more nitrogen than the same aged control plot.

It appears that phosphorus is not only an essential mineral element, but it is also limiting for production. If it is available in excessive amounts in the soil its uptake increases much more than is necessary for the production of biomass. In this case, phosphorus is a good example of luxury consumption of nutrients, which has been discussed by Duvigneaud and Denaeyer De-Smet (1964).

The higher fixation of nutrients in fertilized plots is increased by the higher productivity of biomass and the greater yearly litter fall. The litter is also richer in bioelements in the fertilized plot. For all of these reasons the uptake of nutrients from the soil is substantially greater in the fertilized stand than in the control plot.

In conclusion, fertilization produces a higher quantity of biomass, accompanied by higher amounts of nutrients in the different compartments and pathways of the nutrient cycle. The influence of fertilization, however, seems be limited in time and can induce an over-consumption of some nutrients. The latter problem

Table 4. Losses of nutrients by various harvesting methods in a control and a fertilized stand of *Pinus laricio* (g m^{-2}). F – Fertilized plot; C – Control plot; I – Input

		N	P	K	Ca	Mg
Input with fertilizers		9.2	16.3	13.3	57.1	
Atmospheric inputs		0.5–1.0	0	0.2	0.2–0.5	0.1
Total harvesting (boles, branches, roots and leaves)	F	49.5	5.1	34.6	35.2	7.4
	F-I	40.3	(+11.2)	21.3	(+21.9)	7.4
	C	20.9	1.2	15.1	11.6	2.8
Harvesting total aerial biomass	F	46.1	4.8	32.5	32.8	6.7
	F-I	−36.9	(+11.5)	19.2	(+24.3)	6.7
	C	19.3	0.9	13.8	10.6	2.8
Boles harvested (without bark)	F	5.1	0.9	6.7	5.2	1.5
	F-I	(+ 4.1)	(+15.2)	(+ 6.6)	(+51.9)	1.5
	C	2.6	0.1	3.6	1.9	0.5

needs further study, to determine exactly the amounts of the different nutrients that should be added to the soil as fertilizers to produce the largest amounts of biomass with the lowest investment of fertilizers.

3.5.5 Removal of Forest Products

In contrast to forest fertilization, which involves an input of nutrients into the system, forestry practices have involved the increased utilization of forest products by harvesting all of the biomass. This management practice increases the output from the system.

Foresters have long respected the biological processes and mineral requirements of the forest ecosystems, and have used only a fraction of the biomass, such as boles or big branches, most often excluding leaves, small branches, bark and stumps.

Some utilization of forest litter as fertilizers of agricultural soils has been made in the past in Bavaria. After a few years, the productivity of the forests decreased so much that this method was prohibited.

Some recent projects for exploiting forest products have attempted to harvest not only the aerial part of forest, but also the root system, which can make up an appreciable part of the forest production. To study the effects of such practices, Ranger (1981) used the data of the stand discussed above to analyze the effect of different harvesting processes on the mineral budget. He compared total harvesting of aerial and root biomass, harvesting of the total aerial part and harvesting, by the old methods, of the boles, without bark and branches.

Table 4 indicates that on the control plot, the effect of the three harvesting procedures is always on the output of nutrients from the ecosystem, and that output could possibly be matched for some nutrients, on a several year basis, by inputs from the atmosphere, by rain and by dry and wet deposition. However, when total input and output are considered, the final result is always a loss for succeeding plantations.

Fertilization prevents impoverishment of the ecosystem, especially when the old harvesting methods are used. On the other hand, with the new methods of harvesting of total biomass as well as with total harvesting of aerial parts, fertilization did not make good the loss. Furthermore, Ranger's results did not take into account losses of nutrients by leaching through the rooting zone.

Ranger's study suggests that if we want to preserve the future fertility of forest ecosystems, harvesting must be limited to the boles, excluding leaves, small branches and roots. In the future, new studies could give clear indications of the impact of these new management systems, and it may be possible to design practices which allow total biomass removal.

3.5.6 The Mineral Budget and Plant Succession

In the south of France, as all around the Mediterranean, agriculture began a few thousand years ago and continued unaltered in its methods until 1900. Fires which were used to improve pasture produced a sclerophyllous shrub community in the place of forest. These communities include the garrigue around Montpellier and the maquis on non-calcareous soils.

In this region forest stands are only found in small areas and are not of economic interest. Garrigue can form woodlands with *Quercus ilex* (evergreen oak) and *Quercus pubescens* (deciduous), and on calcareous marl with *Pinus halepensis*. The most common community is a shrub oak community, dominated by *Quercus coccifera* (also evergreen).

A comparison of the nutrient cycling in an old *Quercus ilex* forest (which will be considered for this comparison as a climax formation) with the *Quercus coccifera* garrigue gives some indications of the modification introduced by man on the mineral economy (utilization of the available nutrients) and the organization and adaptation of the biogeochemical cycles. A *Pinus halepensis* woodland, which colonizes calcareous marl soils is also compared.

The aerial perennial biomass of an approximately 150 year old *Quercus ilex* forest amounts to 26.2 kg m^{-2} and includes 525 g m^{-2} of N, P, K, Ca (Rapp 1971; Lossaint and Rapp 1978). Two *Quercus coccifera* garrigues, 17 and 30 years old, had respectively 60.2 and 79.8 g m^{-2} of the four nutrients, for an aerial biomass of 1.95 and 3.15 kg m^{-2} (Rapp 1971).

The *Pinus halepensis* stand, planted 50 years ago on the same soil as the 30 year old garrigue, had an aerial biomass of 15.1 kg m^{-2} and which included 115.9 g m^{-2} of the same mineral elements (Rapp 1974).

If we consider that the three older stands are mature, it appears that with the substitution of the *Quercus ilex* forest by the *Quercus coccifera* garrigue, the annual productivity decreased from 260 g m^{-2} to 110 g m^{-2} (Table 5). This decrease is accompanied by lower amounts of minerals cycling through the system in all pathways: uptake, immobilization in the biomass, restitution to the soil within litter fall.

Table 5. Mineral cycling in mediterranean woodlands

	Community type	Organic matter t ha^{-1}	Nutrients g m^{-2} year^{-1} N	P	K	Ca
Mineral elements in aerial biomass[a]	*Quercus ilex*	262	67.00	2.14	58.30	378.30
	Quercus (1)	19.5	10.50	0.50	5.30	43.90
	coccifera (2)	31.5	15.40	1.00	12.70	50.70
	Pinus halepensis	151	24.10	2.26	8.30	81.20
Retention[b]	*Quercus ilex*	2.6	1.32	0.26	0.89	4.27
	Quercus (1)	1.1	0.62	0.03	0.31	2.58
	coccifera (2)	1.1	0.51	0.03	0.42	1.69
	Pinus halepensis	2.1	0.34	0.03	0.12	1.16
Litterfall[b]	*Quercus ilex*	3.8	3.28	0.28	1.62	6.39
	Quercus (1)	2.3	1.95	0.07	0.90	3.27
	coccifera (2)	2.6	2.35	0.13	0.90	4.68
	Pinus halepensis	3.9	2.80	0.42	0.57	3.95
Uptake[b]	*Quercus ilex*	–	4.60	0.54	2.51	10.66
	Quercus (1)	–	2.57	0.10	1.21	5.85
	coccifera (2)	–	2.86	6.16	1.32	6.37
	Pinus halepensis	–	3.14	6.45	0.69	5.11

(1) 17-year-old stand, (2) 30-year-old stand
[a] g m^{-2}
[b] g m^{-2} year^{-1}

Table 6. Mineral elements required (kg) to build a thousand kilogram a of wood biomass

	Fixation – Immobilization N	P	K	Ca	Total	Absorption – Uptake N	P	K	Ca	Total
Quercus ilex	5.1	1.0	3.4	16.4	25.9	17.7	2.1	9.6	9.6	70.4
Quercus coccifera (1)	5.6	0.3	2.8	23.4	32.1	26.4	1.0	11.6	56.6	95.6
(2)	4.6	0.3	3.8	15.4	27.1	26.0	1.4	12.0	57.9	97.2
Pinus halepensis	1.6	0.1	0.6	5.5	7.8	14.9	2.1	3.3	24.3	44.6

(1) 17-year-old stand, (2) 30-year-old stand

The *Pinus halepensis* stand, in comparison to the garrigue, is characterized by a higher community productivity even though the amount of nutrients required for that productivity is less.

To compare the economy of nutrients in these three communities we can calculate the amount included in a thousand kg of perennial biomass and the amount required (= amount taken up from the soil) to build up this same weight of biomass (Table 6).

Quercus ilex and *Quercus coccifera* communities immobilize nearly the same amount of the four major cations studied (respectively 2.6, 3.2, and 2.4 g m^{-2}). There are however some differences between these elements, particularly phosphorus. The differences between the three communities are more important in the

Table 7. Percent of the uptake of nutrients used to build up biomass

		N	P	K	Ca
Quercus ilex		28.8	47.6	35.4	40.0
Quercus coccifera	(1)	21.2	30.0	24.1	41.3
	(2)	17.7	21.4	31.7	26.6
Pinus halepensis		10.7	4.8	18.2	22.6

(1) 17-year-old stand, (2) 30-year-old stand

uptake of nutrients. The two *Quercus coccifera* garrigues absorb nearly the same amount of nutrients yet require 30% to 40% more nutrients to produce one ton of perennial biomass than does the *Quercus ilex* forest. On the other hand, *Pinus halepensis* uses 30% less nutrients than the evergreen oak forest to build up the same amount of biomass. The coniferous species, even on calcareous soils, are frugal species.

If we calculate the immobilization in the perennial biomass as a percent of the uptake from the soil (Table 7) we note that *Quercus ilex*, often considered as a climax species, used the highest proportion of elements it takes up from the soil in the case of all of the nutrients studied. *Pinus halepensis* immobilized the least, while *Quercus coccifera* occupies an intermediate position. The youngest stand of garrigue used more phosphorus, nitrogen, and calcium than did the 30-year-old garrigue.

From this table we can also see the significant uptake and immobilization of calcium by the evergreen oaks. This is related to the high availability of that cation in the soil and may be an example of luxury consumption. These two oaks (*Quercus ilex* and *Quercus coccifera*) accumulate the highest quantities of calcium that has been noted for all mineral cycling studies undertaken in the IBP (Cole and Rapp 1980). For phosphorus there is a fixation of 47% and 21% of the uptake by the two evergreen oaks, against only 5% by the pine. In the latter case, the metabolism of phosphorus is involved in the retention since the uptake from the soil is the same by *Pinus halepensis* as by *Quercus ilex*. Phosphorus is evidently not limiting in these soils. On the other hand, the low absorption of phosphorus in the garrigue may be the result of a very low availability of phosphorus in soil. It is well known that phosphorus is often immobilized in calcareous soils and that this limits their productivity. This immobilization can be overcome after burning of the garrigue (Trabaud 1980, personal communication), or by biological turnover in the oak forest. In this later case, phosphorus is retained in the biomass. Phosphorus is slowly but regularly restored to the soil by litter fall and litter decomposition, where it can be taken up again by the roots.

The comparison of these three community types indicates clearly the nutrient pathways and pools used by each of them. *Quercus ilex* used the greatest fraction of the nutrients taken up from the soil. The degraded *Quercus coccifera* formation uses more nutrients than the climax vegetation to produce less biomass. Recycling through the garrigue ecosystem is higher than in the *Quercus ilex* stand. *Pinus halepensis* recycles the largest amount of the absorbed nutrients and after a long time reorganizes the soil profile and starts a new pedogenesis.

3.5.7 Conclusions

The studies reviewed in the preceding pages indicate clearly that human interferences on nutrient cycling and nutrient budgets in ecosystems is widely prevalent. The study of these budgets needs special attention, especially in modern management practices, since long term effects on them are not known with accuracy.

The problem of human modification and disturbance of natural ecosystems induces a new field of research including the effects of indirect disturbances, such as atmospheric pollution, the increasing input of nutrients or of heavy metals in the biological cycle of the ecosystems, and the problem of increasing waste production by increasing human activity.

Acknowledgments. The author wishes to express his great gratitude to Dr. Peter M. Vitousek (University of North Carolina), for reviewing and correcting the manuscript.

Résumé

Parmi les nombreux liens qui unissent les différents constituants de l'écosystéme, la répartition et la circulation des éléments minéraux sont parmi les plus importants. Regroupés souvent sous le terme de „cycles biogéochimiques," ils peuvent se schématiser par une série de compartiments d'accumulation et d'immobilisation des éléments biogènes, reliés entre eux par des flux permanents.

Mais l'équilibre de cette organisation est fragile, et de nombreuses interférences, d'origine abiotique, biotique et surtout humaine le perturbent constamment.

Même dans le cas d'écosystèmes dits „naturels," tels les forêts, les interventions humaines sont nombreuses, liées essentiellement aux pratiques et usages sylviculturaux. C'est le cas de la fertilisation en milieu forestier, des récoltes de plus en plus complètes de la biomasse ou encore de la substitution de plantation de résineux, de productivité plus rapide, à des peuplements initialement feuillus.

Cela se retrouve également pour certaines transformations du tapis végétal initial, provoquées par l'homme, lors de l'instauration de pratiques agricoles, ou de l'utilisation de feu comme technique de maintien d'un type de couvert végétal.

A ces influences directes l'on peut ajouter l'action indirecte dont le meilleur exemple est celui de l'eutrophisation de certains milieux, ou de l'apport d'éléments minéraux à partir de l'atmosphère, polluée par des rejets industriels émis dans la basse atmosphère. Présentement, il ne sera tenu compte que des interventions humaines directes, sans envisager les interférences indirectes.

Les perturbations qui font l'object de cette étude sont le feu, le reboisement en résineux, la fertilisation en forêt, l'enlèvement des produits forestiers.

L'effet du feu a été étudié par Villecourt dans une savane de Côte d'Ivoire régulièrement incendiée en fin de saison sèche; le résultat est une perte de 0.99 g m^{-2} d'azote, 0.07 g m^{-2} de phosphore, 0.55 g m^{-2} de potassium; il ne semble pas que, dans ce cas, le système racinaire bénéficie d'une augmentation de son stock d'éléments nutritifs, ce qui amortirait un peu le bilan du feu.

Il est généralement admis que les résineux immobilisent dans leurs parties aériennes nettement moins d'éléments minéraux que les feuillus, mais que la lenteur de la décomposition des aiguilles ralentit le recyclage de ces éléments; par exemple, la masse d'azote stockée dans la litière peut, à la suite d'un reboisement en résineux, être multipliée par trois. Sous climat tempéré, ce type de reboisement peut entraîner une podzolisation qui provoque des pertes sensibles (1 kg m^{-2} de K_2O et 0.42 kg m^{-2} de Na_2O).

Un apport d'engrais en forêt entraîne souvent une sur-consommation d'éléments nutritifs; cette sur-consommation apparaît aussi bien si l'on regarde le total des éléments prélevés que si l'on mesure dans les tissus végétaux les concentrations en éléments minéraux. Ainsi, une tonne de bois produite dans une parcelle fertilisée contient respectivement 122%, 33%, 24%, et 8,5% de phosphore, de calcium, de magnésium et d'azote en plus de la quantité trouvée dans une parcelle voisine non fertilisée.

L'enlèvement des produits forestiers se limitait autrefois à l'exploitation des troncs et des branches. Certaines techniques „modernes" préconisent la récolte de la totalité de la phytomasse aérienne et même d'une part importante de la phytomasse souterraine. Ranger (1981) dans une étude détaillée montre qu'il faut craindre que les techniques n'entraînent rapidement un appauvrissement de la station.

Dans la région méditerranéenne française, une forêt de *Quercus ilex* (feuillage sempervirent) avait produit 26.2 kg m^{-2} (dont 525 g m^{-2} de N, P, K, et Ca) en 150 ans. Des garrigues de *Quercus coccifera* (feuillage sempervirent) qui représentent souvent un stade de dégradation de la forêt initiale, avaient des biomasses aériennes de 1.95 et 3.15 kg m^{-2} aux âges respectifs de 17 et 30 ans. Ces biomasses incluaient 60 et 80 g m^{-2} des quatre éléments déjà signalés (N, P, K, et Ca). Ceci correspond donc à une diminution de la productivité dans cette formation ligneuse basse de substitution de la forêt avec un recyclage plus faible de bioéléments.

Une forêt artificielle de *Pinus halepensis*, âgés de 50 ans, possède une phytomasse aérienne de 15.1 kg m^{-2} des mêmes quatre éléments essentiels. Ces chiffres soulignent la „frugalité" du Pin d'Alep, qui produit assez rapidement une phytomasse aérienne plus importante que les deux essences de chênes sempervirents, tout en contenant et en consommant quatre fois moins d'éléments minéraux que le chêne vert.

References

Bonneau M, Brethes A, Lelong F, Levy G, Nys C, Souchier B (1979) Effets de boisements résineux purs sur l'évolution de la fertilité du sol. Rev For Fr 31:198–207

Cole DW, Rapp M (1980) Elemental cycling in Forest Ecosystems. In: Reichle DE (ed) Dynamic properties of forest ecosystem. IBP 23. Cambridge Univ Press, Cambridge, pp 341–409

Cole DW, Gessel SP, Turner J (1978) Comparative mineral cycling in red alder and Douglas fir. In: Briggs DG, Bell de DS, Atkinson WA (eds) Utilization and management of alder. Pac NW For Rang Exp Stn. Portland Oreg, pp 327–336

Duvigneaud P, Denaeyer De-Smet S (1964) Le cycle des éléments biogènes dans l'écosystème forêt. Lejeunia 28:3–147

Leon R (1968) Balance d'eau et d'azote dans les prairies à litière des environs de Zürich. Ver Geobot Inst Eidg Tech Hochsch Stift Ruebel Zürich 41:2–67

Lossaint P, Rapp M (1978) La forêt méditerranéenne de chêne vert. In: Lamotte M, Bourliere F (eds) Problèmes d'ecologie: Ecosystèmes terrestres. Masson, Paris, pp 129–185

Nihlgard B (1972) Plant biomass, primary productivity and distribution of chemical elements in beech and planted spruce forest in South West Sweden. Oikos 23:69–81

Nyss C (1977) Influence d'une plantation d'épiceas sur un sol granitique du Plateau des Millevaches (France). Proc Symp: Soil as a site factor for forests of the temperate and cool zone, Zvolen, Tchecoslovaquie, pp 142–151

Ranger J (1981) Etude de la minéralomasse et du cycle biologique dans deux plantations de Pin Laricio de Corse dont l'une a été fertilisée à la plantation. Ann Sci For 38:127–158

Rapp M (1971) Contribution à l'étude du bilan et de la dynamique de la matière organique et des éléments minéraux biogènes dans les écosystèmes à chêne vert et chêne kermès du midi de la France. Caractéristiques pédologiques en climat méditerranéen et tempéré. CNRS, Paris, pp 22–184

Rapp M (1974) Le cycle biogéochimique dans un bois de pins d'Alep. In: Pesson P (ed) Ecologie forestière. Gauthier-Villars, Paris, pp 75–97

Rham de P (1971) L'azote dans quelques forêts, savanes et terrains de culture d'Afrique tropicale humide. Thèse, Fac Sci, Lausanne

Ulrich B, Steinhardt U, Muller-Sour A (1973) Untersuchungen über den Bioelementgehalt in der Kronentraufe. Goettinger Bodenkundl Ber 29:133–192

Villecourt P, Roose E (1978) Charge en azote et en éléments minéraux majeurs des eaux de pluie, de pluviolessivage et de drainage dans la Savane de Lamto (Côte d'Ivoire). Rev Ecol Biol Sol 15:1–20

Villecourt P, Schmidt W, Cesar J (1979) Recherche sur la composition chimique (N, P, K) de la strate herbacée de la savane de Lamto (Côte d'Ivoire). Rev Ecol Biol Sol 16:9–15

3.6 Mechanisms of Ion Leaching in Natural and Managed Ecosystems

P. M. Vitousek

3.6.1 Introduction

Interest in mineral cycling and loss in terrestrial ecosystems has intensified in recent years. Impressively detailed studies of a number of temperate zone ecosystems have been completed (e.g., Rapp 1971; Likens et al. 1977; Bormann and Likens 1979; Edmonds 1982). The approach is currently being extended to tropical (Bernhard-Reversat 1977; Herrera and Jordan 1981) and boreal (Van Cleve and Viereck 1981) ecosystems with great success. These studies have focused primarily upon relatively natural ecosystems, but some studies include experimental and comparative examinations of managed ecosystems (Bormann and Likens 1979; Swank and Waide 1980).

A number of syntheses of this rapidly accumulating body of information have been attempted. For example, Likens et al (1977) collected the input-output budgets of a number of watershed ecosystems, Cole and Rapp (1980) reviewed information on within-system nutrient cycling and input-output budgets in 32 forest sites, Ellenberg (1977) summarized measurement of in situ nitrogen mineralization in 120 sites, and Vitousek and Melillo (1979) reviewed the effects of destructive disturbance (especially clearcutting) on nitrogen loss in 30 forest ecosystems.

Measurement of mineral transfers in a range of ecosystems is the essential basis for any advancement in this area, and summaries of such measurements are invaluable. Nonetheless, comparisons of measured fluxes in different ecosystems are unlikely to yield the predictive capability needed by resource managers or the more general understanding sought by terrestrial ecologists. Many more potentially important variables exist than there are (or are likely to be) ecosystem studies, and thus simple comparisons among systems do not suffice to isolate the controlling factors. Such controlling factors include climatic properties, soils, vegetation, and site history. Differences in research methods, especially where international comparisons are made, add a further level of complexity. This problem is exacerbated when managed and natural ecosystems are compared, since many of the variants of land management practices could affect mineral cycling pathways.

Further understanding of mineral fluxes in terrestrial ecosystem is thus dependent on understanding the mechanisms controlling these fluxes. A process-based understanding can be generalized more readily across sites. Moreover, only such an understanding can allow the further development of mechanistic models of production and mineral cycling which can respond more realistically to perturbation effects (Aber et al. 1978).

Here I examine the mechanisms controlling leaching losses of elements from terrestrial ecosystems and consequent gains to downstream aquatic ecosystems.

Relatively large amounts of essential plant nutrients are cycled through the soil solution annually (released by decomposers and taken up by the biota), and substantial quantities of cations are usually held in available form on the soil cation exchange complex. Only a small proportion of these elements are lost to streamwater or groundwater annually despite substantial water movement through most forest soils. I will show why leaching losses of ions are low from natural forests despite large within-system fluxes. Further, the effects of management practices on leaching will be evaluated briefly. This discussion will center upon forest ecosystems, but the ways that other biomes differ from forests will be examined briefly.

Leaching is not the only, and often not the most important, pathway for losses of elements from natural or managed ecosystems. Losses by erosion, volatilization, and harvest can also be substantial. Studies of particulate losses in a range of different ecosystems are accumulating (cf. Bormann et al. 1974) and one outstanding analysis of the geomorphological processes controlling short- and long-term element losses in a forested watershed has been prepared (Swanson et al. 1982a). A systematic process-based synthesis across a range of sites, incorporating managed as well as natural sites, may soon be possible. Volatilization losses are more difficult to measure – only recently have techniques for measuring dentrification potential in situ been developed (Smith and Tiedje 1979), and they are still being evaluated. Harvest losses are of course relatively easily measured (Likens et al. 1978), although their long-term impact on mineral fluxes in a site are less well understood.

3.6.2 Leaching of Anions and Cations

3.6.2.1 Measurement of Leaching Losses

Leaching losses from terrestrial ecosystems have been measured in two major ways – through the small watershed approach (Likens et al. 1967, 1977) and through the use of tension lysimeters placed below the rooting zone (Cole 1958, 1968). Both methods have been widely used in a range of sites, but they measure different fluxes and may yield different results (Cole and Rapp 1980). Each method has distinct advantages and disadvantages, depending on the system under study and the research question to be answered.

The watershed approach offers the advantage that where its basic requirements (especially water-tight bedrock) are satisfied, measurement of both water flux and element concentration provides a clear, unambiguous way to determine the *amount* (rather than just the concentration) of elements lost. It further provides a direct measurement of the integrated (in space) output of minerals from a terrestrial ecosystem as a whole. Where the major emphasis of a study is on land–water interactions, whole landscape units, or the impact of management practices on downstream ecosystems, there is no substitute for the use of the watershed approach.

The fact that the watershed approach does integrate losses from a larger area can be a disadvantage when other research goals are pursued. Where different soils and/or vegetation types exist within a watershed, element outputs cannot be apportioned among the components of the system. For example, identical element losses were assigned to four quite different plant communities within one watershed at Walker Branch, Tennessee (Cole and Rapp 1980). The actual losses from each component could have been quite different. Similar problems are almost certainly present in any natural watershed. Another limitation is that losses are measured after water has: (1) passed through the rooting zone; (2) interacted with lower soil horizons where weathering, adsorption/exchange, and oxidation-reduction reactions can take place; (3) passed through the riparian zone (in some systems), which can be functionally very different from the rest of the watershed (Swanson et al. 1982b); and (4) been processed in the stream (Webster and Patten 1979; Meyer and Likens 1980). In a study of the losses of nutrients required for forest growth following intensive harvesting, for example, only the first is of major interest. Dynamics at the level of the forest stand could be obscured by other processes lower down in the profile or downstream.

The use of tension lysimeters allows the measurement of element concentrations at any point within a system, so that measurements immediately below the rooting zone (or anywhere else) are feasible. Moreover, a smaller, more homogeneous area can be evaluated, and less land area is required for experimental studies. While the integration of outputs from a larger area is prohibitively difficult, adequate replication within a well-defined area is possible using lysimeters. Lysimetric studies are therefore preferable for stand- or plot-level measurements of element losses, or for the many investigations where a knowledge of changes in the soil solution with depth are useful.

One major disadvantage is that while lysimeters allow the measurement of element concentrations anywhere in the soil solution, they only allow the measurement of water flux under the most restrictive, and for samples below the rooting zone probably unrealistic, conditions. Consequently, some other measure of water flux is required – generally a rather complex hydrological model coupled with meteorological measurements. Moreover, results from lysimeters cannot be used directly to interpret nutrient loading to downstream ecosystems.

A logical research solution in suitable sites is to apply both techniques simultaneously, since they do provide rather different and complementary information. At present there is a paucity of such comparative studies, but several investigations applying both methods to the same site are now in progress.

3.6.2.2 Leaching Mechanisms–Anion Mobility

The processes controlling ion movement in the soil solution are relatively well understood, but this understanding has only recently been applied to ecological problems. Basically, the supply of mobile anions controls the leaching of both cations and anions through soils. This concept was developed by Nye and Greenland (1960), and elaborated for particular anions by McColl and Cole (1968) and Likens et al. (1969). It was recently summarized superbly by Johnson and Cole (1980). To illustrate this mechanism, I will start by describing a soil without any anion adsorption or anion exchange capacity.

Most available cations in the soil are exchangeably bound to stationary negative charges on the surfaces of clays and organic colloids. Upon the addition of a dilute electrolyte solution (such as bulk precipitation or throughfall), cations equilibrate between the solution and the exchangeably bound phases. Under the conditions defined here, though, all of the anions remain in the solution phase. The sum of the cations present in solution phase is therefore determined by the amount of anions supplied, since the electrochemical neutrality of the soil solution must of course be maintained. Only the ions present in the soil solution are subject to downward leaching or movement in mass flow to plant roots.

Leaching of a particular cation (as opposed to total cations) is controlled by the total amount of that cation present and its relative affinity for the exchange complex. Losses of a cation can be altered by either a significant change in the fraction of the exchange complex occupied by that cation (in which case there must be a compensatory change in the leaching of another cation), or by a significant change in the (generally small) amount of mobile anion present.

In practice, anion mobility is far from complete – anion adsorption, plant uptake, decomposition (of organic acids), and reduction of anions to volatile forms can all reduce soil solution concentrations and thus leaching potential to levels below those set by anion supply. Nonetheless, an understanding of leaching losses of all cations and anions requires an understanding of the processes controlling anion supply and turnover in forest ecosystems. In order to evaluate the effects of management practices on ion leaching, an analysis of management effects on anion fluxes is both necessary and sufficient (Cole et al. 1975; Johnson and Cole 1980).

3.6.2.3 The Major Anions

The major anions in both natural and managed forest ecosystems are bicarbonate, sulfate, nitrate, chloride, and organic acids. In many sites (especially those with siliceous parent material), these anions are supplied largely from the atmosphere (mostly in bulk precipitation) or the biota – little or none is supplied by rocks or soils. Silicate, the major anion in crystalline rocks, is not dissociated, but remains as H_4SiO_4 at the conditions of temperature, pressure, and pH characteristic of most soil solutions. The hydrous oxides of aluminium and iron which are characteristic of highly weathered soils are nearly insoluble and thus cannot provide anions to the soil solution either. Externally supplied anions are thus essential to the mobility and eventual loss of cations released from silicate minerals or held on exchange sites in the soil (Gorham et al. 1979).

In calcareous areas, weathering can yield both cations and anions ($CO_3^{=}$, which will combine with H^+ to form HCO_3^- at the pH of most soil solutions), but even here the supply (generally by mobile anions) of hydrogen ions to the minerals greatly increases the rate of weathering. In special cases metal sulfides, gypsum ($CaSO_4$) and similar minerals, or even halite (NaCl) can contribute both cations and anions. Above the soil horizon where free carbonates are present (if indeed they are present anywhere in the profile), anions are largely supplied by the biota or by the atmosphere. Minerals supplying both a cation and a mobile anion can often contribute substantially to watershed-level ecosystem budgets (Henderson et al. 1978) without exerting a strong influence within the rooting zone.

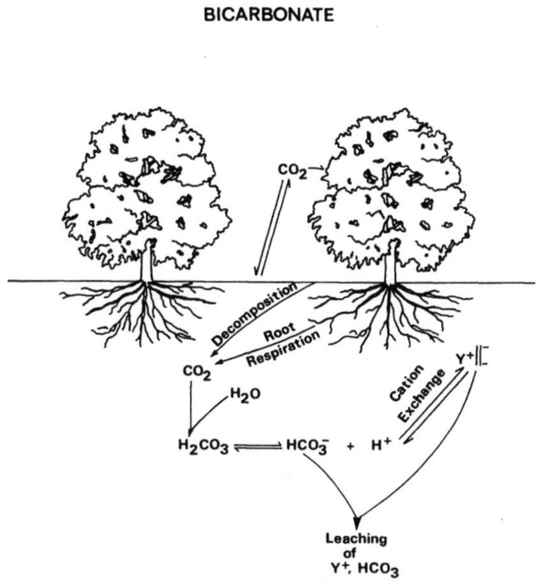

Fig. 1. Pathways of bicarbonate formation and loss in a forested ecosystem. Y^+ represents mineral cation; *double vertical lines on the right* an exchange site

Regardless of whether they are externally or biotically supplied, net anion input or production is always associated with cation input or production, as electrochemical neutrality is always maintained. The cations supplied can then equilibrate with other cations on the soil exchange complex. Where the added cation is a hydrogen ion and the anion leaches from the system, the net result is the acidification of the exchange complex and the removal of base cations.

The major anions, their pathways of delivery, and their associated cations are described briefly below. More detail can be found in Johnson and Cole (1980), upon which this treatment is in part based.

Bicarbonate. The generation and loss of bicarbonate is summarized in Fig. 1. Carbonic acid is supplied to the soil solution by the hydration of CO_2, which enters the soil by decomposition and root respiration:

$$CO_2(g)+H_2O \leftrightarrow H_2CO_3\,, \quad [H_2CO_3]=K1^*pCO_2\,, \tag{1}$$

where pCO_2 = partial pressure of CO_2 (in atmospheres) and $K1 \cong 3.4 \times 10^{-2}$. Carbonic acid can then dissociate to bicarbonate and hydrogen ion:

$$H_2CO_3 \leftrightarrow H^+ + HCO_3^-\,, \quad [HCO_3^-]=\frac{K2^*[H_2CO_3]}{[H^+]}\,, \tag{2}$$

where $K \simeq 4 \times 10^{-7}$. Bicarbonate concentrations in soil are controlled solely by the partial pressure of CO_2 (which is generally elevated well above atmospheric pCO_2 by decomposition and root respiration) and the pH of the soil solution. Below pH 4.5, the contribution of bicarbonate to the anion strength of the soil solution is

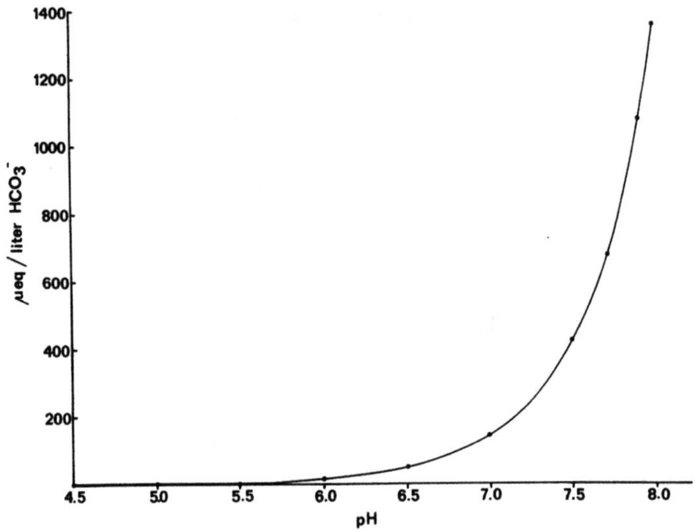

Fig. 2. Bicarbonate concentrations as a function of pH at a $pCO_2 = 1 \times 10^{-3}$ atm, about three times atmospheric

unimportant. Bicarbonate concentrations increase an order of magnitude with each unit increase in pH (Fig. 2).

Hydrogen ions produced by the dissociation of carbonic acid (or by other processes) equilibrate with cations held on the cation exchange complex. For *small* changes in ion activity and short time intervals this can be represented by:

$$H^+ + Y^+X \leftrightarrow Y^+ + H^+X, \quad Y^+ = \frac{Ks[H^+][Y^+X]}{[H^+X]}, \tag{3}$$

where Y^+ stands for any cation, X for the negatively-charged exchange complex, and K_s for the relative affinity of the exchanger for any cation relative to hydrogen ion. The net effect of H^+ production is the removal of some Y^+ from the exchange complex to the soil solution, where it can be leached from the system in association with bicarbonate.

Bicarbonate is a major anion in the soil solution in all circumneutral and basic forest soils (McColl and Cole 1968; Johnson et al. 1977), contributing 436 µEq HCO_3^- 1^{-1} of mobile anion strength at pH 7.5 and a pCO_2 of 1×10^{-3} atmospheres (Fig. 2).

Sulfate. The inputs and outputs of sulfate are summarized in Fig. 3. Sulfur can enter forest ecosystems as sulfate ion in bulk precipitation, as sulfate aerosols, or as SO_2 gas. It is delivered to the soil by throughfall and litterfall, and is rapidly oxidized there to sulfate, if it is not already in that form. A portion of the sulfate is taken up by plants, but sulfur is not commonly limiting to plants in the field, and the amount cycled is generally not large relative to annual inputs. In areas where anthropogenic sulfur inputs have made bulk precipitation a dilute solution of sulfuric acid (Likens and Bormann 1974), both a mobile anion and hydrogen ions are

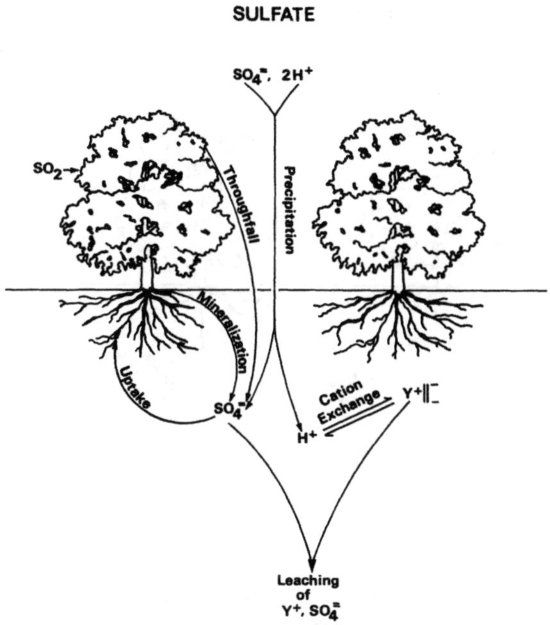

Fig. 3. Pathways of sulfate input and loss in a forested ecosystem. Symbols as in Fig. 1

supplied to a forest ecosystem. The sulfate can then leach through the soil solution accompanied by cations released from the exchange complex (Cronan et al. 1978). However, once it is in the soil solution, sulfate does not necessarily leach through the profile to groundwater or streamwater. Johnson et al. (1980) showed that sulfate adsorption on iron and aluminum sesquioxides in the B horizon can substantially reduce sulfate and cation losses.

The importance of sulfate in the soil solution is very largely dependent on atmospheric inputs of sulfur. Sulfate is now the predominant anion in the soil solution in large areas of eastern North America (Cronan et al. 1978; Cronan and Schofield 1979) and presumably northern Europe. Once concentrated by evapotranspiration (Vitousek 1977), it can be present in concentrations of 150–250 μEq $SO_4\ l^{-1}$.

Organic Acids. Soluble organic acids (Fig. 4) can be leached from live or decomposing plant parts, produced by decomposers, or exuded by roots (Graustein et al. 1977). A wide range of organic acids are found in forest soil solutions (Bruckert and Jacquin 1969). They can dissociate by:

$$OA \leftrightarrow O^- + H^+, \quad [O^-] = \frac{KA*[OA]}{[H^+]}, \tag{4}$$

where OA stands for an organic acid, O^- for the organic anion, and KA for the dissociation constant of the acid.

Soluble organic acids are generally present in relatively low concentrations. Where soils are highly acid due to parent material, geological age and leaching in-

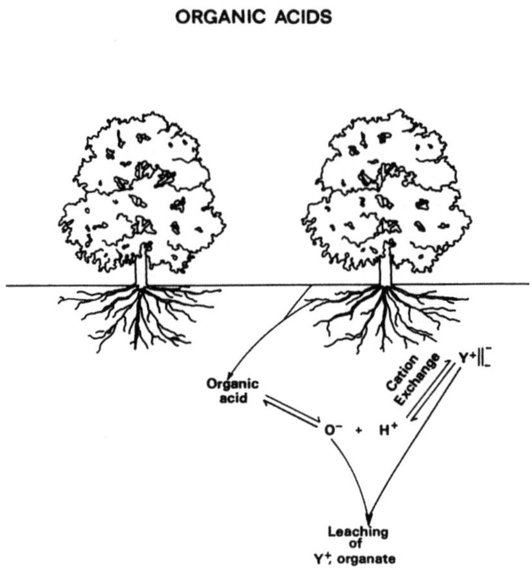

Fig. 4. Pathways of organic acid supply and loss in a forested ecosytem. O^- dissociated organic anion; otherwise, symbols as in Fig. 1

tensity, or insoluble organic acids, bicarbonate is, however, nearly absent. If such sites do not receive substantial inputs of other anions in precipitation, organic acids can dominate the soil solution. Apparently, larger amounts of organic acids are produced in cold regions (Johnson et al. 1977). Organic anions differ from the other anions discussed in that they can either decompose or precipitate in the soil, thus reducing anion mobility.

Nitrate. The nitrogen cycle is complex, in that nitrogen can enter a site either by nitrogen fixation or bulk precipitation, it can exist in the soil as either a cation (NH_4^+) or an anion (NO_3^- or to a much lesser extent NO_2^-), and it can leave either through leaching or volatilization (ammonia volatilization or denitrification). Ammonium is released upon the decomposition of N-containing organic matter (Fig. 5). It can either be taken up in that form by microbes or plants, or be oxidized by specialized autotrophic bacteria to nitrite and then nitrate. The overall oxidation reaction is:

$$NH_4 + 2O_2 \rightarrow 2H^+ + H_2O + NO_3^-. \quad (5)$$

Nitrate can be taken up by organisms, reduced to N_2 or N_2O, or leached through the soil accompanied by the cations displaced by the hydrogen ions produced in (5).

Many forests are limited by nitrogen supply, and most cycle far more nitrogen than they gain or lose annually (Rosswall 1976; Cole and Rapp 1980). Consequently, leaching losses of nitrate are generally very small or confined to the non-growing season (Vitousek 1977). Even the nitrate added in acid precipitation is generally absorbed within a site and contributes only indirectly to the leaching of cations.

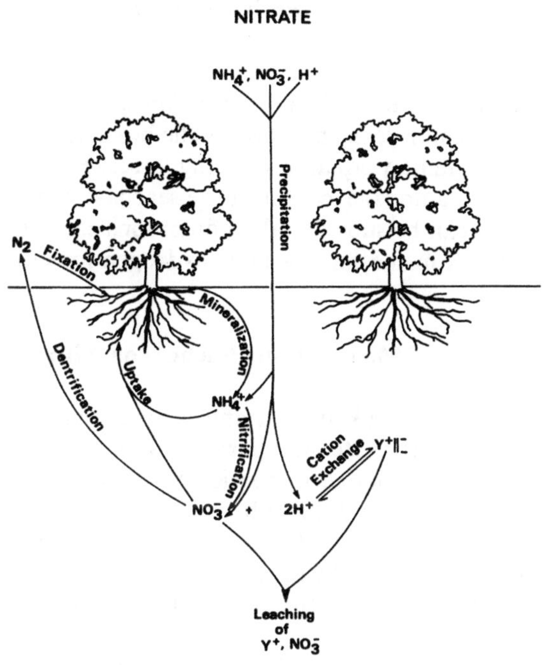

Fig. 5. Pathways of nitrate formation and loss in a forested ecosystem. Symbols as in Fig. 1. Most nitrogen cycles within the forested system; usually only small amounts are lost in natural forests

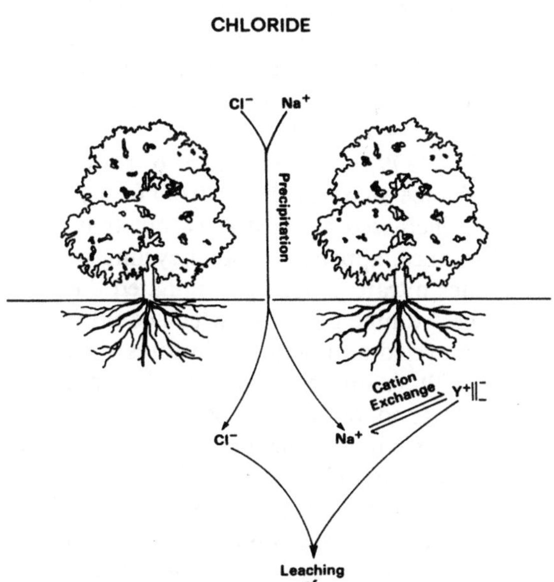

Fig. 6. Chloride inputs and losses in a forested ecosystem. Symbols as in Fig. 1

As discussed below, however, nitrate can become very important in disturbed or managed ecosystems.

Chloride. The primary source of chloride in forest soils is bulk precipitation (Fig. 6). Relatively little chloride is absorbed by the biota, and total inputs in bulk precipitation are approximately equal to leaching losses. Both inputs and outputs are low (relative to other anions) except near the seacoast, where chloride and associated sodium inputs can be substantial. Sodium ions can replace other cations on the exchange complex in the same way as hydrogen ions do as shown in Eq. (3).

3.6.2.4 Effects of Management Practices on Nitrate Fluxes

The anions which are most likely to change substantially in managed ecosystems are organic acids, bicarbonate, and particularly nitrate. Organic acids could change substantially with a change in species composition, but details of such changes are poorly understood and their absolute magnitude is probably generally small. The effects of various forest site manipulations on bicarbonate fluxes are very well described by Johnson and Cole (1980). I will focus here upon changes in nitrate production and loss, and on the effects of nitrate production on fluxes of the other anions.

Despite its relative unimportance in leaching in aggrading forests, nitrate is the single most important anion in the soil solution in many, perhaps most, agricultural systems (Raney 1960; Nye and Tinker 1977) and in some disturbed forests (Nye and Greenland 1960; Likens et al. 1969; Johnson and Edwards 1979). It can achieve this importance because of the large amounts of nitrogen cycled annually within forest ecosystems (Rosswall 1976; Cole and Rapp 1980). While plant uptake is temporarily reduced or prevented by land clearing, net nitrogen mineralization is increased (Stone et al. 1978; Bormann and Likens 1979). Nitrogen can then accumulate in available form in the soil, and, if oxidized to nitrate, be leached from the system or volatilized as N_2 or N_2O.

Vitousek and Melillo (1979) reviewed studies of the effect of clearcutting and other destructive disturbance on nitrate losses from forest ecosystems. They found that nitrate losses were increased in most sites, but that only rarely did annual losses, following disturbance, approach annual cycling within the forest prior to disturbance. In such sites nitrate probably dominated the soil solution following disturbance. They suggested nine processes which could prevent or delay the loss of most of the nitrogen cycled annually within forests prior to disturbance.

Vitousek et al. (1979) examined some of the mechanisms which could prevent or delay nitrate production and loss in a range of forest ecosystems. They concluded that nitrogen immobilization and lags in nitrification could delay nitrate production, especially in infertile sites, while nitrogen uptake by regrowing vegetation would eventually reestablish the soil – plant – microorganism cycle within sites and prevent further nitrate losses. In fertile sites, both immobilization and lags in nitrification are minimized, and substantial nitrate losses can be observed immediately following disturbance (Wiklander 1980). Where regrowth is prevented, as in the trenched plots used by Vitousek et al. (1979) and the herbicide-

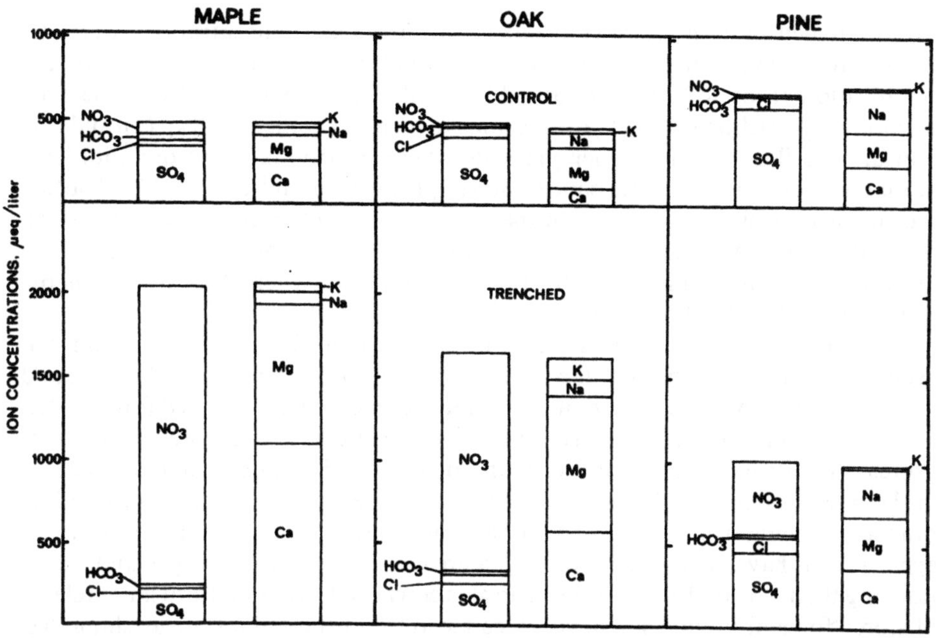

Fig. 7. The effects of root trenching and the prevention of vegetation regrowth on cation and anion leaching in three forests in southern Indiana. *Upper bars* mean concentrations of anions and cations in control tension lysimeters; *lower bars* concentrations in trenched-plot lysimeters. All ions are reported in $\mu Eq l^{-1}$

treated watershed examined by Bormann and Likens (1979), nitrate losses equivalent to or greater than annual cycling should eventually occur.

The potential magnitude of nitrate losses and their impact on cation and other anion losses is illustrated in Fig. 7. The ionic charge balance for the soil solution below the rooting zone is presented for the mean of ten control and ten trenched-plot (Vitousek et al. 1979) lysimeters in three southern Indiana forests. Collections were made and analyzed weekly for 14 months.

The soil solutions at all three sites are dominated by sulfate under control conditions. Following trenching, however, nitrate losses increased rapidly to very high levels in the maple site, more slowly but to similar levels in the oak, and much more slowly and to lower levels in the pine site (Vitousek et al. 1979). The increase in nitrate caused an increase in losses of the non-hydrogen cations in all sites. The actual magnitude of the increase in nitrate and cation losses is understated in Fig. 7, since water movement through the trenched plots was greater than that through the control plots.

The trenching treatment also had a substantial impact on other anion fluxes. Organic acids were not determined, but (due to the closeness of the charge balances) they must have been minor. Sulfate decreased in concentration, due at least in part to dilution by greater water flow. High nitrate concentrations may also affect sulfate oxidation or adsorption (Bormann and Likens 1979). Chloride was unaltered or slightly increased in concentration, so some increase in chloride output from trenched plots must have occurred. Bicarbonate concentrations were decreased following trenching. The decrease could be explained simply, since the acidification of the soil solution caused by nitrification could have led to the reassociation of bicarbonate and hydrogen ions to form carbonic acid [Eq. (2)] and carbon dioxide [Eq. (1)]. In higher pH soils where bicarbonate was quantitatively more important before disturbance, nitrate production and loss could substitute nitrate ions for bicarbonate ions in the soil solution up to a point, without necessarily altering the total anion or cation strength of the soil solution very much. The same process could be important for those organic acids where the soil solution pH was near or slightly above the pKA of the acid [Eq. (4)].

Experimental trenched plots can in no way be considered equivalent to management practice, but they do illustrate the magnitude, potential significance, and possible interactions with other anions of nitrate leaching in disturbed forests. Relative to natural forest ecosystems, most management systems involve a long-term decrease in nitrogen storage in living and dead organic matter, often an increase in the rate of cycling of the nitrogen present, and periodic disturbances caused by practices ranging from plowing to harvests of forest plantations. Thus most managed systems have, at least periodically, a greater potential for nitrogen and cation leaching than natural forests. When forest is converted to agricultural land, release of stored forest floor and soil nitrogen can lead to substantial nitrate leaching. Replacement of that nitrogen by nitrogen fertilizers maintains and reinforces the predominance of nitrate in the soil solution (Woodwell 1979).

An examination of the effects of specific management practices on nitrate anion flux in particular kinds of sites can be of considerable interest. For example, immobilization of nitrogen is one of the major delays in nitrate production and loss in infertile forests. Use of the recent "complete forest removal" harvesting techniques, which involves pulling and utilizing stumps and large woody roots as well as tree stems and branches, removes what is potentially one of the most important immobilizing substrates from disturbed forests, and thus it could increase nitrate and cation leaching substantially (Vitousek 1980).

It should be stressed that disturbance to forest vegetation is not only a human-caused process (White 1979). Moreover, natural old-growth forests are likely to be a mosaic of patches of different sizes and different times since natural disturbance while managed forests are likely to be aggrading in biomass and nutrient content more or less synchronously. Consequently, long-undisturbed natural forest ecosystems may have nitrate losses systematically elevated above those of aggrading forests or forest plantations (Leak and Martin 1975; Vitousek and Reiners 1975; Bormann and Likens 1979). Moreover, some management practices, such as prescribed burning, may actually decrease cation losses by volatilizing nitrogen and sulfur which could otherwise be lost by leaching.

3.6.2.5 Leaching Losses in Other Biomes

Few hydrologically based input–output budgets of sites in which the natural vegetation is not forest have been made. Woodmansee (1978) suggested that volatilization (especially from animal urine and feces) is the most important pathway of nitrogen losses from shortgrass prairie ecosystems, and that increasing the intensity of grazing (through management) increases the losses. West and Skujins (1978) suggested that volatilization dominates nitrogen losses from deserts as well, though episodic surface runoff can also be important (Fletcher et al. 1978). It is reasonable that other pathways are more important than leaching in drier systems, since water movement through such systems in episodic and infrequent or even absent. When leaching does occur in such sites, anion availability should be very high due to precipitated carbonate, sulfate, and (in the most arid areas) chloride salts present in the profile. Studies of leaching mechanisms in wetter nonforest systems such as tallgrass prairie and tundra would be of considerable interest.

3.6.3 Conclusions

Studies of element losses from terrestrial ecosystems have been carried out in a wide range of sites, and it is clear that leaching losses of major nutrient elements below the rooting zone are much less than the amounts cycled within forests or held in available form in the soil. Two techniques for measuring such losses are in widespread use. Watershed studies effectively measure losses from large, often heterogeneous landscape units, while lysimeter studies measure element concentrations immediately below the rooting zone – or elsewhere as desired. Watershed studies are more suitable for examinations of land–water interactions, while lysimeter studies are preferred for stand-level nutrient cycling measures.

The supply and mobility of anions control anion and cation leaching losses. The major anions in humid-zone forest soils are: bicarbonate, which derives from the solution and dissociation of carbon dioxide from decomposition and root respiration; sulfate, largely from atmospheric sources; organic acids, from decomposition or root or fungal exudation; nitrate, from the atmosphere or from the oxidation of ammonium released upon decomposition; and chloride, from the atmosphere. In a very real sense, the leaching of all ions through forest soils is controlled by externally or biologically supplied anions.

Management practices can alter the supply or mobility of all of these anions. Nitrate in particular is often increased greatly in disturbed or managed ecosystems, at times becoming the predominant anion in the soil solution.

Résumé

Les cycles biogéochimiques commencent à être assez bien connus, aussi bien dans les régions tempérées que dans les régions méditerranéennes, tropicales et boréales. L'influence de l'Homme sur ces cycles est moins bien connue et la comparaison des nombreux travaux réalisés dans le monde entier est souvent difficile, en particulier parce que les méthodes sont loin d'être normalisées.

L'un des points critiques de ces cycles est la perte d'ions par lessivage; elle peut être mesurée soit par l'étude de bassins versants, soit par des mesures en cases lysimétriques; ces deux approches sont très complémentaires et l'auteur discute les avantages et les inconvénients de l'une et de l'autre.

Les anions mobiles contrôlent l'ensemble de l'équilibre ionique de tout écosystème terrestre, et l'apport d'anions mobiles est la source majeure de l'ensemble des sorties de cations et d'anions. Or, la plus grande partie des anions est fournie par les précipitations et par les végétaux. Dans les trois exemples forestiers cités (Fig. 7), l'ion $SO4^{2-}$ est de beaucoup le plus abondant, et il est véhiculé par les pluies; l'ion HCO_3^- résulte de la dissolution du gaz carbonique, et sa concentration dépend directement du pH des solutions du sol; les acides organiques tiennent une place assez faible; l'ion Cl^- reste en proportions assez constantes; les ions issus de d'azote sont en faible quantité dans ces trois forêts.

Quand la forêt est supprimée, les concentrations en ions SO_4^{2-}, Cl^-, acides organiques et HCO_3^- diminuent un peu, mais la concentration en ions NO_3^- augmente considérablement, et entraîne une forte mobilisation des cations (en particulier du magnésium et du calcium). Il semble ainsi que la forêt naturelle protège ses réserves en azote, et que celles-ci sont dilapidées quand l'Homme détruit la forêt.

Dans d'autres milieux, la volatilisation de l'azote peut être une source de „sorties" importantes, en particulier quand les herbivores laissent à l'air libre des déjections riches en azote.

Les techniques d'aménagement doivent prendre en compte cette sensibilité des systèmes écologiques.

References

Aber JD, Botkin DB, Melillo JM (1978) Predicting the effects of different harvesting regimes on forest floor dynamics in northern hardwoods. Can J For Res 8:308–316

Bernhard-Reversat F (1977) Recherches sur les variations stationelles des cycles biogéochemiques en fórêt ombrophile de Côte d'Ivoire. Cah. ORSTOM Ser Pedol 15:175–189

Bormann FH, Likens GE (1979) Pattern and process in a forested ecosystem. Springer, Berlin Heidelberg New York, 253 p

Bormann FH, Likens GE, Siccama TG, Pierce RS, Eaton JS (1974) The export of nutrients and recovery of stable conditions following deforestation at Hubbard Brook. Ecol Monogr 44:255–277

Bruckert S, Jacquin F (1969) Interaction entre la mobilité de plusiers acides organiques et de divers cations dans un sol à mull et dans un sol à mor. Soil Biol Biochem 1:275–294

Cole DW (1958) An alundum lysimeter. Soil Sci 85:293

Cole DW (1968) A system for measuring conductivity, acidity, and rate of water flow in a forest soil. Water Resour Res 4:1127–1136

Cole DW, Rapp M (1980) Elemental cycling in forested ecosystems. In: Reichle DE (ed) Dynamic properties of forest ecosystems. Cambridge Univ Press, Cambridge, pp 341–409

Cole DW, Crane WJB, Grier CC (1975) The effect of forest management practices on water chemistry in a second-growth Douglas-fir ecosystem. In: Bernier B, Winget CF (eds). Laval Univ Press, Quebec, pp 195–207

Cronan CS, Schofield CL (1979) Aluminum leaching response to acid precipitation: effects on high elevation watersheds in the northeast. Science 204:304–306

Cronan CS, Reiners WA, Reynolds RC, Lang GE (1978) Forest floor leaching: contributions from mineral, organic, and carbonic acids in New Hampshire subalpine forests. Science 200:309–311

Edmonds RL (ed) (1982) Analysis of coniferous forest ecosystems in the western United States. Dowden, Hutchinson Ross, Publishing, Co. Stroudsburg, Pa

Ellenberg H (1977) Stickstoff als Standortsfaktor, insbesondere für mitteleuropäische Pflanzengesellschaften. Oecol Plant 12:1–22

Fletcher JE, Sorenson DL, Porcella DB (1978) Erosional transfer of nitrogen in desert ecosystems. In: West NE, Skujins JJ (eds) Nitrogen in desert ecosystems. Dowden, Hutchinson and Ross, Stroudsburg Pa, pp 171–181

Gorham E, Vitousek PM, Reiners WA (1979) The regulation of chemical budgets over the course of terrestrial ecosystem succession. Annu Rev Ecol Syst 10:53–84

Graustein WC, Cromack K, Sollins P (1977) Calcium oxalate: occurrence in soils and effect on nutrient and geochemical cycles. Science 198:1252–1254

Henderson GS, Swank WT, Waide JB, Grier CC (1978) Nutrient budgets of Appalachian and Cascade region watersheds: a comparison. For Sci 24:385–397

Herrera R, Jordan CF (1981) Nitrogen cycle in a tropical rain forest of Amazonia: the case of the Amazon Caatinga of low mineral nutrient status. Ecol Bull (Stockholm) 33:493–505

Johnson DW, Cole DW (1980) Anion mobility in soils: relevance to nutrient transport from forest ecosystems. Environ Int 3:79–90

Johnson DW, Edwards NT (1979) The effects of stem girdling on biogeochemical cycles within a mixed deciduous forest in eastern Tennessee. II. Soil nitrogen mineralization and nitrification rates. Oecologia 40:259–271

Johnson DW, Cole DW, Gessel SP, Singer MJ, Minden RB (1977) Carbonic acid leaching in a tropical, temperate, subalpine, and northern forest soil. Arct Alp Res 9:329–343

Johnson DW, Hornbeck JW, Kelly JM, Swank WT, Todd DE (1980) Regional patterns of soil sulfate accumulation: relevance to ecosystem sulfur budgets. In: Shriner DS, Richmond CR, Lindberg SE (eds) Atmospheric sulfate deposition: Environmental impacts and health effects. Ann Arbor Sci, Ann Arbor Mich, pp 501–520

Leak WB, Martin CW (1975) Relationship of stand age to streamwater nitrate in New Hampshire. USDA For Serv Res Note NE-211. US Dep Agric, Upper Darby Pa, 4 p

Likens GE, Bormann FH (1974) Acid rain: a serious regional environmental problem. Science 184:1176–1179

Likens GE, Bormann FH, Johnson NM, Pierce RS (1967) The calcium, magnesium, potassium, and sodium budgets for a small forested ecosystem. Ecology 48:772–785

Likens GE, Bormann FH, Johnson NM (1969) Nitrification: importance to nutrient losses from a cutover forested ecosystem. Science 163:1205–1206

Likens GE, Bormann FH, Pierce RS, Eaton JS, Johnson NM (1977) Biogeochemistry of a forested ecosystem. Springer, Berlin Heidelberg New York, 146 p

Likens GE, Bormann FH, Pierce RS, Reiners WA (1978) Recovery of a deforested ecosystem. Science 199:492–496

McColl JG, Cole DW (1968) A mechanism of cation transport in a forest soil. Northwest Sci 42:134–140

Meyer JL, Likens GE (1979) Transport and transformations of phosphorus in a forest stream ecosystem. Ecology 60:1255–1269

Nye PH, Greenland DJ (1960) The soil under shifting cultivation. Commonw Bur Soils Tech Bull 51:156 p

Nye PH, Tinker PB (1977) Solute movement in the soil–root system. Univ California-Press, Berkeley, 342 p

Raney WA (1960) The dominant role of nitrogen in leaching losses from soils of humid regions. Agron J 52:563–566

Rapp M (1971) Cycle de la matière organique et des elements minéraux dans quelques ecosystèmes méditerranéens. CNRS, Paris

Rosswall T (1976) The internal cycle between vegetation, microorganisms, and soil. Ecol Bull (Stockholm) 22:157–167

Smith MS, Tiedje JM (1979) Phases of denitrification following oxygen depletion in soil. Soil Biol Biochem 11:261–267

Stone EL, Swank WT, Hornbeck JW (1978) Impacts of timber harvest and regeneration on stream flow and soils in the eastern deciduous region. In: Youngberg CT (ed) Forest soils and land use. Colorado State Univ Press, Fort Collins, pp 516–535

Swank WT, Waide JB (1980) Interpretation of nutrient cycling research in a management context: evaluating potential effects of alternative management strategies on site productivity. In: Waring RH (ed) Symp For: Fresh perspectives from ecosystem analysis. Oregon State Univ Press, Corvallis, pp 137–158

Swanson FJ, Fredriksen RL, McCorison FM (1982a) Material transfer in a western Oregon forested watershed. In: Edmonds RL (ed) Analysis of coniferous forest ecosystems in the Western United States. Dowden, Hutchinson and Ross, Stroudsburg Pa, pp 233–266

Swanson FJ, Sedell JR, Triska FJ (1982b) Land-water interactions: the riparian zone. In: Edmonds RL (ed) Analysis of coniferous forest ecosystems in the western United States. Dowden, Hutchinson and Ross Stroudsburg Pa, pp 267–291

Van Cleve K, Viereck L (1981) Forest succession in relation to nutrient cycling in the boreal forest. In: West D, Shugart HH, Botkin DB (eds) Forest succession: Concept and application. Springer, Berlin Heidelberg New York, pp 185–211

Vitousek PM (1977) The regulation of element concentrations in mountain streams in the northeastern United States. Ecol Monogr 47:65–87

Vitousek PM (1981) Clearcutting and the nitrogen cycle. Ecol Bull (Stockholm) 33:631–642

Vitousek PN, Melillo JM (1979) Nitrate losses from disturbed forests: patterns and mechanisms. For Sci 25:605–619

Vitousek PM, Reiners WA (1975) Ecosystem succession and nutrient retention: a hypothesis. BioScience 25:376–381

Vitousek PM, Gosz JR, Grier CC, Melillo JM, Reiners WA, Todd RL (1979) Nitrate losses from disturbed ecosystems. Science 204:469–474

Webster JR, Patten BC (1979) Effects of watershed perturbation on stream potassium and calcium dynamics. Ecol Monogr 49:51–72

West NE, Skujins JJ (1978) Nitrogen in desert ecosystems. Dowden, Hutchinson and Ross. Stroudsburg Pa, 307 p

White PS (1979) Pattern, process, and natural disturbance in vegetation. Bot Rev 45:229–299

Wiklander G (1981) Rapporteur's comment on clearcutting. Ecol Bull (Stockholm) 33:642–647

Woodmansee RG (1978) Additions and losses of nitrogen in grassland ecosystems. BioScience 28:448–453

Woodwell GM (1979) Leaky ecosystems: nutrient fluxes and succession in the pine barrens vegetation. In: Forman RTT (ed) Pine Barrens, ecosystem, and landscape. Academic Press, London New York, pp 333–343

Section 4 Species Physiological Characteristics

4.1 The Determinants of Plant Productivity-Natural Versus Man-Modified Communities

H. A. MOONEY and S. L. GULMON

4.1.1 Introduction

There is great variation in the terrestrial primary productivity of different climatic regions of the earth as well as within a given climatic region dependent on local habitat properties. This variability is a result of the resources available to the biota – water, light, and nutrients – as well as the prevailing temperatures which regulate resource availability. In addition, of course, site productivity is determined by the particular properties of the resident plants. Here we explore the interaction between habitat resources and plant potential in determining site productivity.

Man has greatly influenced habitat productivity, both intentionally, through agriculture and forestry, and unintentionally, through a variety of cultural practices. If we are to assess the impact of man on the earth's productive capacity, we must understand more fully resource–plant interactions and how man may be altering these interactions.

Here we present a specific hypothesis relating habitat resources to site productivity which has implications in the management of landscapes. The hypothesis is based on the proposition that primary productivity of a habitat is determined by the resources available at the site. We propose that this relationship is independent of the particular species present, assuming they are adapted to conditions at the site, and that evolution has produced community aggregations which fully utilize the available resources of the habitat. The data to support these latter propositions are few. It is our purpose here to explore these relationships since, even if only usually true, they have important implications in the management of resources to maximize biomass yields.

One example of the potential importance of these relationships is the use of plant biomass as fuel or as substrate for fuel production. One of the primary economic criteria needed to evaluate the feasibility of this technology is the rate of production of biomass per unit of land. Much of the promise in the biomass fuels approach rests on the assumption that very high biomass yields may be obtained on a sustained basis from certain crops or crop combinations. This assumption has resulted in a search for particularly promising plants for biomass farms. If there are intrinsic limits to plant productivity for a given resource base, which have already been realized through the process of evolution, then examination of productivities of natural systems will provide a good indication of the attainable limits of biomass production in any combination of climate and soil type. Also, this means that exotic species or species combinations would offer no greater biomass accumulation potential than native plants. If, however, the above proposition is not true, then understanding the basic limits to the production process will provide the most promising directions for plant breeding programs to maximize biomass production.

4.1.2 Comparisons of Productivity

Surprisingly, we cannot examine the evolutionary limit proposition directly because of the lack of data to make meaningful comparisons – for example, it is very difficult to compare the potential productivity of differing land surfaces. Although there are abundant data on the productivity of natural vegetation, crops, and managed forests, these data are not directly comparable. Crop data usually apply to the yields of a particular plant part, such as seeds, and those for plantations to marketable timber. In both cases, below ground productivity is excluded. Even in ecological studies of plant production, below ground productivity is often not noted.

Even if the total biomass accumulation was known for various natural and cultivated community types in a given climatic region, direct comparisons still could not be made, since the communities are growing on soils with differing nutrient resources available. That is, the cultivated crops are heavily fertilized, the plantations possibly lightly so, and the natural vegetation not at all. Furthermore, the cultivated crops are generally grown on the deeper, more naturally fertile soils, and the natural vegetation remains on slopes with shallow soils. Therefore, our analysis of the potential productivity of different sites must be somewhat indirect.

4.1.3 The Components of Plant Productivity

4.1.3.1 The Biotic Component

There are four principal components to productivity: Leaf area index (LAI), net photosynthetic rate (P_s), root-to-shoot ratio (R/S), and the duration of a leaf crop (L.D.). Each has a separate effect on carbon gain, but each also affects the magnitudes of the other three. It is a consequence of this interdependence that productivity can be maximized, and does not increase indefinitely. In addition to the four principal biological components of production, the costs of producing and maintaining tissue are also important and vary among plant types. These relationships are not discussed here (see Penning de Vries et al. 1974 for an analysis of these costs).

The photosynthetic rate, P_s, is the most fundamental component of productivity. It defines the theoretical upper limit for the rate of biomass increase since a plant cannot gain weight any faster than its leaves can fix new carbon. P_s rates of over 59 $\mu mol\, m^{-2} s^{-1}$ have been measured (Mooney et al. 1976), but these rates were sustained only in full sunlight. This precludes any significant degree of biomass buildup, since self shading among leaves in a canopy reduces photosynthetic rates considerably. For broadleaf forest trees with a full canopy, P_s rates range from 16 μmol $m^{-2}s^{-1}$ at the top of the canopy to 3 μmol $m^{-2}s^{-1}$ for leaves at the canopy bottom (Larcher 1980).

The leaf area index (LAI) is closely allied to the photosynthetic rate in defining limits of primary production. It is the total area of leaf exposed per unit area of ground. Carbon gain (weight of carbon fixed/unit time) per unit ground area is equal to the LAI times the mean P_s of the leaves. The limit to the LAI is controlled by light penetration through the canopy. When there is insufficient light at the bottom of a canopy to sustain photosynthesis, addition of leaves will result in a net decrease of total carbon gain. The maximum benefit of high LAI depends on arrangement of the leaves to expose the largest leaf surface to the highest average P_s rate. The highest leaf area indices occur in canopies with vertical leaves or in plants with small needlelike leaves. In natural forests LAI varies from about 3 to over 11 (Kira 1975).

The annual rate of carbon gain is equal to the mean P_s rate times the LAI times the period of leaf duration, L.D. A year-long L.D. is potentially the most productive, but deciduous leaves are mandated in many vegetations by climatic factors. Where both evergreen and deciduous types ço-occur, it appears that deciduous leaves have higher mean P_s rates, which offset their shortened duration; productivities of the two types can be similar (Miller and Mooney 1974).

The final component of production is the root-to-shoot ratio. The effect of root growth versus new leaf growth on productivity is best understood by considering new leaf production analogous to compounding interest, and other production to withdrawing interest; the former mode results in the highest total productivity. However, two other factors must be considered. First, where an inadequate supply of nutrients, water, or light is limiting photosynthesis, additional root or stem growth may increase the P_s rate by increasing the supply of the limiting factor to the leaves. Second, if the maximum LAI has already been attained, additional leaf growth will not increase productivity.

4.1.3.2 Environmental Influences on the Biological Components of Productivity

The relationship of a number of these biological productivity components to availability of the three basic resources, light, water, and nutrients, is indicated in Fig. 1. As can be seen, photosynthetic capacity generally increases asymptotically with an increase in either nutrients, water or light. The resource level at which saturation occurs is related to the availability of other limiting resources as well as to both growth conditions and the plant's evolutionary history (Mooney and Gulmon 1979). An increase in resource level also leads to a high leaf area index, a component with a high productivity. This curve is also asymptotic due to light extinction in the canopy.

Leaf duration is broadly constrained by climatic factors since many plants shed the entire canopy during periods of extreme drought or cold. Within this context, longevity is generally inversely related to productivity, although cause and effect are intertwined (Fig. 1). Productivity will be low if either light, water, or nutrients are severely limiting to photosynthesis (see above), or if the growing season is

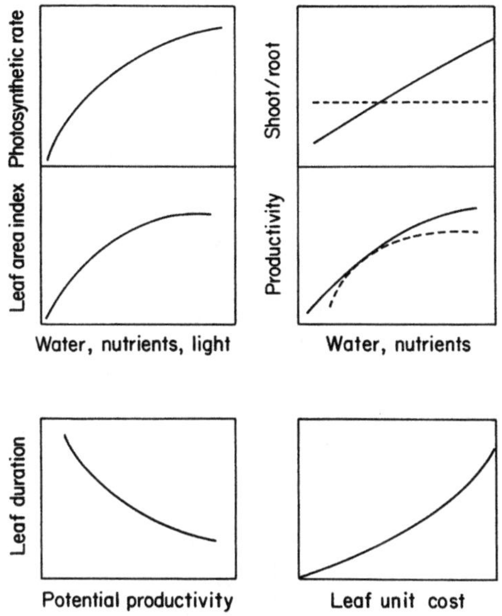

Fig. 1. The relationship between the biological components of primary productivity (photosynthetic rate, leaf area index, shoot/root ratio, leaf duration, and leaf cost) and resource level. All axes increase in value in the positive direction. *Dotted curves* indicate the effect on productivity of increasing water or nutrients when shoot/root allocation remains constant. *Solid curves* in the same pair of graphs indicates the gain in productivity when shoot/root allocation changes in response to resource level

limited by low temperature or drought. Under such conditions, leaves are generally longer-lived and justify the investment of carbon in their construction (Grime 1979). Bound by the constraints of limited resource availability, productivity will be enhanced by longer-lived leaves (Schulze et al. 1977; Ewers and Schmid 1981). By contrast, when resource availability is high, high rates of carbon gain permit rapid growth, and the consequent self-shading necessitates more rapid leaf turnover. In this case, productivity is enhanced by redeploying carbon and nitrogen to the outside or top of the canopy and shedding older leaves.

Although longer-lived leaves may be considered a response to low potential productivity, they are generally more costly to produce (Fig. 1) (Mooney and Gulmon 1982), and are generally more sclerophytic with lower concommitant photosynthetic rates (Medina 1981). Thus, peak productivity may be lower with longer-lived leaves even though yearly productivity is enhanced.

Allocation to roots increases as either water or nutrients become more limiting relative to light (Fig. 1). As discussed earlier, such allocation patterns result in partial compensation for the reduction in these resources. That is, although increased allocation to roots results in decreased production of photosynthetic tissue, more resources are captured for that photosynthetic tissue which is produced (Fig. 1).

In most environments, availability of some or all of the basic resources is not constant. Disturbed habitats, for example, are generally characterized by a large pulse of nutrient release (Vitousek and Reiners 1975), and light intensity in forest understory may vary from full sunlight to deep shade within minutes. The plant

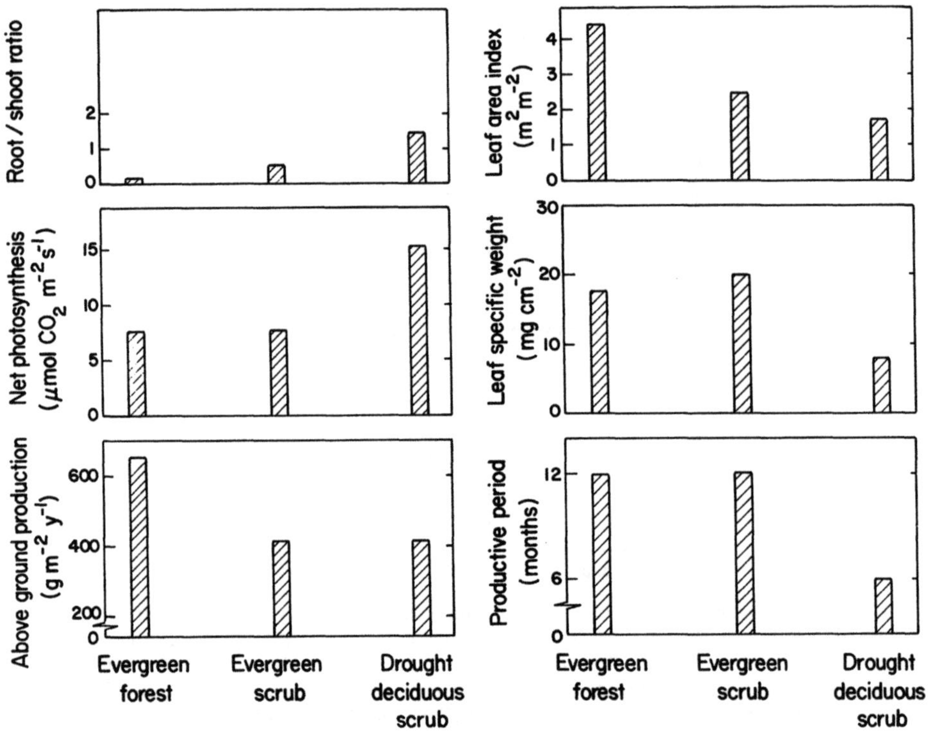

Fig. 2. Productivity characteristics of a series of mediterranean-climate communities. (Ehleringer and Mooney 1982)

photosynthetic response to variable resource levels depends on the periodicity of the variation and the capacity to respond to peaks in resource availability. Disturbed habitats are colonized by species which utilize the nutrient flush through a very high photosynthetic rate and high initial biomass production, especially of foliage (Bazzaz 1979). In general, however, this high productivity is not sustained over the long term.

4.1.3.3 Interactions of Productivity Components and Resource Level in Natural Communities

The limiting value of stand productivity is defined by the limits to the four component processes (photosynthetic rate, leaf area index, leaf duration and root/shoot ratio). The interrelationships between these components in a series of natural communities is indicated in Fig. 2.

Progressing from moist to dry sites in mediterranean-climate regions, communities change from evergreen forest to evergreen scrub and then to a drought-deciduous scrub. As would be expected, the evergreen forest, as contrasted with the evergreen scrub, has the highest productivity, since it has the lowest root-to-shoot

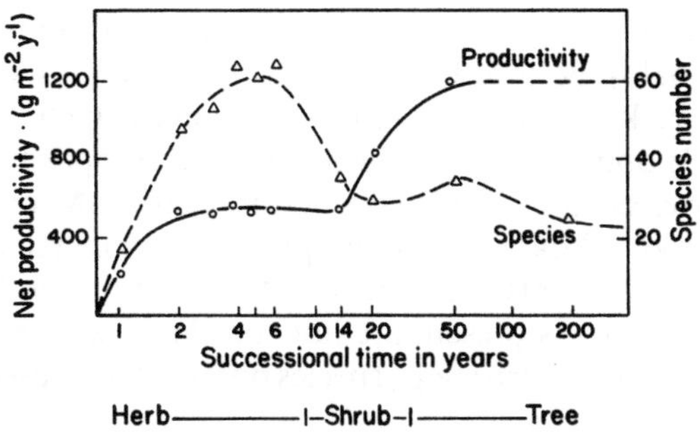

Fig. 3. Net primary productivity and numbers of native species in a successional sequence in Long Island, New York. (Whittaker 1975, data of Holt and Woodwell)

ratio and the highest leaf area index, although photosynthetic capacities and productive periods are equal.

In contrast, the evergreen scrub and drought-deciduous scrub communities have identical above-ground productivities, but apparently this is attained in quite different manners. Factors that lead to low production in the drought-deciduous community, such as a short productive period, a high root-shoot ratio and a low leaf area are compensated for by a low allocation to leaf material and, most important, a high inherent photosynthetic capacity.

With this background we now explore further the proposition of the environmental and evolutionary limits to plant productivity. For this we first view the process of succession.

4.1.4 Succession and Plant Productivity

The succession process is normally characterized by an initial increase in the number of species occupying the same piece of landscape. If each new addition to the community invaded it by exploiting a new resource, then it could be predicted that site productivity would increase. Is this the case? Remarkably, data on this fundamental point are generally lacking.

Odum's (1960) classic study of old field succession in South Carolina covered a time sequence of only seven years. During this period species number changed from five initial dominant annual and perennial herbs to a community of nearly 20 dominants in which the perennial *Andropogon* species contributed the greatest biomass. The highest annual productivity in the sequence occurred during the first two years; this was attributed to residual effects of fertilizer from the previous agricultural use of the fields. In subsequent years, although species numbers increased, annual productivity remained fairly constant. Odum predicted that as succession proceeded to dominance by tree forms, annual productivity would increase. Thus he envisaged productivity increasing in a stepwise fashion with time, with each step representing a new major life form which could tap a resource not available to the previous members of the community. Woodwell (1974) indicates that this is indeed the case in the oak-pine forests of coastal New York (Fig. 3). He found that annual productivity of the early herb stage was about 800 g $m^{-2}y^{-1}$, and the 50-year-old oak-pine forest about 1,200 g $m^{-2}y^{-1}$. Woodwell predicted that the climax oak-hickory forest would have an even higher net productivity. Ovington et al. (1963) obtained similar results from a comparison of net primary productivity in prairie, savanna, and oakwood in Central Minnesota. Whittaker (1975), citing the Woodwell data, notes that productivity increases with time with "increasing use of environmental resources by the community." He further indicates that the species numbers present are not correlated with net production. As Odum found in the perennial herb stage, species numbers increased but net productivity did not.

Peet (1982) reviewed available data on productivity during secondary succession in temperate forests, Brown (1980) considered biomass accumulation in tropical forests, and Shugart and West (1981) used computer simulation models to generate biomass accumulation patterns in various forest types. The general pattern described by these data shows an initial rapid rise in primary productivity with canopy closure followed by stabilization, or a slow decline in productivity with time. There is very little evidence that new species in later successional stages result in higher productivity. Peet (1982) cites several examples in which the climax tree species are less productive than the early successional forest.

These limited data indicate that within growth forms there is a division of the resource pool in the habitat, and new resources become available only with the entry of a completely different major growth form. This would explain the lack of relationship between species diversity and productivity (Grime 1979; Drury and Nisbit 1973) (Fig. 3).

4.1.5 Succession Anomalies

Succession appears to provide two anomalies to the proposed relationship between productivity and the resource base. As noted earlier, disturbed sites are often characterized by release of a pulse of nutrients and are colonized by species with high photosynthetic rates. This should result in high net productivity, but the first successional state, composed of annuals and herbaceous perennials, is much less productive than later woody stages. This occurs because the annual species deplete the nutrients rapidly and then switch completely to reproduction. Allocation of nitrogen is diverted from leaves to reproductive parts, and the leaves "self-destruct" (Sinclair and de Wit 1976). There is no living biomass carry-over to the following year, except for seeds, which precludes cumulating the effects of exponential growth. Thus, long term productivity is sacrificed for higher short term productivity (Fig. 4).

Another reason for this anomaly is that the initial production of woody stems and larger root systems reduces productivity, but over an extended time period both the light environment and the water resource base which can be exploited are expanded. Also, nutrients can be stored in permanent structures and recycled (Ryan and Bormann 1982); both these factors lead to high productivity in the longer-term (Fig. 5).

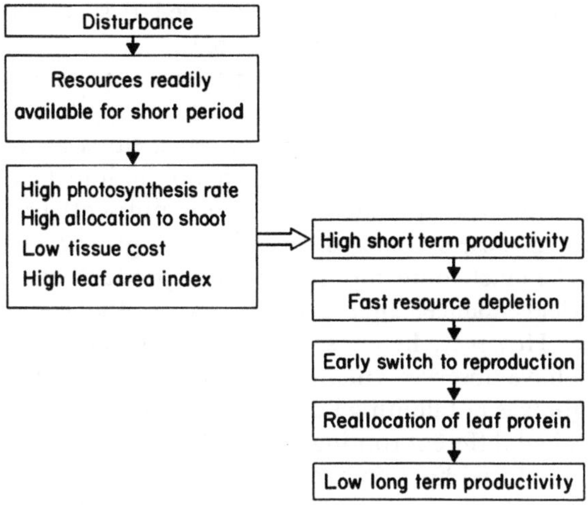

Fig. 4. Productive characteristics of plants occupying disturbance habitats

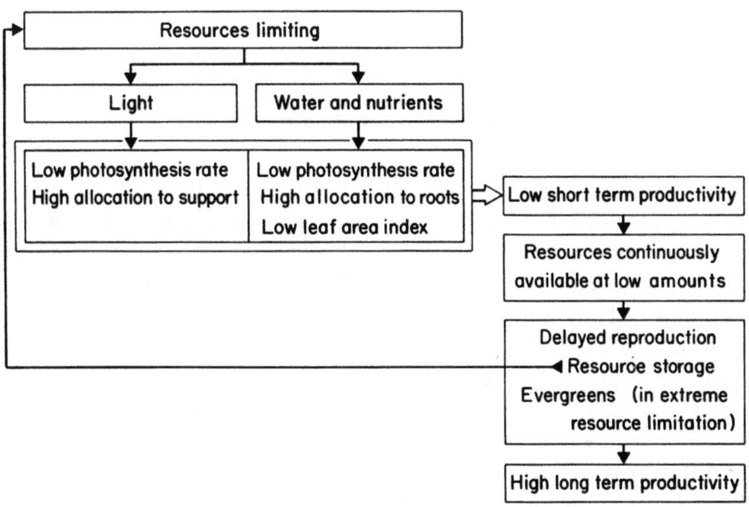

Fig. 5. Productive characteristics of plants occupying later successional stages

The second anomaly is why there appears to be no increase in productivity with the addition of new species within a growth form, if it is assumed that new species enter by exploiting a new or expanded resource base.

There are three possible explanations for this: (1) new species do increase productivity, but the increment cannot be detected given the high variability of productivity measurements in a natural system; (2) the resource base is expanded by newly entering species, but the cost of additional non-photosynthetic components (root, stem, or storage tissue) needed to mine these resources offsets the potential gain in productivity; (3) new species do not expand the resource base, but simply displace existing species. An example of (3) might be late successional trees which are able to reproduce in the shade of existing trees.

A final consideration in the relationship between successional sequence and productivity concerns the primary limiting resource in the system. The studies cited have all referred to essentially closed canopy systems, either at the herbaceous or arboreal level, in which it appears that light is a major limiting factor. In other systems, such as deserts or serpentine soil (or other low-nutrient) grasslands, the canopy does not close. In these systems, differential root growth or nutrient uptake among species may be much more important to total productivity. Trenbath (1974) has observed that yield enhancement in mixtures has generally not been experimentally demonstrated. However, Berendse (1981), Bebawi and Naylor (1981), and Gulmon (unpubl.) have observed enhancement of yield by a mixture. In the first and third cases, light was not limiting because of low nutrient availability, and the component species had different rooting depths. In the second case, frequent clipping probably also reduced light competition.

It is clear that further measurements are needed, particularly in more successional sequences, if we are to establish the universality of the constancy of productivity on a site within a growth form. Further, we need a quantitative test of the proposals given above.

4.1.6 Convergence in Productivity

The resource limitations to productivity were illustrated above. Further, some evidence was given that, for a given resource base, productivity is independent of species numbers, at least for a given growth form. What of the proposition that evolution has produced aggregations of species that utilize all of the resources of the habitat? The only evidence we can cite for this is indirect. In different regions of the world comparable community types exist where climates are similar. The major biomes, evergreen and deciduous forests, grasslands, etc., are found on all the continents, though they are composed of entirely different species aggregations. The homoclimates in which these analogs are found can be considered to represent similar resource areas. If we assume that productivity is resource limited and that all of the resources of the habitat are being used, then we would expect comparable productivities in the homoclimatic regions.

Whittaker and Woodwell (1971) have noted that the net primary productivities of climax temperate forests of the world converge at a value of 1,200 to 1,500 g $m^{-2}y^{-1}$. Further, the dimension analyses of these forests are comparable. Low-elevation temperate forests on mesic sites generally have a basal area of 50 to 64 m^2ha^{-1}. Mean tree height ranges from 20 to 35 m, and above-ground biomass from 400 to 600 t ha^{-1}. Above-ground net productivity is between 1,000 and 1,200 g $m^{-2}y^{-1}$, and foliage production between 320 and 420 g $m^{-2}y^{-1}$.

As with the temperate forests, there is also convergence in structural and functional attributes of mature mediterranean-climate scrub communities. Standing biomass ranges from 20 to 40 t ha^{-1}. Annual biomass accumulation is around 150 g m^{-2} and litter fall 200 g m^{-2}. Above-ground annual production averages

400 g m^{-2} (Mooney 1977). Vegetations of this type form a closed canopy and attain an average height of less than 2 m. Leaf area indices range between 2 and 3 (Mooney et al. 1977).

These observations indicate that under similar climates, climax vegetations that are geographically isolated and have entirely different species compositions will still have comparable productivities and structures.

4.1.7 Agricultural Versus Natural Community Productivity

As noted earlier, few data exist to enable comparisons between the productivities of managed and natural ecosystems to be made. Of course, it is of little interest to the agriculturalist if the native vegetation is more or less productive than crops, since it is only the latter that are of immediate economic value. This is not the case, however, if biomass for energy conversion is the product of interest.

The high yield of crops has been achieved principally through increasing the resource base of the habitat through fertilization (de Wit 1968) and irrigation. In an arid region in the northern Negev Desert, van Keulen (1975) found that the productivity of the annual grasses was independent of species composition. Further, the growth rates of the native annuals when fertilized were similar to that of the wheat variety utilized locally (as well as to growth rates of wheat in the Great Plains). These data would indicate, again, that it is the resources of the habitat rather than the plant species that are the determinants of primary productivity.

4.1.8 Conclusions

The productive potential of a site is determined primarily by the available water, light, and nutrients. Herbaceous growth forms all have similar resource-gathering capacities and hence potential productivities. Such plants have fast growth rates and efficiently use the declining resource base of disturbed or ephemeral habitats. As resources decline to a certain level, such plants switch to reproduction, foregoing further growth. Although native herbaceous species of a given habitat may vary in their particular resource-gathering strategies, a balance exists between costs and benefits of resource acquisition such that community productivity is not enhanced.

Plants with characteristics that lead to lower short-term growth can persist in habitats with lower resource availabilities at any given instant. Woody plants, by storing resources, delaying reproduction, and building resource-gathering structures which last many years, can have higher long-term productivity than herbaceous plants in spite of lower short-term growth. However, it appears again that the productivity of a site of a given resource level dominated by woody plants may be independent of species composition.

There are many implications of these possibilities. One is that attempts to increase the yield of a site must of course be accompanied by an increase in the re-

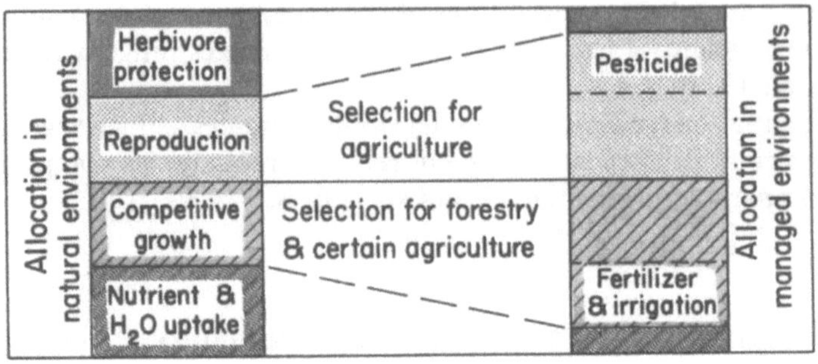

Fig. 6. Influence of selection on allocation

source level of the site. Improving yields of biomass farms by utilizing species with fast closure times is in essence utilizing species with rapid resource use. Subsequent harvests will decline unless additional resources are added. This would be true of slower growing species also, but the amounts of resources needed would be less. There is evidence of a decline in yields of second rotation forests in Australia (Ovington 1979).

The productive components of plants are leaf duration, photosynthetic capacity, leaf area index, allocation pattern, and tissue costs. The balance of these components is suited to the resource base of the habitat in which a given plant has evolved. The balance can be shifted by breeding, but there will be consequences on the resource-gathering capacity of the plant. For example, plants bred for components such as fast growth rate will generally have higher allocation to leaf tissue. This shift may come at the expense of allocation to resource-gathering roots, which then must be compensated for by the addition of these resources by management practices (Fig. 8). Also, plants bred for a greater palatability may have a reduction in allocation to herbivore defense, which could lead to a greater yield if the carbon were allocated to leaf instead. But, here again, energy from pesticides will have to be applied through management programs to realize this potential increased yield.

As we learn more about the productive components of plants and of their interactions with the resources available in the habitat, we should be able to rationally manage landscapes to strike the desired balance between biomass yield and resource use.

Résumé

L'une des rares hypothèses écologiques générales que l'on pourrait essayer de tester sur tout la gamme des milieux du globe est que, dans chacun de ces milieux, la biocénose qui s'établit spontanément est celle qui „produit" annuellement la biomasse maximale, indépendamment de la composition floristique.

Les quatre principales composantes de la production sont l'indice foliaire, le taux de phytosynthèse, le ratio tiges/racines et la durée de la saison de végétation. Les trois premières augmentent quand les ressources de base (eau, éléments nutritifs, lumière) augmentent; les auteurs pensent que, à l'opposé, la durée de vie des feuilles a tendance à être inversement proportionnelle à la fertilité. En fait, dans les communautés naturelles, les phénomènes sont plus complexes, en particulier parce que la plupart des climats comportent des saisons „défavorables", et la Fig. 2 montre les résultats atteints par trois types de communautés.

Au cours d'une succession, la production vraie a tendance à augmenter, mais elle se stabilise (ou même quelquefois décroît) dans les stades les plus mûrs. Ce schéma général est interrompu par les „perturbations", qui accroissent temporairement les flux internes du système, avant qu'il retrouve sa tendance initiale.

Dans les cultures, l'Homme modifie les composantes du système, et l'une de ses actions les plus fréquentes est d'ajouter des éléments fertilisants. Il serait nécessaire de faire des comparaisons précises entre le fonctionnement de la végétation naturelle et celui de la végétation cultivée, en calculant les „coûts et bénéfices" respectifs dans les deux cas, car les „bénéfices" apparents obtenus par exemple par le choix de cultivars utilisant rapidement les ressources disponibles sont souvent compensés par la perte d'autres qualités.

References

Bazzaz FA (1979) The physiological ecology of plant succession. Ann Rev Ecol Syst 10:351–372

Bebawi FF, Naylor REL (1981) Performance of pure and mixed stands of forage grasses at the establishment phase. I. Two and three species mixtures. New Phytol 89:347–356

Berendse F (1979) Competition between plant populations with differing root depths. I. Theoretical considerations. Oecologia 43:19–26

Berendse F (1981) Competition between plant populations with different rooting depths. II. Pot experiments. Oecologia 48:334–341

Brown S (1980) Rates of organic matter accumulation and litter productions in tropical forest ecosystems. In: Brown S, Lugo AE, Liegel B (eds) The role of tropical forests in the world carbon cycle. US Dep Energy, Gainesville Fla, pp 118–139

Drury WH, Nisbit ICT (1973) Succession. J Arnold Arbor 54:331–368

Ehleringer J, Mooney HA (1982) Photosynthesis and productivity of desert and mediterranean-climate plants. In: Lange OL, Nobel PS, Osmond CB, Ziegler H (eds) Physiological plant ecology, Vol 12D. Springer, Berlin Heidelberg New York

Ewers FW, Schmid R (1981) Longevity of needle fascicles of *Pinus longaeva* (bristlecone pine) and other North American pines. Oecologia 51:107–115

Grime JP (1979) Plant strategies and vegetation processes. Wiley, New York, p 222

Keulen van H (1975) Simulation of water use and herbage growth in arid regions. PUDOC Cent Agr Publ Doc, Wageningen, p 176

Kira T (1975) Primary production of forests. In: Cooper JP (ed) Photosynthesis and productivity in different environments. Cambridge Univ Press, Cambridge, pp 5–40

Larcher W (1980) Physiological plant ecology, 2nd edn. Springer, Berlin Heidelberg New York, p 303

Medina E (1981) Nitrogen content, leaf structure and photosynthesis in higher plants. A report to the UNEP study group on photosynthesis and bioproductivity

Miller PC, Mooney HA (1974) The origin and structure of American arid zone ecosystems. The producers: interactions between environment, form, and function. In: Proc 1st Int Congr Ecol. The Hague, Netherlands

Mooney HA (1977) The carbon cycle in mediterranean-climate evergreen scrub communities. In: Mooney HA, Conrad CE (eds) Proc Symp. The environmental consequences of fire and fuel management in mediterranean ecosystems. USDA For Serv Gen Tech Rep WO-3. US Dep Agric, Washington DC, pp 107–115

Mooney HA, Gulmon SL (1979) Environmental and evolutionary constraints on the photosynthetic characteristics of higher plants. In: Solbrig OT, Jain S, Johnson GB, Raven PH (eds) Topics in plant population biology. Columbia Univ Press, New York, pp 316–337

Mooney HA, Gulmon SL (1982) Constraints on leaf structure and function in reference to herbivory. Bio Science 32:198–206
Mooney HA, Ehleringer J, Berry HA (1976) High photosynthetic capacity of a winter annual in Death Valley. Science 194:322–324
Mooney HA, Kumerow J, Johnson AW, Parsons DJ, Keeley S, Hoffmann A, Hays RI, Giliberto J, Chu C (1977) The producers – their resources and adaptive responses. In: Mooney HA (ed) Convergent evolution in Chile and California mediterranean climate ecosystems. Dowden, Hutchinson and Ross, Stroudsburg Pa, pp 85–143
Odum ED (1960) Organic production and turnover in old field succession. Ecology 41:34–39
Ovington JD (1979) Some considerations of forest use and energy flow. In: Boyce SG (ed) Biological and sociological basis for a rational use of forest resources for energy and organics. USDA For Serv, Southeast. For Exp Stn, Asheville NC, pp 23–26
Ovington JD, Heitkamp D, Lawrence DB (1963) Plant biomass and productivity of prairie, savanna, oakwood, and maize field ecosystems in Central Minnesota. Ecology 44:52–63
Peet RK (1982) Changes in biomass and production during secondary forest succession. In: Shugart HH, Botkin DF, West D (eds) Forest succession: Concept and application. Springer, Berlin Heidelberg New York
Penning de Vries FWT, Brunsting AHM, Laar HH (1974) Products, requirements, and efficiency of biosynthesis: a quantitative approach. J Theoret Biol 45:339–377
Ryan DF, Bormann FH (1982) Nutrient resorption in northern hardwood forests. Bio Science 32:29–32
Schulze E-D, Fuchs M, Fuchs MI (1977) Spatial distribution of photosynthetic capacity and performance in a montane spruce forest of Northern Germany. III. The significance of the evergreen habit. Oecologia 30:239–248
Shugart HH Jr, West DC (1981) Long-term dynamics of forest ecosystems. Am Sci 69:647–652
Sinclair TR, Wit de CT (1976) Analysis of the carbon and nitrogen limitations to soybean yield. Agron J 68:319–324
Trenbath B (1974) Biomass productivity of mixtures. Adv Agron 26:177–260
Vitousek PM, Reiners WA (1975) Ecosystems succession and nutrient retention: a hypothesis. Bio Science 25:376–381
Whittaker RH (1975) Communities and ecosystems, 2nd edn. MacMillan, New York, p 385
Whittaker RH, Woodwell GM (1971) Measurement of net primary production of forests. In: Duvigneaud P (ed) Productivity of forest ecosystems. Proc Brussels Symp Oct 1969. UNESCO, Paris, pp 159–175
Wit de CT (1968) Plant production. Misc Pap Landbouwhogesch Wageningen 3:25–50
Woodwell GM (1974) Success, succession and Adam Smith. Bio Science 24:81–87

4.2 Plant Growth and its Limitations in Crops and Natural Communities

B. Saugier

4.2.1 Introduction

The growth of a plant is limited by the scarcity of resources in its environment and by its capacity to gather these resources and to use them in its metabolism. Changes in the environment occur naturally or by man's action, selecting adapted plant species and genotypes.

The present work describes a few examples of such adaptation to a specific feature, man's needs in the case of wheat, or drought in the case of *Dactylis glomerata* and *Pinus* sp. It then compares the use of energy and raw materials in several crops and natural communities and attempts to define what is meant by available resources at the community level.

Basic metabolic processes appeared early in evolution and are thus common to most higher plants. This implies a relative constancy in the elemental composition of active tissues and a certain balance between the various fluxes of matter absorbed by the plant. These fluxes must also be adjusted to the growth capacities of the plant. Relations between growth and carbon balance are briefly discussed in the next section as an introduction to specific examples of adaptation.

4.2.2 Plant Growth Parameters

The growth of a plant is reflected in its carbon balance, i.e., the balance between carbon gains by photosynthesis and carbon losses by respiration during both day and night. At a given time carbon uptake is determined by the leaf area of the plant, by the photosynthetic capacity of its various leaves and by the existing environmental conditions.

4.2.2.1 Photosynthesis

Leaf photosynthesis is dependent on several environmental variables; absorbed light, concentration of oxygen and carbon dioxide in the air, temperature, water, and mineral status of the leaf (Fig. 1).

To account for these variations of leaf photosynthesis with the environment, and also for variations among species or genotypes, various models of leaf photosynthesis have been proposed (e.g., see Chartier and Prioul 1976). Emphasis has been given to the understanding of light and CO_2 responses. Photosynthesis at low light intensities is affected by various leaf parameters; radiation absorptance, maximum quantum yield, and the photorespiration/photosynthesis ratio.

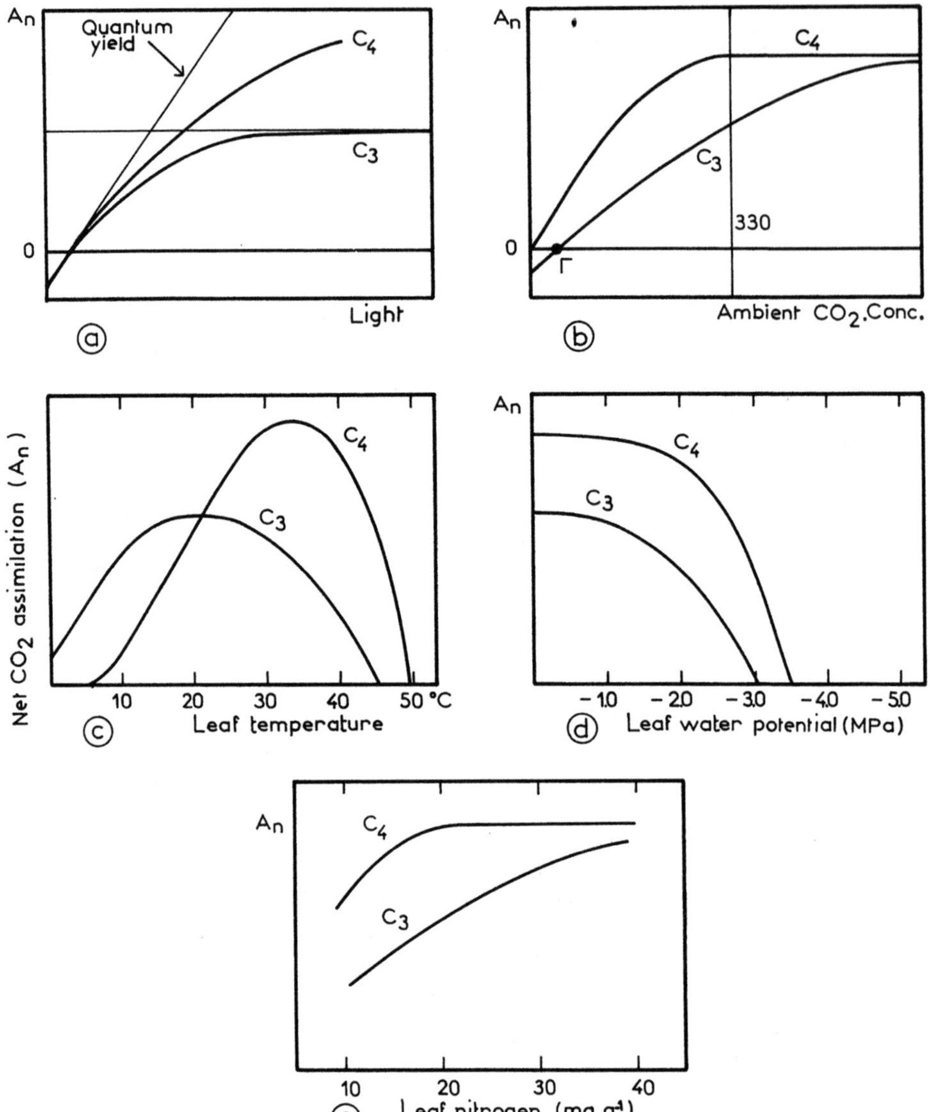

Fig. 1a–e. Patterns in variations of leaf photosynthesis with various limiting factors; **a** light; **b** CO_2; **c** temperature; **d** leaf water potential; **e** leaf nitrogen

At high light intensities, photosynthesis becomes limited by either CO_2 diffusion to the chloroplasts or by the enzymatic capacities of the photosynthetic apparatus to utilize the available CO_2. Since the work of Gaastra (1959), CO_2 diffusion is usually treated by analogy with the circulation of an electrical current through a series of resistances opposed respectively by the boundary layer and by the epidermis (stomates and cuticle) in gas phase, and by the mesophyll in liquid (dissolved) phase. When CO_2 is limiting, the carboxylation process is simply mimicked by an additional, so-called carboxylation resistance. When it is in excess, then photosynthesis is artificially limited to a maximum value which can be quantified (Chartier and Prioul 1976). Improving on this empirical treatment, a recent model (Farquhar et al. 1980) incorporates enzymatic activities as measured in vitro in a more mechanistic model of photosynthesis and photorespiration in C_3 plants. This model emphasizes the significance of the concentration of RuBP carboxylase, which often represents half of the total soluble leaf protein pool. Thus the effects of a nitrogen shortage on photosynthesis may be simulated via a change in RuBPcase level.

The various characteristics determining leaf photosynthesis in a given environment are not as rigidly fixed as are the physical properties of molecules. Photosynthetic capacity is a result of the expression of the genotype in a particular environment. Thus when comparing photosynthetic capacities of various species one distinguishes between the possible variations caused by genotypic diversity within species as well as by the plasticity of the phenotype.

Photosynthesis may also be regulated by the assimilate requirement of the plant, which varies with phenological stage. The plant appears to achieve a balance between carbon production and carbon utilization (e.g., Ho 1979), although knowledge on the underlying mechanism is still fragmentary.

4.2.2.2 Respiration

Dark respiration of plant tissues has been conveniently separated into a growth component corresponding to the elaboration of new plant material, and a component corresponding to the maintenance of the functional structure of the plant and similar to the basal metabolism of the animals.

Growth respiration has been evaluated from experimental results using various techniques (Mc Cree 1974; Ryle and Powell 1974) and found to be in reasonably good agreement with the theoretical calculations of Penning de Vries et al. (1974) based on ATP and energy requirements for the synthesis of the various plant materials using known metabolic pathways.

Maintenance respiration seems more difficult to deal with. Although the main expenses of maintenance energy appear to be protein turnover and maintenance of ionic gradients across membranes, theoretical evaluation of this component does not fully explain the large variability exhibited in measurements made in various plant species (El Aouni 1980).

It is however possible to express total respiration of a plant as the sum of the two components using the classical expression of Mc Cree (1970),

$$R = kP + cW,$$

where P represents the gross assimilation rate (in mol C m^{-2} day^{-1}), W the dry weight to the plant in mol C m^{-2}, and k and c two parameters that vary with chemical composition of plant tissue and thus with the development stage.

4.2.2.3 Other Growth Processes

Other growth processes include the various transformations of assimilates into plant material; translocation, mineral uptake, and assimilation, metabolic syntheses and organ growth. We can measure the results of growth – dry weights of the various plant organs and their evolution through time and assimilate partitioning by using ^{14}C techniques. We have little understanding, however, of what starts or stops growth and differentiation.

We still have little general understanding of how the plant controls the growth of its various organs according to the incoming fluxes of matter, carbon, water, and minerals, but there is little question that it does achieve such a control. For instance, a lack of water or nitrogen increases the root/shoot ratio in what Brouwer (1963) has been called a functional balance. Such general ideas should help focus on some optimizing principle that acts as a constraint at the whole plant level.

4.2.3 Comparison of Cultivated and Wild Species

Carbon balance of a wild species is oriented towards a reproduction ensuring genetic continuity in a given environment, whereas in crops man has attempted to maximize harvestable yield, acting both on the vegetation and on the environment. This has resulted in very different patterns in leaf photosynthesis, assimilate partitioning and dry matter production.

The history of breeding for more productive species and varieties is well documented for at least a few species and is especially interesting because of the light it throws on the factors which are limiting to crop production.

4.2.3.1 The Case of Wheat

As with most cereals, man first looked for a wheat with a less fragile rachis, allowing the ear to be harvested (Harlan and Zohary 1966). Subsequently ears with large grains were selected. Large grains led to larger leaf area of seedlings. There was also a close relation between the area of the largest leaf on the main stem and the weight of the ears and of the individual grains. Thus parallel increases in leaf and grain size occurred in the evolution of wheat, but at the expense of a decrease in leaf photosynthetic rate from 40 to 22 $\mu mol\ CO_2\ m^{-2}s^{-1}$ (Evans and Dunstone 1970). This seemingly unfavorable character was more than compensated by the larger size of the flag leaf in modern wheats (the predominant source of assimilates to the ear) and by an increase of its active lifespan, attributed by Evans and Dunstone to the increased demand for assimilates of the larger ear. Tillering was also decreased, as well as the fraction of assimilates directed to stem and root growth. As a result of these changes, grain weight at harvest increased roughly from 10% of shoot weight in the primitive diploid *Aegylops speltoides* to 50% and more in modern hexaploid varieties of *Triticum aestivum* (Evans and Dunstone 1970).

Thus evolution of wheat has essentially increased the movement of assimilates outside of the leaves (mainly of the flag leaf) towards the growing ear. This has been accompanied by an increase in the cross-sectional area of phloem tissue at the top of the stem (Evans et al. 1970).

To summarize, the photosynthetic capacity of the flag leaf has decreased through the evolution of wheat but it has been more than compensated for by a larger size and a longer lifespan, leading to a total dry weight of modern wheats similar to that of their wild ancestors. Thus the main change has been in the distribution of assimilates, with a five-fold increase in the harvestable fraction.

We find another interesting result when expressing Evans and Dunstone's data on a per plant basis. The total number of grains produced per plant has decreased during wheat selection, following the reduction in tiller number, although total grain weight was increased many times. Apparently wild wheat with large numbers of tillers, and of grains per plant, and a consequently high reproductive output were competitive in natural environments, whereas our modern wheats which require seeding, fertilizers, and pesticides to give an adequate harvest would presumably be less so.

Evans and Dunstone (1970) noted that leaf photosynthetic rate had not limited wheat yield so far, but could well limit further evolution.

Agriculture in recent years has become a monoculture of a few species in environments with a high level of available resources. The increased cost of fertilizers may lead to a search for varieties making the best use of sub-optimal nutrition levels. To that end, natural species may give us valuable hints on the adaptive possibilities in limiting environments.

4.2.3.2 Adaptation of Natural Species to a Given Level of Resources

Adaptation to water stress has been a recurrent theme in the activities of the ecophysiology group of CNRS in Montpellier since its beginning in 1966. Two recent studies, one on several populations of a grass species, the other one on two species of pine trees, are illustrative of adaptation to varying resource levels.

Adaptation to Water Stress in Dactylis glomerata (Roy 1980). *Dactylis glomerata* is a very common grass found in an extensive area ranging from cold and humid sites to hot and dry environments. Nearly all individuals are tetraploid. Among them, some have a winter dormancy partly induced by the cold (*glomerata* type), whereas mediterranean types have a summer dormancy during the drought (*hispanica* type). Plants coming from four different regions have been transplanted and grown in pots under natural conditions at Montpellier under three irrigation regimes and adequate fertilization. Plants were brought to the laboratory and irrigated once. Photosynthesis, transpiration, and water potential were then monitored on the same leaf during a drought cycle (Roy 1980). The differences observed in the well-irrigated treatment among the four populations when submitted to a drying period of about 5 days, which was enough to decrease net photosynthesis by a factor of more than 5, are reported here.

Table 1 compares various parameters related to photosynthesis of well-watered plants. Among the four populations only one was a *glomerata* type, coming from a mountainous location in Massif Central (Laqueuille, elevation 1,050 m). The others were *hispanica* types.

Table 1. Comparison of photosynthetic parameters of populations of *Dactylis glomerata* originating from different localities. (After Roy 1980)

Location	Laqueuille	La Jasse		Zaïo
Region	Massif Central	30 km North of Montpelier		Northern Morocco
Special features	Elev. 1,050 m	Deep soil	Shallow soil	Arid zone
Annual precipitation (mm)	1.271	1.090		200
Average annual temperature (C)	8.2	12.3		18
Quantum yield (mol CO_2/einst.)	0.044	0.047	0.039	0.043
Max. photosynthesis (μmol CO_2 $m^{-2}s^{-1}$)	20.7 (±1.8)	17.7 (±0.9)	14.1 (±1.4)	22.5 (±2.0)
Stomatal density (mm^{-2}, upper side)	100	115	224	217
Stomatal density ratio (upper/lower side)	1.2	1.1	1.66	3.8
Stomatal conductance (mol H_2O $m^{-2}s^{-1}$)	0.39 (±0.09)	0.54 (±0.27)	0.20 (±0.05)	0.56 (±0.25)

Two populations were located 50 m apart in the same station, La Jasse, about 30 km North of Montpellier in a relatively humid climate with an average annual precipitation of 1,090 mm, compared to 750 mm in Montpellier. The so-called dry location was on shallow soil whereas the "wet" one was on deeper soil. These two populations obviously had genetic interchange but electrophoretic studies of enzymatic polymorphism had shown a gradient in the frequency of the allozymes on two genes (acid phosphatase and glutamate-oxaloacetate transaminase) moving from wet to dry habitats (Lumaret, pers. comm.; Lumaret and Valdeyron 1978). In each population representative genotypes were chosen for physiological studies.

The plants originating from northern Morocco near Dar-Driouch were grown from seeds, so they cannot be considered as representative of the natural population, although presumably all the seeds retain some adaptive characters of their parents.

Quantum yield of photosynthesis was about the same for the four populations (Table 1). Observed differences in the maximum net photosynthetic rate (at 1,500 μE $m^{-2}s^{-1}$, 320 vpm CO_2, 25 °C) cannot be easily explained. Higher photosynthesis of Laqueuille and Zaïo plants could be a result of their short growing period (limited by the cold or by the drought), but this possibility requires further investigation.

Stomatal density of the upper side, as well as the ratio of the densities on the two sides of the leaf, increase as one goes from wet to dry sites. Maximum stomatal conductance, however, is highly variable and does not exhibit a specific trend. The length of the guard cells is about the same in the four populations (data not shown). It is thus possible that well-watered plants coming from dry sites always keep their stomates partly closed, although further work is needed to verify this possibility.

Plant water potential was measured with an in situ leaf hygrometer located within the leaf chamber used for photosynthesis and transpiration measurements.

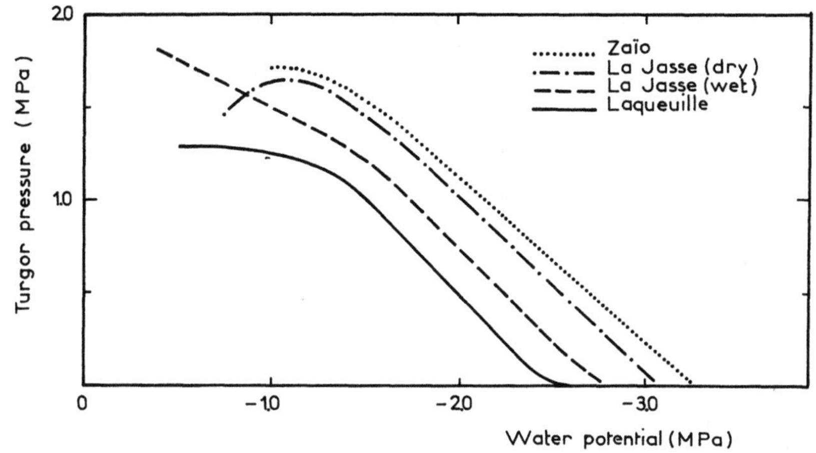

Fig. 2. Relations between turgor pressure and water potential in leaves of different genotypes of *Dactylis glomerata*. (After Roy 1980)

After studying variations of leaf photosynthesis against water potential during drought in the four populations it became clear that the main differences observed among them were related to differences in their water relationships (Fig. 2). *Dactylis* plants originating from dry habitats have higher turgor pressures at low water stress (thus causing hydropassive stomatal closure) and can maintain a positive turgor at lower leaf water potentials than those coming from wet places. Plants of the former populations also have thicker cell walls and, according to Roy (1980), can develop higher water potential differences between the apoplasm and the symplasm, thus resulting in a certain isolation of the cytoplasm from the drying apoplasm. These plants may be considered to a certain extent as drought endurers (as opposed to avoiders), although the cell apparently manages to escape at least part of the stress by increasing its membrane resistance to water circulation.

The maintenance of turgor pressure at high water stress must be associated with a certain energy cost. Knowledge of this associated cost would allow us to quantify the competitive advantage (or disadvantage) of such plants on a given site. Another interesting question arises from differences observed in curves of Fig. 2. We know that a given genotype may change its response curve when grown under water stress. Is this phenotypic plasticity higher in plants grown in dry places and if so, can we calculate the adaptive significance of an increased plasticity? These questions cannot be answered by physiologists alone, but require close cooperation with ecological geneticists, and population biologists.

Comparison of Pinus halepensis and Pinus pinea (El Aouni 1980). Pinus halepensis and *P. pinea* have similar distributions in the Mediterranean Basin but *P. pinea* grows at low elevation, relatively humid locations close to the sea, whereas *P. halepensis* grows on dry, hilly sites. Natural stands of *P. halepensis* in Tunisia have a timber production with less than 2 m^3 $ha^{-1}yr^{-1}$, a low figure within the productive *Pinus* genus. Yet the maximum photosynthetic rate of its needles is similar to that of other pine-trees with about 6 μmol CO_2 $m^{-2}s^{-1}$. This consideration led El Aouni to study the various factors that could limit the productivity of *P. halepensis*.

Table 2. Comparison of productivity factors in two species of pines. (After El Aouni, 1980)

Parameter	*Pinus pinea*	*Pinus halepensis*	Unit
Leaf absorptance (400–700 nm)	0.88	0.86	–
Quantum yield	0.051	0.045	mol CO_2/abs. einstein
Hill reaction	400	250	µmol Ferric. $(mgCHl)^{-1}h^{-1}$
Stomatal conductance	0.4 –0.8	0.07–0.11	mol CO_2 $m^{-2}s^{-1}$
Internal conductance	0.03–0.06	0.02–0.03	mol CO_2 $m^{-2}s^{-1}$
Max. photosynthetic rate	6–10	4–8	µmol CO_2 $m^{-2}s^{-1}$
Photorespiratory rate	1.2	1.2	µmol CO_2 $m^{-2}s^{-1}$
Critical water potential for stomatal closure	−1.2	−2.0	MPa
Water saturation deficit at −2.0 MPa water potential	25	13	%
Dark resp./Net assim. (from ^{14}C studies)	–	50–55	%
Root/shoot ratio (2 year seedlings)	–	about 1	

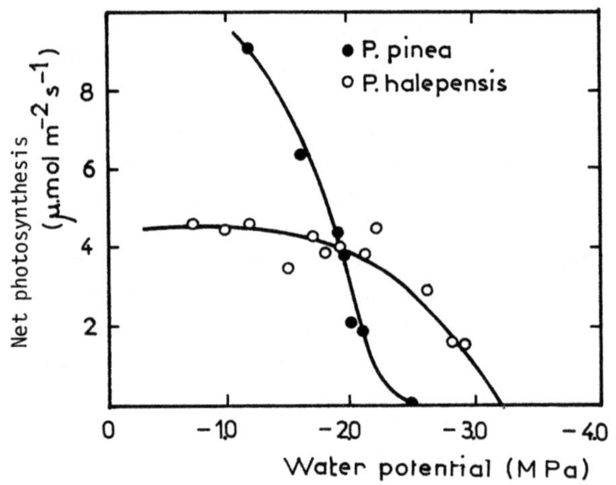

Fig. 3. Influence of water potential on net photosynthesis of needles of *Pinus pinea* and *Pinus halepensis*. (After El Aouni 1980)

He examined various parameters affecting photosynthesis and its response to water stress in both species. He also estimated the cost of respiration in mature trees of *P. halepensis* using ^{14}C techniques, and the relative production of roots and shoot in young plants of the same species. Some of his results are summarized in Table 2. Leaf absorptance and quantum yield were similar in the two species and comparable to the values obtained by Mc Cree (1972) on field grown plants of various species. Hill activity of chloroplasts was higher in *P. pinea*, as was the maximum stomatal conductance. Internal conductance, maximum photosynthetic rate and photorespiration rate were similar in the two species. Their response to water stress was very different as shown by the values of critical leaf water potential for stomatal closure (Table 2, Fig. 3). Also *P. halepensis* loses far less water than *P.*

pinea when subjected to the same leaf water potential of −2.0 MPa. As it was the case with *Dactylis*, adaptation to water stress does not seem to stem from a more resistant photosynthetic apparatus, but rather from a better protection of productive tissues against desiccation.

The cost of growth respiration (50% of the assimilated carbon) was found to be very high in *P. halepensis* as compared to 20%–30% in herbaceous species (Mc Cree 1974). This is due in part to the high carbon synthesis cost of lignin and lipid that make about 1/3 of the total biomass (El Aouni 1980). Other pines, however, also generally contain high amounts of lignin and lipids, yet they have a high timber production. In *Pinus halepensis*, a high fraction of assimilates goes into root production; 45% in young trees, 27% in adult trees, against 7% in adult individuals of *P. pinea* (El Aouni 1980). This allocation pattern in *P. halepensis* may be an adaptive character to permit growth in dry, nutrient-poor soils. Lastly, the cost of reproduction is high in *P. halepensis* since these trees produce numerous cones and large amounts of seeds. These seeds are naked, have limited carbohydrate reserves and are frequently eaten by small mammals and birds. A large production of seeds is evidently necessary to insure some successful establishment.

When all factors are taken together, the result is that in young seedlings of *P. halepensis* less than 10% of the assimilated carbon goes into wood formation in favorable (no water or mineral stress) conditions. Further studies in the field would be necessary to estimate the length of the growing period in natural stands, which is severely limited by drought. Preliminary investigations have shown that plant water potentials lower than critical values required for stomatal closure are not uncommon.

Summing up, a detailed investigation has helped to pinpoint the main factors limiting timber productivity of *P. halepensis;* high respiration rate, high reproductive cost, high root production, limitation of growing period by recurring droughts and likely limitation of growth by a shortage of nutrients. This species maintains stomatal opening and photosynthesis at greater water stresses than *P. pinea*, but its active growth period is certainly much shorter since it occurs on much drier sites.

4.2.4 Crops Versus Natural Communities

Carbon uptake may be studied on a single leaf, on a whole plant, or at the community level. Details of the mechanisms of CO_2 assimilation are best understood at the leaf level, but a comprehensive picture of primary production as related to available resources is more readily obtained at the vegetation level, since resources such as light, water, and minerals are more easily defined per unit area of ground than per unit area of leaf or per plant.

Eckardt et al. (1977) studied the primary production processes of four community types: a stand of *Quercus ilex* (phanerophyte), a salt marsh with *Salicornia fruticosa* (chamephyte), a stand of the perennial grass *Arrhenaterum elatius* and a sunflower crop (Fig. 4). Leaf specific weight increased from 41 in *Arrhenaterum elatius* to 160 g m^{-2} for the xerophytic leaves of *Quercus ilex*. Total above ground biomass varied greatly, being highest in the forest stand. The slope and the plateau of the light response curve of leaf photosynthesis was very different for the four

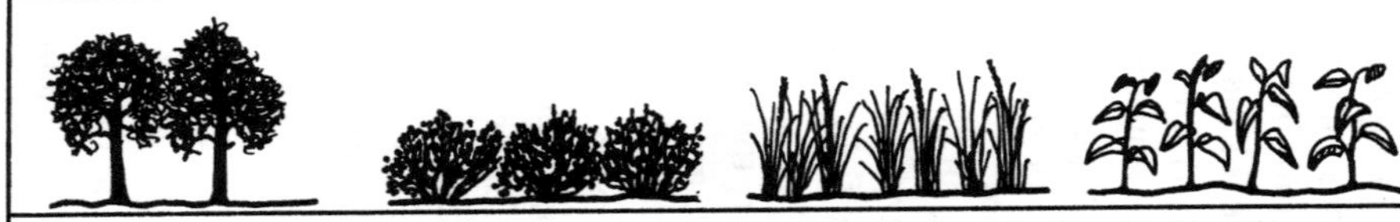

	Quercus ilex	*Salicornia fruticosa*	*Arrhenatherum elatius*	*Helianthus annuus*	
Distribution of Leaf area index	12 m 4.4 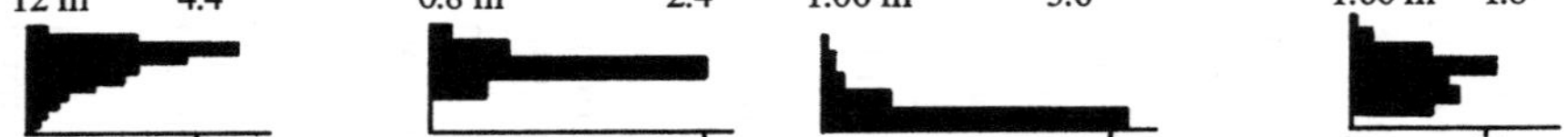 0 20	0.8 m 2.4 0 55	1.00 m 3.6 0 70	1.60 m 1.8 0 25	%
Leaf specific weight	160	125	41 (lamina)	90	$g\ m^{-2}$
Aerial biomass	27	2–4	0.4	0.9	$kg\ m^{-2}$
Leaf biomass	0.7	0.3–0.4	0.15	0.16	$kg\ m^{-2}$
Leaf area duration	12	7	10–12	2	months
Maximum leaf photosynthesis	6.5	7.4	10.4	29	$\mu mol\ CO_2 m^{-2} s^{-1}$
Quantum yield	0.021	0.019	0.021	0.059	mol CO_2 per incident einstein
Yearly above-ground production	0.65	0.5	1.1	0.9	$kg\ m^{-2}\ yr^{-1}$
Yearly energetic yield	0.23	0.16	0.35	0.28	% of incident global rad.
Maximum daily energetic yield	0.67	0.62	1.6	1.6	% of incident global rad.

Fig. 4. Compared figures for production and associated factors in four vegetation types growing under mediterranean climate near Montpellier: a forest stand of *Quercus ilex*, a salt marsh with *Salicornia fruticosa*, a forage crop of *Arrhenaterum elatius* and a sunflower crop. (After Eckardt et al. 1977)

species, as was the photosynthetic period which varied from just over 2 months (sunflower) to 12 months for the evergreen tree.

In spite of all these differences, annual dry matter production is relatively similar among the four types, with a range of 0.5 to 1.1 kg $m^{-2}yr^{-1}$. When expressed on an energy basis as a percentage of incident global radiation, it varies between 0.16 and 0.35%, whereas maximum yields on a daily basis vary between 0.6% and 1.6%. *Arrhenaterum elatius* and *Helianthus annuus* have relatively high leaf photosynthetic rates but short active growth periods (the 10–12 months period for *A. elatius* includes many months with very little growth). In contrast, *Quercus ilex* and *Salicornia fruticosa* have low rates of leaf photosynthesis that are somewhat compensated by longer growing periods, allowing these species to take advantage of the minerals that are released at a small rate throughout the year.

The comparison between these four vegetation types is not completely valid because the two crops (grass and sunflower) were irrigated and fertilized, which may account for their higher productivities. Further, the productivity of the *Salicornia fruticosa* community was likely depressed by the presence of salt in soil water, that is, the plants have to invest a certain amount of carbon in order to decrease water potential to absorb soil water (salt is absorbed and accumulated in the vacuoles and is actively excluded from the cytoplasma). Midday leaf water potential decreases from −3.5 to −8.0 MPa through the summer drought, leading to stomatal closure below −5.0 to −6.0 MPa (Berger in Eckardt et al. 1977).

If these differences in resource availability are taken into account they probably explain most of the observed differences in primary production.

The last example is taken from the work of Bernhardt-Reversat et al. (1978) and Lemée (1978) on two forests of different climates (Table 3). The first is a tropical rainforest of the Ivory Coast. It receives high yearly global radiation and is not limited by water since its water consumption is close to potential evaporation. Its productivity may, however, be limited by a shortage of nutrients, since the biomass contains 93% of the calcium and 90% of the exchangeable potassium present in the whole ecosystem. This means that the community has built biomass until it has virtually exhausted the supply of these nutrients. An equilibrium results in which uptake of these nutrients by the roots is compensated by the decomposition of the small amount of organic matter and possibly by chemical weathering of the rocks. There is here a precise adjustment of the system to the level of available resources.

The other forest is a natural preserve dominated by *Fagus silvatica* in the Fontainebleau forest 50 km south of Paris, France. It has been unmanaged for at least three centuries and is an approximation of a "natural" ecosystem. In this case the soil is the main storage place of the exchangeable nutrients which are evidently not limiting productivity. It appears here that water is the limiting factor. Annual precipitation balances annual potential evaporation, but precipitation is distributed equally throughout the year in contrast to evaporation which is of course maximum in summer. Since the sandy soil of the forest has a small water holding capacity, the trees endure in summer a certain amount of water stress that is reflected in the small annual water consumption.

Thus we have two climax forests under different climates with productivity limited by different resources. How can we determine:

Table 3. Comparison of two climax forests under tropical and temperature climates

Country	Ivory Coast	France	
Location	Banco (5°23′N, 4°2′W)	La Tillaie, Fontainebleau (50 km South of Paris)	
Dominant species	Numerous	*Fagus silvatica* (high forest)	
Basal area	30	30 to 35	$m^2\ ha^{-1}$
Biomass	56.2	34.2	$kg\ m^{-2}$
Leaf Area Index	8.3	6.6	
Specific leaf weight	100	55	$g\ m^{-2}$
Productivity	1.7	0.95	$kg\ m^{-2}\ yr^{-1}$
Photosynthetic period	12	5 to $5^1/_2$	months
Annual global radiation	5,680	4,140	$MJ\ m^{-2}\ yr^{-1}$
Total precipitation	1,800	697	$mm\ yr^{-1}$
Potential evaporation	1,220	660	$mm\ yr^{-1}$
Water consumption	1,145	475	$mm\ yr^{-1}$
N in ecosystem	787	913	$g\ m^{-2}$
Fraction in biomass	0.17	0.16	
K in ecosystem	79	215	$g\ m^{-2}$
Fraction in biomass	0.88	0.23	
Ca in Ecosystem	153	373	g m¯2
Fraction in biomass	0.90	0.21	
Source	Bernardt-Reversat et al. (1978)	Lemée (1978)	

1. To what extent minerals or water are limiting?
2. What would be the production of the community if radiation was the only limiting factor?

The second question has been answered theoretically by Loomis and Williams (1963) who computed the upper limit of productivity of a vegetation in which all leaves would photosynthesize with the maximum quantum yield. Some highly productive crops such as sugar cane may approach this upper limit of productivity when abundantly supplied with water and nutrients. But can we work backwards and estimate the degree of scarcity of the resources by comparing the actual with the potential production? We need not consider the species composing the vegetation if we assume that the species growing in a particular environment have developed to take advantage of that environment and of its resources. Perhaps we would improve our estimate of potential production by including physiological characteristics of the actual species such as photosynthetic parameters. This may tell us something on how to manage the actual vegetation to get an improved production but it will not tell us about the ultimate level of available resources.

4.2.5 Towards an Estimate in the Level of Available Resources

The concept of availability raises immediately the question, available to what? We may consider a plant organ (leaf, root), a whole plant, a population or a whole community. It is perhaps easier to estimate resources at the community level, and to express them per unit ground area. The three main resources, light, water, and nutrients, are discussed separately.

4.2.5.1 Light

The amount of light received by a horizontal surface may be measured easily and expressed on a daily or yearly basis. Light is intercepted by the vegetation according to its structure and LAI. If we know the development in the vegetation architecture and the light intensity throughout the year, we can compute a close estimate of the fraction of incident light that is absorbed by the vegetation during one year. Such an approach has been developed successfully for crops, e.g., by Varlet Grancher and Bonhomme (1979). They define a trapping efficiency ε_i as the ratio of absorbed to incident photosynthetically active radiation. On a yearly basis, they found $\varepsilon_i = 0.72$ for sugarcane which maintains a nearly continuous crop cover, 0.13 for one crop of cowpea (*Vigna sinensis*) in French West Indies. In France values ranged from 0.32 for maize to 0.57 for alfalfa. ε_i would be close to 0.8 for an evergreen forest and about 0.6 for a deciduous beech forest.

4.2.5.2 Water

Annual precipitation is the primary measure of water availability but run-off and drainage must be subtracted from this amount to obtain available water. This may be done by taking regular measurements of the soil water profile using a neutron probe or by the gravimetric method. When such measurements are not available one may use a model of soil water balance to predict this profile. Then we need to know how a reduction in this reserve affects: (1) real evapotranspiration and (2) growth and productivity.

In French agronomic research, the degree of water stress has often been defined as the ratio at actual (T) to maximal (T_m) evapotranspiration taken on a daily basis or averaged over the growing period.

When this ratio is decreased below 1, dry matter (and harvestable matter) is also reduced, more or less according to the sensibility to drought of the particular species. Figure 5 shows some results obtained by Puech and Maertens (1974) in the south-west of France near Toulouse. Although C_4 species such as maize and sorghum have a greater water-use efficiency in terms of dry matter produced per unit of transpired when water is freely available they show under water stress a relative reduction in production that is equal or greater to that observed in C_3 species. This is understandable since water stress usually affects stomatal resistance, r_s, more than the inner resistance, r_i, to CO_2 diffusion, and since the r_s/r_i ratio is higher for C_4 species. Similar results have been obtained by Stewart et al. (1973) as reported by Feddes et al. (1978, p. 55) for sorghum and maize.

Summing up, we may first define a growing season for a given community as being determined by the temperature regime through the year. The result is a certain variation with time of the potential leaf area index of the community LAI (t),

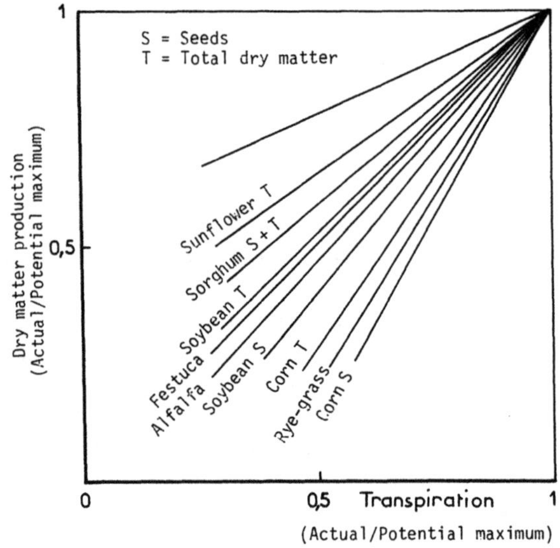

Fig. 5. Relations between dry matter production, P, and transpiration, T, for various crops. P_m and T_m are the maximal values of P and T when water is not limiting. (After Puech and Maertens 1974)

i.e., of the leaf area that would be present if there were no shortage of water or nutrients.

This, together with the variation of the photosynthetically available radiation PAR (t), allows a computation of the ratio ε_i of absorbed to incident PAR. If we know from climatic data the variations $E_0(t)$ in free water evaporation, we can compute from LAI(t) and $E_0(t)$ a maximum evapotranspiration Tm(t). We may then calculate a simple index of water availability knowing incident precipitation Pr(t) as $\varepsilon_W = \int \min [Pr(t), Tm(t)]\, dt / \int Tm(t) dt$. This index may of course be somewhat refined by introduction of a simple model of soil water balance.

4.2.5.3 Nutrients

Nutrients are perhaps the most difficult resource to assess. Since variations in mineralization rate are difficult to measure accurately, we may define an index, ε_N, of say nitrogen availability as the ratio of actual nitrogen uptake by the plants of the community (obtained from a measurement of the dry matter produced and of its nitrogen content) to the potential nitrogen uptake that would occur under ideal conditions, i.e., when production is limited only by intercepted radiation and CO_2 concentration (de Wit et al. 1978) and when nitrogen concentration in plant tissue is at its optimum.

This general approach is very preliminary and is given mainly to stimulate research for a better definition of this rather vague concept of resource availability, a necessary step if we are to progress in the understanding of man's action on ecosystems as a perturbation in the level of available resources.

4.2.6 Conclusions

Most plant ecophysiologists have concerned themselves with only one aspect of plant production and have developed models for this particular aspect. Such models include, crop photosynthesis in relation to radiation penetration, CO_2 and microclimate, growth, and maintenance respiration, water circulation in the soil–plant–atmosphere system, mineral nutrition and so on. But these models have not yet been integrated into a coherent picture of total plant functioning. To achieve this goal we will have to simplify the description of each individual process, yet retaining its major aspects. Such an approach will be rewarding if we want to understand and compare the adaptive strategies of various vegetation types, or species, or genotypes.

We have shown in *Dactylis* that adaptation to drought results both from the existence of different genotypes occurring with varying frequencies within a population, and from the plasticity of the physiological response to water stress. Such studies combining genetic and physiological approaches though as yet incomplete may lead the way to a better understanding of the fitness of natural populations.

At the community level, comparison of actual and potential productivity is a first approach to the study of resource limitations. Adaptation of the community to the scarcity of these resources may be understood when studying the resource requirements of each individual species. Then perhaps we may one day be able to measure how close an ecosystem is to optimal use of the environment.

Résumé

La croissance d'une plante est limitée par les ressources disponibles dans son milieu et par sa capacité à capter ces ressources et à les utiliser dans son métabolisme.

L'homme intervient dans les processus naturels en modifiant la végétation et le niveau des ressources, et en orientant à son profit la production végétale.

Le travail ci-dessous décrit brièvement l'action des facteurs du milieu sur la photosynthèse et présente quelques exemples d'adaptation à un facteur donné: selection du blé par l'homme, adaptation à la sécheresse chez *Dactylis glomerata*, *Pinus halepensis* et *Pinus pinea.*

Il compare ensuite l'efficacité d'utilisation de l'énergie et des matières premières par diverses cultures et communautés naturelles et cherche à préciser le niveau des ressources disponibles au niveau de l'écosystème.

References

Bernhardt-Reversat F, Huttel C, Lemée G (1978) La forêt sempervirente de la basse Côte-d'Ivoire. In: Lamotte M, Bourlière F (eds) Problèmes d'ecologie: Structure et fonctionnement des écosystèmes terrestres. Masson, Paris, pp 313–345

Brouwer R (1963) Some aspects of the equilibrium between overground and underground plant parts. Jarrb IBS Wageningen 1963:31–39

Chartier P, Prioul JL (1976) The effects of light, carbon dioxide, and oxygen on the net photosynthetic rate of the leaf: a mechanistic model. Photosynthetica 10:20–24

Eckardt FE, Berger A, Methy M, Heim G, Sauvezon R (1977) Interception de l'énergie rayonnante, échanges de CO_2, régime hydrique et production chez différents types de végétation sous climat méditerranéen. In: Moyse A (ed) Les processus de la production végétale primaire. Gauthier-Villars, Paris, pp 1–75

El Aouni MH (1980) Processus déterminant la production du pin d'Alep (*Pinus halepensis* Mill.): photosynthèse, croissance et répartition des assimilates. Thèse de doctorat d'etat, Univ Paris 7, 164 pp + 61 pp annexes

Evans LT, Dunstone RL (1970) Some physiological aspects of evolution in wheat. Aust J Biol Sci 23:725–741

Evans LT, Dunstone RL, Rawson HM, Williams RF (1970) The phloem of the wheat stem in relation to requirements for assimilate by the ear. Aus J Biol Sci 23:743–752

Farquhar GD, Caemmerer Von S, Berry JA (1980) A biochemical model of photosynthetic CO_2 assimilation in leaves of C_3 species. Planta 149:78–90

Feddes RA, Kowalik PJ, Zaradny H (1978) Simulation of field water use and crop yield. PUDOC, Simulation Monogr, Wageningen, 188 pp

Gaastra P (1959) Photosynthesis of crop plants as influenced by light, carbon dioxide, temperature, and stomatal diffusion resistance. Meded Landbouwhogesch Wageningen 59:1–68

Harlan JR, Zohary D (1966) Distribution of wild wheats and barley. Science 153:1074–1080

Ho LC (1979) Regulation of assimilate translocation between leaves and fruits in the Tomato. Ann Bot (London) 43:437–448

Lemée G (1978) La hêtraie naturelle de Fontainebleau.In: Lamotte M, Bourlière F (eds) Problèmes d'ecologie: Structure et fonctionnement des écosystèmes terrestres. Masson, Paris, pp 75–128

Loomis RS, Williams WA (1963) Maximum crop productivity: an estimate. Crop Sci 3:67–72

Lumaret R, Valdeyron G (1978) Les glumatate-oxaloacétate transaminases du dactyle (*Dactylis glomerata* L.): génétique formelle d'un locus. CR Acad Sci Ser D 287:705–708

Mc Cree KJ (1970) An equation for the rate of respiration of white clover plants grown under controlled conditions. In: Setlik I (ed) Prediction and measurement of photosynthetic productivity. Proc IBP/PP Tech Meet. Trebon PUDOC, Wageningen, pp 221–229

Mc Cree KJ (1972) The action spectrum, absorptance, and quantum yield of photosynthesis in plants. Agric Meteorol 9:191–216

Mc Cree KJ (1974) Equations for the rate of dark respiration of white clover and grain sorghum, as functions of dry weight, photosynthetic rate, and temperature. Crop Sci 14:509–514

Penning De Vries FWT, Brunsting AHM, Laar Van HH (1974) Products requirements and efficiency of biosynthesis: a quantitative approach. J Theoret Biol 45:339–377

Puech J, Maertens C (1974) Efficience de l'eau consommée de quelques cultures placées dans différentes conditions écologiques. Agrochimica 18:223–230

Roy J (1980) Comportement photosynthétique et hydrique de la feuille chez *Dactylis glomerata* L. Adaptation phénotypique et génotypique à la sécheresse. Thèse de spécialité, Univ Montpellier, 118 pp + 153 pp annexes

Ryle GJA, Powell CE (1974) The utilization of recently assimilated carbon in graminaceous plants. Ann Appl Bot 77:145–158

Stewart JI, Hagan RM, Pruitt WO, Hall WA (1973) Water production functions and irrigation programming for greater economy in project and irrigation system design and for increased efficiency in water use. Report 14-06-D-7 329. Univ California, Davis, 145 pp

Varlet Grancher C, Bonhomme R (1979) Application aux couverts végétaux des lois de rayonnement en milieu diffusant. II. Interception de l'énergie solaire par une culture. Ann Agron 30:1–26

Wit de CT et al. (1978) Simulation of assimilation, respiration, and transpiration of crops. PUDOC, Simulation Monogr, Wageningen, 140 pp

4.3 Patterns of Nutrient Absorption and Use by Plants from Natural and Man-Modified Environments

F. STUART CHAPIN, III

4.3.1 Introduction

Man's activities in natural ecosystems generally alter nutrient availability, thus affecting nutrient absorption and allocation by plants. In this way man has substantially changed plant nutrient relations on more than 75% of the continental United States (Table 1). Prior to 1940 when new lands were being developed most rapidly, the predominant effect of European man in America was to reduce the productive potential of the land, in part through depletion or loss of available nutrients. Declining soil fertility as a result of human exploitation has played a major role in the changing sociology and history of the United States (e.g. southeastern and midwestern states) and is playing a similar role today in many of the developing nations of the world.

Land-use practices such as the introduction of agriculture, logging (Likens et al. 1978), and grazing (Dean et al. 1975) generally cause a short-term increase in nutrient availability above that of undisturbed ecosystems through reduced uptake by plants, increased organic matter mineralization rates (Gorham et al. 1979; Vitousek et al. 1979), and fertilization. In contrast, mining and other activities that remove or bury surface soil horizons reduce nutrient availability. Long-term effects of these land uses often differ from short-term effects and depend upon the balance between increased weathering and organic matter mineralization on the one hand and losses due to plant removal, leaching, and erosion on the other. The scale at which man is altering nutrient availability throughout the world makes it imperative that we develop a predictive understanding of the changing patterns of plant

Table 1. American land use patterns and effects on nutrient availability. (Clawson 1972; Peone et al. 1978)

	Area (% of total)	Land use effects on nutrient availability		
		1700–1940	Current practices	
			Short-term	Long-term
Pasture and rangeland	34	↓	↑?	↓?
Forest and Woodland	32	↓	↑	–
Cropland	23	↓	↑	↑
Recreation	3	–	–	–
Urban land	1	↓	↓	–
Mining	0.2–2	↓	↓	↓
Other	5 –7	–	–	–

nutrient use resulting from altered nutrient availability. In this paper I describe the differences in patterns of nutrient use between plants from fertile and infertile environments and discuss examples in which man's alteration of nutrient availability has led to changes in nutrient use by plants. These general considerations suggest that many of the current revegetation practices cannot succeed and should be revised.

4.3.2 General Patterns of Nutrient Use

Abandoned fields, deforested areas, pastures, and similar man-modified environments generally have more light and higher nutrient availability than unmodified climax vegetation on the same parent material. The species that predominate in such man-modified high-resource habitats typically grow rapidly (Rorison 1968; Marks 1974; Grime and Hunt 1975; Chapin 1980). They have high rates of photosynthesis (Bazzaz 1979) and nutrient absorption (Clarkson 1967; Rorison 1968; Chapin 1980) which provide the precursors necessary for their rapid growth (Chapin 1980). Photosynthesis declines with leaf age (Orians and Solbrig 1977), and nutrient absorption declines with root age (Clarkson 1974), so that high rates of both leaf and root turnover (at least annual) are requisite to the rapid growth (high-resource) strategy. Generally 40%–60% of the maximum nutrient (nitrogen, phosphorus, and potassium) content of leaves is lost to the plant at abscission, so that the rapid tissue turnover of the high resource strategy is dependent upon a continuous replenishment of tissue nutrient reserves, and this is possible only in relatively fertile soils. Such plants grow as rapidly as possible under given conditions of photosynthetic carbon and nutrient supply, so they respond to nutrient addition by greatly accelerated growth rate rather than by increasing nutrient stores. They thus maintain a competitive advantage under favorable conditions. However, given a low nutrient supply, these rapidly growing species deplete internal nutrient reserves, cease growing and may be vulnerable to disease and other environmental stresses (Rorison 1968; Chapin 1980).

At the opposite extreme, plants typical of soils with relatively low nutrient availability are characterized by slow growth rates (Clarkson 1967; Rorison 1968; Grime and Hunt 1975; Chapin 1980) and consequently with relatively low annual requirements for carbon and nutrients. Such species typically have somewhat slower rates of photosynthesis and nutrient absorption that can be supported even with slow root and leaf turnover (Chapin 1980). Slow leaf and root turnover in turn entail lower annual nutrient requirements. Therefore, slowly growing species from infertile environments can grow for a longer period of time on a given nutrient supply and are more likely to survive periods of low nutrient availability. These species are not highly responsive to nutrient addition, so that, when grown in fertile soils, they accumulate nutrient reserves (luxury consumption) but do not show a large increase in growth rate. These generalizations are discussed and documented elsewhere (Chapin 1980).

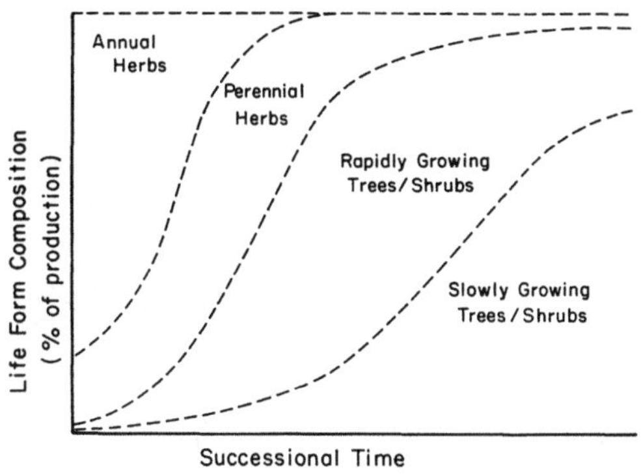

Fig. 1. Representative change in life-form composition through succession in a forested zone

4.3.3 Successional Changes in Nutrient Use

In a general sense, man most frequently alters vegetation by altering normal patterns of disturbance and subsequent successional development, e.g. by clearing land, by changing fire frequency, or by preventing succession in agricultural fields and pastures. The nutrient relations of plants change through succession due both to changing species and growth form composition and to changing nutrient availability in the soil (Gorham et al. 1979; Chapin and Van Cleve 1981). The general pattern of change in growth-form composition is from short-lived herbaceous species (generally annual) with large reproductive allocation to longer-lived herbaceous species and finally to long-lived species (generally woody) with large storage reserves and relatively small reproductive output (Fig. 1). Each of these growth forms has a distinct pattern of nutrient utilization, so the change in growth form composition has profound effects upon community patterns of nutrient use.

Annual plants are frequently important colonizing species. Their seasonal patterns of nutrient accumulation are determined primarily by a sequence of phenological events, modified secondarily by patterns of nutrient availability. Annuals have small seeds (Harper et al. 1970) and consequently small nutrient reserves, so their growth is directly and immediately dependent upon absorption from soil. Under either constant or unlimiting nutrient supply, there is an exponential increase in the standing crop of plant nutrients during vegetative growth due to exponential increase in root biomass (i.e., constant relative growth rate) (Fig. 2) (Williams 1948; Milthorpe and Moorby 1974). Thus annual plants extract nutrients from soil most rapidly when they have the greatest root biomass. Flowering and seed development are supported in large part by nutrients translocated out of leaves and roots. Removal of nutrients from leaves and roots coincides with a decline in photosynthesis (Murata 1969; Bouma 1970; Nàtr 1975) and nutrient absorption (Williams 1948, 1955), respectively. Thus rates of nutrient accumulation are tightly keyed to a phenological calendar and are highest late in the vegetative growth stage. In the autumn the plant dies, and all nutrients return to the soil.

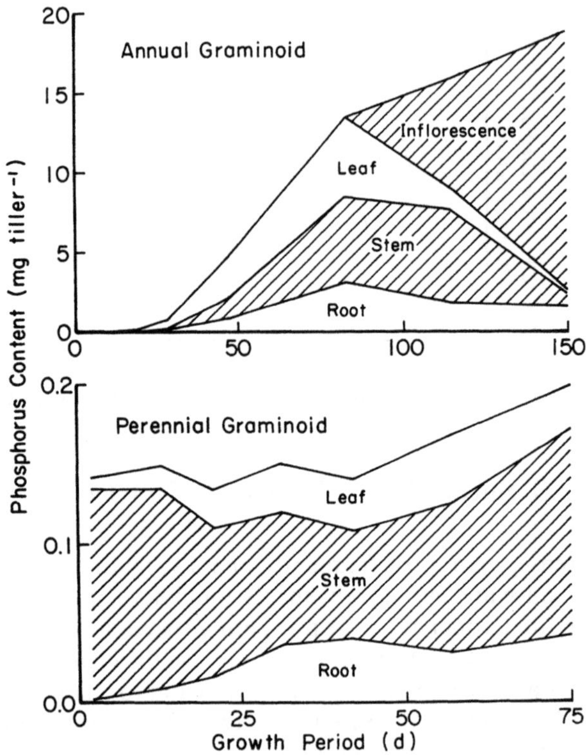

Fig. 2. Seasonal changes in the phosphorus content of various plant parts of an annual graminoid (oats; Williams 1948) and a perennial graminoid (*Eriophorum vaginatum;* Chapin et al. 1980)

Under natural conditions, if nutrients are at all limiting, this seasonal pattern of growth and nutrient accumulation is affected by seasonal changes in nutrient availability.

In contrast, perennial species have nutrient and carbohydrate stores that support rapid early spring growth. Moreover, many late successional species (e.g., oak) have large seeds with large nutrient reserves so that early seedling growth may not depend upon absorption from soil for several months (Specht and Groves 1966). Due to storage in seeds or stems the dependence of growth upon soil nutrients is buffered. For example, *Eriophorum vaginatum*, a tundra sedge, showed no net phosphorus accumulation during the first half of the growing season, and shoot and root growth were supported entirely by phosphorus translocation from rhizome stores (Fig. 2). Because of stored reserves, nutrient accumulation by perennial plants is not closely tied to a phenological calendar but depends upon seasonal patterns of both nutrient availability and root production. In species without a large perennial root biomass, absorption may be greatest late in the growing season (Fig. 2). However, in the more common circumstance of a large perennial root biomass, absorption closely follows nutrient availability (Chapin and Bloom 1976; Mooney and Rundel 1979). In most non-agricultural soils there are bursts of nutrient availability in spring and sometimes in autumn and winter, due to leaching

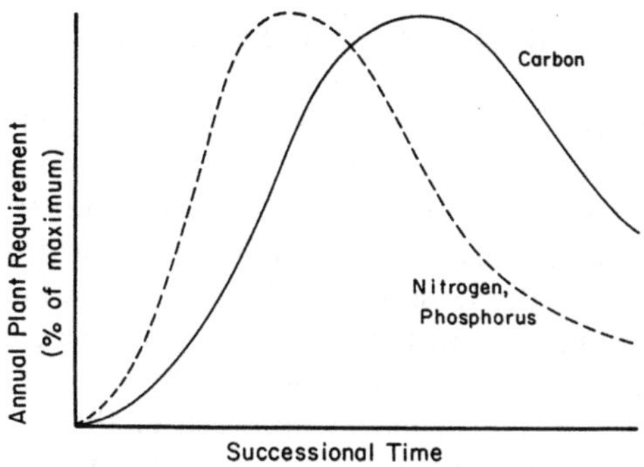

Fig. 3. Successional changes in annual plant requirement of nitrogen, phosphorus, and carbon by vegetation

of new litter, lysing of microbial cells, etc. (Biederbeck and Campbell 1971; Gupta and Rorison 1975; Chapin et al. 1978; Harrison 1979). These nutrient pulses would be more effectively exploited by perennial than by annual plants, in part explaining the competitive advantage of perennial species in established vegetation. In contrast to annuals, perennial plants retranslocate nutrients from leaves in autumn primarily to storage tissues rather than to reproductive structures.

In summary, as succession proceeds, seasonal patterns of nutrient absorption by vegetation become less dependent on plant phenology and more dependent on seasonal patterns of availability in soil. Absorption during non-growing seasons can be important in communities like the chaparral where winter is an important time of mineralization (Mooney and Rundel 1979).

4.3.4 Nutritional Patterns Related to Disturbance

4.3.4.1 Abandoned Fields

Secondary succession in fertile habitats (e.g., abandoned fields) occurs through a gradual transition from annuals and short-lived perennials to more slowly growing perennial species (Fig. 1) (Grime and Hunt 1975; Bazzaz 1979). The nutritional patterns of plants in this successional sequence are similar to those described above, changing from, (a) immediate dependence of growth upon nutrients absorbed from soil to, (b) increased importance of stored nutrients to support growth with uptake serving primarily to replenish stored reserves. Total nutrient uptake by the vegetation increases through succession (Odum 1969; Vitousek and Reiners 1975) due to increases in root biomass and increased capacity of the vegetation to exploit pulses of nutrient release in spring and autumn. In late succession, the change to slowly growing species (Horn 1974; Connell and Slatyer 1977) with low annual nutrient requirements is partially responsible for the decline in nutrient accumulation by ecosystems (Fig. 3) (Vitousek and Reiners 1975; Gorham et al. 1979).

4.3.4.2 Post-Fire Succession

Changing patterns of nutrient use following fire are similar to those occurring in abandoned agricultural fields. Fire prevention programs allow high fuel accumulation, so that, when fires do occur they tend to be hotter, kill more perennial vegetation, and mineralize more nutrients than during normal fire cycles. During initial years following fire, the seasonality of plant activity has an important influence upon seasonal patterns of nutrient absorption and leaching loss. In later successional stages, dominated by perennials with large root biomass, the seasonality of nutrient absorption depends primarily upon seasonality of nutrient availability, and leaching loss is less important (Likens et al. 1978). In areas where fire does not kill all the perennial vegetation, plants recover rapidly from below-ground parts and are quite effective in absorbing fire-released nutrients and preventing leaching (Stark 1977; Stark and Steele 1977; Chapin and Van Cleve 1981).

During post-fire succession the capacity of roots to absorb nutrients (g^{-1} root) and soil nutrient availability decline, whereas root biomass continues to increase with successional time (Chapin and Van Cleve 1981). The net result is that annual nutrient accumulation by vegetation is most rapid during mid-succession, at approximately the average fire return time for a given ecosystem (Chapin and Van Cleve 1981). Thus, nutrient accumulation by vegetation is most rapid several months after fire in a grassland (Steward and Ornes 1975), 10 years after fire in shrub heath (Chapman 1967), and 60 years after fire in a pine forest (MacLean and Wein 1977).

Fire and cutting of vegetation are often viewed as alternative methods of bringing a community to an early successional stage. However, these two practices have quite different effects upon nutrient use by plants and have different management implications. Fire causes immediate release of most organically bound nutrients, whereas nutrient release from organic matter is more gradual after cutting, particularly if nutrient-rich leaves are removed, as in whole-tree logging (Likens et al. 1978). Burning favors rapidly growing species with high nutrient concentrations that may provide highly palatable forage for wildlife. In contrast, species resprouting or invading a cut-over area may have lower nutrient requirements, grow more slowly, and be less palatable to wildlife than species invading a burn (Chapin and Van Cleve 1981).

4.3.4.3 Tundra Disturbance

Human impacts upon established communities alter nutrient availability and directly or indirectly cause changes in community composition. These effects are particularly evident in arctic and alpine tundras, where low temperature greatly reduces decomposition rates, and most of the nitrogen and phosphorus in the system are bound in soil organic matter (Babb and Whitfield 1977; Chapin et al. 1978). A general effect of disturbance by vehicle passage, hiking, skiing, oil spills, or partial vegetation removal is to increase decomposition rates and nutrient availability, in part through decreased insulation and warmer soil temperatures (Bliss and Wein 1972; Challinor and Gersper 1975; Haag and Bliss 1973).

Following such tundra disturbances there is generally an increased abundance of rapidly growing species, particularly certain graminoids, and decreased abundance of slowly growing species, e.g., evergreen shrubs (Bliss and Wein 1972; Hernandez 1973; Chapin and Chapin 1980). Plants in such disturbances have greater biomass and higher nutrient concentrations than the same species outside the disturbance, indicating considerably higher nutrient absorption (Bliss and Wein 1972; Chapin and Shaver 1981).

Disturbance by frost action, fire, and animal activity is a normal part of tundra ecosystems and man's impact is primarily a change in the scale of disturbance. On natural disturbances such as animal burrows there is a comparable increase in abundance of rapidly growing species with high nutrient concentrations (Batzli and Sobaski 1980). Thus in both natural and man-caused disturbances there is a change in community composition toward species with a high resource requirement. Such species have high nutrient absorption rates and high rates of nutrient return in litter (Bliss and Wein 1972; Chapin and Shaver 1981).

4.3.4.4 Disturbances Causing Reduced Nutrient Availability

All of the disturbances discussed above are associated with increased nutrient availability, and the species that invade naturally have rapid growth rates and high rates of nutrient absorption, thus generally minimizing nutrient losses from the ecosystem. Vegetation establishment on such sites is a relatively minor problem, because species from the normal secondary succession are well adapted to colonize and effectively exploit this high resource environment (Bazzaz 1979). Restoration of such sites becomes a problem primarily when man is interested in successful establishment of certain species (e.g., trees) that may have slower growth rates and be out-competed by normal early successional high-resource species. In cases where man uses herbicides to prevent establishment of the normal early successional flora, increased nutrient loss can result (Likens et al. 1978).

However, the bulk of revegetation research has centered on the more difficult situation where disturbance reduces nutrient availability below levels present in the unmodified community, requiring a vegetation development similar to that in primary succession. Primary succession is a slow process, requiring tens to thousands of years, a time period generally considered unacceptable for community restoration. Disturbances causing reduced nutrient availability include mine spoils, rock quarries, road cuts, and other situations where developed soil horizons are removed or buried. Surface soils can also be lost by wind or water erosion following overgrazing, as in Africa and the Middle East, improper dryland agriculture, as in the midwestern dust bowl, deforestation of tropical rain forests (Nye and Greenland 1960), or of steep hillsides (Likens et al. 1978). In such subsoils the difficulties of low nutrient availability may be compounded with problems of extreme pH, heavy metal toxicity, and changes in soil structure (Antonovics et al. 1971; Bradshaw et al. 1978; Mays and Bengtson 1978).

Mining disturbances resulting in greatly reduced nutrient availability comprise 0.2% of the land area in the United States, an area larger than the state of Connecticut (Paone et al. 1978). Such disturbances are becoming increasingly common as population pressures increase the exploitation of lands in third world nations and as mining, particularly for coal, expands in developed nations. These lands can be "reclaimed," setting in motion the normal successional processes, if proper attention is paid to nutritional characteristics of soils and plants used in restoration efforts. Most successful restoration efforts involve improvement of soil physical and chemical properties to the point that the soil has sufficient nutrient-supplying capacity to meet the high nutritional requirements of rapidly growing grasses or shrubs that are sown or planted. The situation is then similar to that at the beginning of secondary succession, and normal successional development can be expected, as described above. Methods of soil improvement in restoration of disturbed lands are beyond the scope of this paper but include, (a) direct nutrient addition through use of fertilizer, stockpiled topsoil, or sewage sludge (Halderson and Zenz 1978; Bradshaw et al. 1978), (b) improvement of soil pH (Mays and Bengtson 1978), or, (c) increase in cation exchange capacity through addition of organic matter or fine-particle soil.

Frequently, however, economic constraints and inadequate incentives limit the extent to which soil nutrient availability is improved over the long term in drastically disturbed sites. Under infertile conditions rapidly growing species are unlikely to survive for long periods because of their inherently high nutrient requirement, as discussed above. Thus rapid revegetation of infertile disturbances is unlikely (Elias and Chadwick 1979). For example, if tundra disturbances are seeded with rapidly growing exotic grasses without refertilization, most plants die within 2–5 years (Nordmeyer et al. 1974; Van Cleve 1977; Chapin and Chapin 1980). Repeated nutrient addition prolongs survival of these grasses (e.g., along the Trans-Alaska Pipeline route), but most plants die within a few years if fertilizer additions are stopped, so that further rapid recovery of the community is unlikely.

Similarly, reseeding of mine spoils with rapidly growing plants is largely unsuccessful without repeated fertilization. Such repeated fertilization generally ceases after a mining company is released from its performance bond (Mays and Bengtson 1978). It is therefore important to consider alternative means of reclaiming infertile sites:

1. Infertile soils have been successfully revegetated by seeding species with low relative growth rates and correspondingly low nutrient requirements. Such sites include mine spoils (Bradshaw et al. 1978; Bennet et al. 1978; Smith and Bradshaw 1979), areas devoid of topsoil (Nordmeyer et al. 1974; Chapin and Chapin 1980), and tropical lateritic soils following deforestation (Eiten 1969). The low productivity can be an advantage in these sites because of low nutrient requirements for establishment and reproduction (Elias and Chadwick 1979). Moreover, in the case of mine spoils, the rapid transpiration associated with rapid growth results in greater accumulation of toxic heavy metals in the plant (Antonovics et al. 1971). In most cases fertilization aids initial establishment (Bradshaw et al. 1978; Bennet et al. 1978), but repeated heavy fertilization is unnecessary and could result in competitive elimination of low-nutrient-adapted species by rapidly growing species. Unfortunately, there have been few commercially available seed sources for such

plants (Sutton 1975; Elias and Chadwick 1979), although such varieties are now becoming increasingly available (Smith and Bradshaw 1979).

2. The establishment and growth of plants on infertile soils can often be greatly enhanced by innoculation with mycorrhizae (Daft and Hacskaylo 1976; Vogel and Curtis 1978), because mycorrhizae increase the rates of nutrient acquisition by plants from infertile soil. Mycorrhizae are important in transfer to roots of ions that diffuse slowly in soil ($PO_4 \ll NH_4 < K \ll NO_3$) and so will be particularly important in low-phosphorus sites. Many mycorrhizal fungi have differentiated into a series of ecotypes, each of which may have narrow tolerance limits (Trappe and Fogel 1977). The mycorrhizal innoculum should therefore be collected from a site similar to that where it will be used. In certain cases, soils thought to be toxic to plants were actually toxic to mycorrhizal fungi, resulting in decreased phosphorus supply to plants (Trappe et al. 1973), again pointing to the importance of selection of appropriate mycorrhizal species and strains for a given site. In general, ectomycorrhizae have proven more successful than endomycorrhizae on extreme sites such as coal mine spoils (Schramm 1966; Trappe and Fogel 1977). Heavy phosphorus fertilization reduces mycorrhizal infection and in some cases actually impedes the long-term establishment of vegetation.

3. Legumes, alders, and other plants with symbiotic nitrogen-fixing organisms have been used successfully to build up nitrogen capital of a disturbed site (Bradshaw et al. 1978; Bennet et al. 1978; Palaniappan et al. 1979). Nitrogen is commonly one of the most strongly limiting nutrients on an undeveloped soil profile, because, (a) it comes primarily from biotic fixation rather than from weathering of rocks, and, (b) nitrogen is the mineral element required in greatest quantities by most plants (Epstein 1972). In normal primary succession, dominance by nitrogen-fixing vegetation commonly precedes and may be a requirement for normal further successional development (Crocker and Major 1955; Van Cleve et al. 1971). If such nitrogen-fixing vegetation can be included in the revegetation effort, it speeds later succession (Palaniappan et al. 1979).

Nitrogen fixation has a high phosphorus and energy requirement. Light energy is generally not a major limitation in early stages of revegetation, but phosphorus limitation can be important on many parent materials. In New Zealand and Australia where soils are generally phosphorus-deficient, there has been detailed research on the quantity of phosphorus that must be supplied to support legume growth (e.g., Nordmeyer et al. 1974). Such legume growth then provides the nitrogen necessary to support growth of other species in the revegetation effort.

In summary, present information suggests that attempts to revegetate disturbed sites with rapidly growing species do not lead to rapid community recovery, except where soil fertility has been maintained or restored. In other cases planting of nitrogen-fixing species, often with phosphorus fertilization, can improve soil nitrogen status. Infertile soils are more effectively colonized by slowly growing species in association with appropriate mycorrhizal fungi, than by the rapidly growing commercial grasses that have received widespread use. However, the most effective solution to recover drastically disturbed lands lies in plans to minimize disturbance in areas where community restoration is difficult (Bradshaw et al. 1978; Mertes 1978).

4.3.5 Conclusions

Man's activities frequently alter soil nutrient availability, thus affecting nutrient absorption and allocation by plants. The changing pattern of plant nutrition following disturbance can be predicted from an understanding of the nutrient relations of plants in diverse natural environments. Plants adapted to infertile environments grow slowly, have a low annual nutrient requirement, and are not highly responsive to nutrient addition. In contrast, plants from more fertile and less stressful environments have high rates of nutrient absorption and growth. Under conditions of nutrient stress, these rapidly growing plants deplete internal nutrient reserves, cease growing, and are highly vulnerable to disease or environmental stress. Such rapidly growing species respond to nutrient addition by greatly accelerated growth rate rather than by nutrient accumulation and thus maintain a competitive advantage under favorable conditions. Human impacts such as agriculture and fire cause an increase in the resources available for plant growth and thus favor plants with rapid rates of nutrient absorption and growth. Mining and other disturbances reduce nutrient availability.

Attempts to revegegate such disturbance areas with rapidly growing plants having high nutrient requirements are generally unsuccessful without massive repeated fertilization. Such sites can be successfully recolonized by either, (1) increasing the nutrient supplying capacity of the soil directly or through plants with symbiotic nitrogen fixation, or, (2) using plants with slow growth rates and low nutrient requirements aided by mycorrhizal innoculation.

Acknowledgments. Research leading to these ideas was supported by the Guggenheim Fondation and the U.S. Army Research Office. I thank S.F. MacLean for critical review of the manuscript.

Résumé

Dans les champs abandonnés, les coupes forestières, les pâturages et autres milieux modifiés par l'Homme, chaque plante peut bénéficier de ressources en lumière et en éléments minéraux plus grandes que celles qu'elle trouverait dans la végétation climacique croissant sur la même roche-mère. Les plantes qui prolifèrent dans ces milieux sur les sols riches ont une croissance rapide, un rendement photosynthétique élevé, et un taux élevé d'absorption des nutrients. Quand la feuille vieillit, son fonctionnement se ralentit, et son contenu en NKP décroît de 40% à 60% du tau initial. Ces plantes à croissance rapide sont très compétitives, tant que le milieu est riche. A l'opposé, les plantes de milieux pauvres fonctionnent lentement mais sont frugales, et profitent peu des apports d'engrais.

Au cours d'une succession, les plantes rapides, à cycle phénologique très contrasté, sont progressivement remplacées par des plantes lentes, dotées de réserves importantes, et moins liées aux cycles saisonniers (Fig. 1); de même, le prélèvement de nutrients par la végétation s'accroît, en particulier parce que les plantes pérennes peuvent mieux profiter des périodes favorables. Dans les milieux périodiquement incendiés, le prélèvement cumulé de nutrients atteint son maximum peu avant l'incendie (au bout de quelques mois dans les formations herbacées incendiées tous les ans, au bout de dix ans dans les landes de Bruyère, et de soixante ans dans une forêt de Pins). Les plantes qui poussent après un incendie ont souvent une meilleure valeur pastorale que celles qui croissent après une coupe rase.

Dans les toundras, les perturbations augmentent aussi les vitesses de décomposition et la disponibilité en nutrients, et elles favorisent aussi les plantes à croissance rapide aux dépens des buissons sempervirents.

Quand l'Homme empêche les espèces "rapides" de coloniser les milieux perturbés, il risque de provoquer une forte perte d'éléments minéraux. Quand il a enlevé (ou laissé éroder, ou trop perturbé) les horizons supérieurs du sol, la colonisation par les plantes "rapides" est difficile, et il faut prévoir l'apport de fertilisants, ou l'installation d'espèces peu exigeantes, eventuellement inoculées de champignons mycorrhiziens (en particulier dans les sols pauvres en phosphore, où un apport de phosphore peut être nécessaire pour la croissance de Légumineuses ou d'Aulnes fixateurs d'azote).

References

Antonovics J, Bradshaw AD, Turner RG (1971) Heavy metal tolerance in plants. Adv Ecol Res 7:1–85

Babb TA, Whitfield DWA (1977) Mineral nutrient cycling and limitation of plant growth in the Truelove Lowland ecosystem. In: Bliss LC (ed) Truelove Lowland, Devon Island, Canada: A high arctic ecosystem. Univ Alberta Press, Edmonton, pp 589–606

Batzli GO, Sobaski S (1980) Distribution, abundance, and foraging patterns of ground squirrels near Atkasook, Alaska. Arct Alp Res 12:501–510

Bazzaz FA (1979) The physiological ecology of plant succession. Annu Rev Ecol Syst 10:351–371

Bennet OL, Mathias EL, Armiger WH, Jones JN Jr (1978) Plant materials and their requirements for growth in humid regions. In: Schaller FW, Sutton P (eds) Reclamation of drastically disturbed lands. Am Soc Agron, Madison, pp 285–306

Biederbeck VO, Campbell CA (1971) Influence of simulated fall and spring conditions on the soil system. I. Effects on soil microflora. Soil Sci Soc Am Proc 35:474–479

Bliss LC, Wein RW (1972) Plant community responses to disturbances in the western Canadian Arctic. Can J Bot 50:1097–1109

Bouma D (1970) Effects of nitrogen nutrition on leaf expansion and photosynthesis of *Trifolium subterraneum* L. 1. Comparison between different levels of nitrogen supply. Ann Bot (London) 34:1131–1142

Bradshaw AD, Humphries RN, Johnson MS, Roberts RD (1978) The restoration of vegetation on derelict land produced by industrial activity. In: Holdgate MW, Woodman MJ (eds) The breakdown and restoration of ecosystems. Plenum Press, New York, pp 249–274

Challinor JL, Gersper PL (1975) Vehicle perturbation effects upon a tundra soil-plant system: II. Effects on the chemical regime. Soil Sci Soc Am Proc 39:689–695

Chapin FS III (1980) The mineral nutrition of wild plants. Annu Rev Ecol Syst 11:233–260

Chapin FS III, Bloom AJ (1976) Phosphate absorption: adaptation of tundra graminoids to a low temperature, low phosphorus environment. Oikos 26:111–121

Chapin FS III, Chapin MC (1980) Revegetation of an arctic disturbed site by native tundra species. J Appl Ecol 17:449–456

Chapin FS III,Van Cleve K (1981) Plant nutrient absorption and retention under differing fire regimes. In: Mooney HA, Bonnickson TM, Christensen NL, Lotan JE, Reiners WA (eds) Fire regimes and ecosystem properties. USDA For Serv Gen Tech Rep US Dep Agric, Washington, pp 301–321

Chapin FS III, Shaver GR (1981) Changes in soil properties and vegetation following disturbance of Alaskan arctic tundra. J Appl Ecol 18:605–617

Chapin FS III, Barsdate RJ, Barèl D (1978) Phosphorus cycling in Alaskan coastal tundra: A hypothesis for the regulation of nutrient cycling. Oikos 31:189–199

Chapin FS III, Johnson DA, McKendrick JD (1980) Seasonal movement of nutrients in plants of differing growth form in an Alaskan tundra ecosystem: implications for herbivory. J Ecol 68:189–209

Chapman SB (1967) Nutrient budgets for a dry heath ecosystem in the south of England. J Ecol 55:677–689

Clarkson DT (1967) Phosphorus supply and growth rate in species of *Agrostis* L. J Ecol 55:111–118

Clarkson DT (1974) Ion transport and cell structure in plants. McGraw-Hill, London, p 350

Clawson M (1972) America's lands and its uses. Johns Hopkins Press, Baltimore, p 166

Connell JH, Slatyer RO (1977) Mechanisms of succession in natural communities and their role in community stability and organization. Am Nat 111:1119–1144

Crocker RL, Major J (1955) Soil development in relation to vegetation and surface age at Glacier Bay, Alaska. J Ecol 43:427–448

Daft MJ, Hacskaylo E (1976) Arbuscular mycorrhizas in the anthracite and bituminous coal wastes of Pennsylvania. J Appl Ecol 13:523–531

Dean R, Ellis JE, Rice RW, Bement RE (1975) Nutrient removal by cattle from a shortgrass prairie. J Appl Ecol 12:25–29

Eiten G (1972) The cerrado vegetation of Brazil. Bot Rev 38:201–341

Elias CO, Chadwick MJ (1979) Growth characteristics of grass and legume cultivars and their potential for land reclamation. J Appl Ecol 16:537–544

Epstein E (1972) Mineral nutrition of plants: principles and perspectives. Wiley, New York, p 412

Gorham E, Vitousek PM, Reiners WA (1979) The regulation of chemical budgets over the course of terrestrial ecosystem succession. Annu Rev Ecol Syst 10:53–84

Grime JP, Hunt R (1975) Relative growth rate: Its range and adaptive significance in a local flora. J Ecol 63:393–422

Gupta PL, Rorison IH (1975) Seasonal differences in the availability of nutrients down a podzolic profile. J Ecol 63:521–534

Haag RW, Bliss LC (1974) Energy budget changes following surface disturbance to upland tundra. J Appl Ecol 11:355–374

Halderson JL, Zenz DR (1978) Use of municipal sewage sludge in reclamation of soils. In: Schaller FW, Sutton P (eds) Reclamation of drastically disturbed lands. Am Soc Agron, Madison, pp 355–377

Harper JL, Lovell PH, Moore KG (1970) The shapes and sizes of seeds. Annu Rev Ecol Syst 1:327–356

Harrison AF (1979) Variation of four phosphorus properties in woodland soils. Soil Biol Biochem 11:393–403

Hernandez H (1973) Natural plant recolonization of surficial disturbances, Tuktoaktuk Peninsula Region, Northwest Territories. Can J Bot 51:2177–2196

Horn HS (1974) The ecology of secondary succession. Annu Rev Ecol Syst 5:25–37

Likens GE, Bormann FH, Pierce RS, Reiners WA (1978) Recovery of a deforested ecosystem. Science 199:492–496

MacLean DA, Wein RW (1977) Nutrient accumulation for postfire jack pine and hardwood succession patterns in New Brunswick. Can J For Res 7:562–578

Marks PL (1974) The role of pin cherry (*Prunus pennsylvanica* L.) in the maintenance of stability in northern hardwood ecosystems. Ecol Monogr 44:73–88

Mays DA, Bengtson GW (1978) Lime and fertilizer use in land reclamation in humid regions. In: Schaller FW, Sutton P (eds) Reclamation of drastically disturbed lands. Am Soc Agron, Madison, pp 307–328

Mertes JD (1978) Criteria for selecting lands that are not to be disturbed. In: Schaller FW, Sutton P (eds) Reclamation of drastically disturbed lands. Am Soc Agron, Madison, pp 205–221

Milthorpe FL, Moorby J (1974) An introduction to crop physiology. Cambridge Univ Press, Cambridge, p 202

Mooney HA, Rundel PW (1979) Nutrient relations of the evergreen shrub, *Adenostoma fasciculatum* in the California chaparral. Bot Gaz 140:109–113

Murata Y (1969) Physiological responses to nitrogen in plants. In: Eastin JD, Haskins FA, Sullivan CY, Bavel van CHM (eds) Physiological aspects of crop yield. Am Soc Agron, Madison, pp 235–259

Nátr L (1975) Influence of mineral nutrition on photosynthesis and use of assimilates. In: Cooper JP (ed) Photosynthesis and productivity in different environments. Cambridge Univ Press, Cambridge, pp 537–555

Nordmeyer AH, Lang MH, Roberts Q (1974) Legume establishment. In: Orwin J (ed) Revegetation in the rehabilitation of mountain lands. For Res Inst Symp No 16, Auckland, pp 9–20
Nye PH, Greenland DJ (1960) The soil under shifting cultivation. Commonw Bur Soils Tech Bull 51:156
Odum EP (1969) The strategy of ecosystem development. Science 164:262–270
Orians GH, Solbrig OT (1977) A cost-income model of leaves and roots with special reference to arid and semiarid areas. Am Nat 111:677–690
Palaniappan VM, Marrs RH, Bradshaw AD (1979) The effect of *Lupinus arboreus* on the nitrogen status of china clay wastes. J Appl Ecol 16:825–831
Paone J, Struthers P, Johnson W (1978) Extent of disturbed lands and major reclamation problems in the United States. In: Schaller FW, Sutton P (eds) Reclamation of drastically disturbed lands. Am Soc Agron, Madison, pp 11–22
Rorison IH (1968) The response to phosphorus of some ecologically distinct plant species. I. Growth rates and phosphorus absorption. New Phytol 67:913–923
Schramm JR (1966) Plant colonization studies on black wastes from anthracite mining in Pennsylvania. Trans Am Philos Soc NS 56:1–194
Smith RAH, Bradshaw AD (1979) The use of metal tolerant plant populations for the reclamation of metalliferous wastes. J Appl Ecol 16:595–612
Specht RL, Groves RH (1966) A comparison of the phosphorus nutrition of Australian heath plants and introduced economic plants. Aust J Bot 14:201–221
Stark NM (1977) Fire and nutrient cycling in a Douglas-fir/larch forest. Ecology 58:16–30
Stark NM, Steele R (1977) Nutrient content of forest shrubs following burning. Am J Bot 64:1218–1224
Steward KK, Ornes WH (1975) The autecology of sawgrass in the Florida Everglades. Ecology 56:162–171
Sutton RK (1975) Why native plants aren't used more. J Water Conserv 30:240–242
Trappe JM, Fogel RD (1977) Ecosystematic functions of mycorrhizae. In: Marshall JK (ed) The below-ground ecosystem: A synthesis of plant-associated processes. Range Sci Dep Sci Ser No 26. Colorado State Univ, Fort Collins, pp 205–214
Trappe JM, Stahly EA, Benson NR, Duff DM (1973) Mycorrhizal deficiency of apple trees in high arsenic soils. Hortic Sci 8:52–53
Van Cleve K (1977) Recovery of disturbed tundra and taiga surfaces in Alaska. In: Cairns J Jr, Dickson KL, Herricks EE (eds) Recovery and restoration of damaged ecosystems. Univ Press Virginia, Charlottesville, pp 422–455
Van Cleve K, Viereck LA, Schlentner RL (1971) Accumulation of nitrogen in alder (*Alnus*) ecosystems near Fairbanks, Alaska. Arct Alp Res 3:101–114
Vitousek PM, Reiners WA (1975) Ecosystem succession and nutrient retention: a hypothesis. BioScience 25:376–381
Vitousek PM, Gosz JR, Grier CC, Melillo JM, Reiners WA, Todd RL (1979) Nitrate losses from disturbed ecosystems. Science 204:469–474
Vogel WG, Curtis WR (1978) Reclamation research on coal surface-mined lands in the humid east. In: Schaller FW, Sutton P (eds) Reclamation of drastically disturbed lands. Am Soc Agron, Madison, pp 379–397
Williams RF (1948) The effects of phosphorus supply on the rates of intake of phosphorus and nitrogen and upon certain aspects of phosphorus metabolism in gramineous plants. Aust J Sci Res Ser B 1:333–361
Williams RF (1955) Redistribution of mineral elements during development. Annu Rev Plant Physiol 6:25–42

4.4 Comparison of Water Balance Characteristics of Plant Species in "Natural" Versus Modified Ecosystems

P.C. MILLER

4.4.1 Introduction

The disturbance of natural ecosystems leads to sites with altered microclimatic and soil conditions and to the establishment of new plant species populations either by deliberate introduction by man or by natural processes of revegetation. These disturbed lands include areas changed by factors such as agricultural activity and grazing, recreational impacts, road, pipeline, and building construction, deposition of mine spoils, and fire (Brown et al. 1978) and can be a sizable fraction of natural ecosystems (Johnston and Brown 1979). Interest in revegetating these lands comes from efforts to reduce soil erosion, siltation, and leaching of acid producing chemicals, heavy metals, and other materials into streams and lakes, as well as to restore the esthetic appeal of the site. Ideally, vegetation reestablishment should produce a self-perpetuating plant cover directly or foster entrapment and germination of native plant seeds which will form a self-regenerating community (Dean et al. 1973). Plant introduction should ideally reduce the undesirable side effects of disturbance while encouraging natural processes of plant succession leading to the establishment of a natural self-perpetuating vegetative cover.

Ecological theories relating to the mechanisms of plant succession usually emphasize competition and resource use by the species populations. The theories imply that early successional species, whether deliberately introduced or naturally invading, should capture enough of the resources, including water, to survive but not enough to exclude later invaders which will form the final self-perpetuating community of native plants (MacArthur 1970, 1972; Connell and Slatyer 1977). Introduced species should not use all resources fully, so that native species can invade. Ecological theories often include concepts of strategies and optimization of resource use (Clements and Shelford 1939; MacArthur 1970, 1972; Orians 1975; Pickett and Bazzaz 1978). One expects to find similar plant characteristics in similar environments because these characteristics optimize some vital functions. These plant characteristics, if found, would aid the land manager in selecting species for revegetation which will be successful. However, competition and resource use should be low in the early stages of recovery of disturbed sites. The lower competitive selection may allow a greater diversity of plant characteristics to be present amongst early successional species. Generalities relating to plant characteristics in early successional species may be more complex than those for later successional species. The diversity of characteristics relates to the generalization that the number of primary associations is always greater than the number of climax associations (Tüxen 1960).

Some of the most dramatic environmental changes in disturbed communities include changes in the heat exchange processes and site water relations which in turn affect transpiration rates and plant water balance. Transpiration rates and water balance affect the survivorship of the population. Increased soil temperatures and high heat loads of disturbed sites have been cited by various authors (Isaac 1938; Schramm 1966; Chadwick 1973; Deely and Borden 1973; Goodman et al. 1973; Schimp 1973; Lee et al. 1975). The increased droughtiness of the soils of disturbed sites has been mentioned even more commonly (Curtis 1973a; Darmer 1973; Goodman et al. 1973; Schimp 1973). Soil moisture and soil temperature are inversely related as are transpiration and plant leaf temperature (Gates 1965; Deely and Borden 1973; Schimp 1973). Plant characteristics modify heat loads and transpiration rates and affect survival (Gates 1965; Taylor 1975). High transpiration rates can produce chemical shock related to accompanying high rates of uptake of chemicals as well as detrimental accumulations of chemicals (R. Jones and Etherington 1971; H.E. Jones 1971; Repp 1973).

In the course of vegetative recovery the site energy and water balance changes because of changes in the soil water relations, soil surface shading, and transpiration. Thus, plant characteristics related to plant water balance may differ between species occurring on disturbed sites and species in natural, undisturbed communities. Characteristics may also differ among species with different modes of invasion or regrowth, whether by seed, rhizome, or resprout.

The objective of this paper is to compare water balance characteristics of plant species of natural and modified ecosystems. The characteristics of importance include those governing the loss and uptake of water by the plant, those describing the state of water in the plant, and those describing the growth and death of the plant in relation to the water status of the plant.

4.4.2 Theoretical Background

4.4.2.1 Heat and Water Exchange Processes

Climate and vegetation structure interact through the physical processes of energy exchange, which include radiation, convection, evaporation, and conduction and which affect the heat and water balance of the plant (Table 1). The plant absorbs solar and far infrared radiation and dissipates this absorbed energy by radiating in the far infrared, by evaporation, and by convection. If the air temperature is above the plant temperature the plant can gain energy by convection. The processes of energy exchange between the environment and the individual leaf or stem are affected by several structural and physiological properties of the plant part, including its absorptance, emittance, width or diameter, inclination to the sun, and conductance to water vapor diffusion. Within the plant canopy the energy exchange is affected additionally by the total leaf and stem area indices, clustering of leaves and of stems, vertical profiles of leaf and stem areas and of leaf and stem inclinations, and density of the foliage with respect to air movement. The canopy will exchange energy with the soil surface by shading the surface from the incoming solar radiation and radiating infrared, and in return will receive reflected solar radiation and far-infrared radiation from the surface. Considering the whole plant,

Table 1. Processes of energy exchange which affect the heat and water balance of the plant. (After Collier et al. 1973)

Process of energy exchange	Environmental factors involved	Coupling factors	Organism properties involved in coupling factor	Organism response
Convection	Wind velocity and direction air temperature	Convection coefficient	Size Shape Texture Orientation	Temperature
Evaporation	Wind velocity and direction humidity	Boundary layer resistance	Size Shape Texture Orientation	Temperature Water loss
		Leaf resistance	Density and size of stomata Resistance of cuticle to water loss	
Radiation	Solar radiation		Absorbance Orientation	Temperature Photosynthesis
	Infrared radiation		Emittance	Temperature
Conduction	Matrix temperature	Conduction coefficient	Conductance	Temperature

the processes of energy exchange will be affected by the absorptive capacity of the roots, because continued evaporation from the plant ultimately depends upon water absorption by roots. The processes of energy exchange influence the plant water balance and are affected, favorably or detrimentally for plant survival, by the morphological and physiological characteristics of the plant.

The processes of energy exchange affect plant survival both directly and indirectly. They result in adjustment of plant temperature so that the energy absorbed equals the energy lost. An equilibrium temperature which exceeds the thermal tolerance of the organism is lethal to it. Even a sublethal equilibrium temperature may affect the plant adversely if it differs from the optimal temperature for photosynthesis or growth. The transpirational loss of water may exceed the rate of water uptake, thus reducing the plant water content to damaging limits. Transpiration is linked to photosynthesis. When transpiration is reduced to conserve water, photosynthesis is generally reduced, although photosynthesis per unit of water loss may be increased. Water stress on a daily or annual basis reduces photosynthesis. Photosynthesis affects the carbohydrate available for additional leaf growth, root growth, and reproduction. The reduced reproductive effort may reduce the population and increase the chance of species replacement.

Plant height above a nearly bare soil surface is important in maintaining cooler plant temperatures. Surface temperatures can be 15°–20 °C higher than temperatures in the freely moving air (Schramm 1966; Deely and Borden 1973; Lee et al. 1975; P.C. Miller et al. 1977). The air close to the surface is similarly heated depending on the turbulent exchange of air near the surface. Stems in contact with the hot soil surface are unable to convect the absorbed heat away because of the still air and high air temperatures near the soil surface, and such stems are subject to heat damage (Schramm 1966).

4.4.2.2 Water Availability and Plant Characteristics

Plant species have been classified with respect to their mechanism of surviving environmental drought as drought escaping, drought avoiding, and drought tolerating (Oppenheimer 1960; Levitt 1972). Plants in the first category complete their life cycle before the environmental drought occurs and pass through the dry conditions in drought-resistant forms, such as seeds. Drought avoiders maintain high water contents and low water stress in spite of generally dry environmental conditions either by tapping deep soil water or by suppressing water loss. Levitt (1972) divided drought avoiders into water spenders and water savers depending on whether the species took up and used water rapidly or conserved water loss (Table 2). Drought tolerators survive low water contents and high water stresses. The classification system does not uniquely categorize a species, and in the semiarid mediterranean regions most species fall into more than one category.

Considering the plant characteristics in terms of the soil-plant-atmosphere continuum (Lemon et al. 1971; Brown 1977) is a simpler organizational framework which also relates plant characteristics functionally with plant water balance (Fig. 1). The plant water content is related to the soil water content by the rates of uptake and loss. The rate of uptake depends directly on the plant water potential, root area or root length density, soil water conductance, and root depth, and inversely on the soil water potential. The rate of loss depends directly on stomatal conductance to water, cuticular conductance, boundary layer air conductance, leaf area, stem area and conductance, and leaf temperature and depends inversely on the humidity of the air. These plant and soil variables are affected by many other environmental conditions. Water savers decrease the conductances to water loss and the areas involved in water loss; water spenders increase the conductances and the areas involved in water uptake.

The soil water balance is intimately associated with the plant water balance and depends on patterns of precipitation, interception in the canopy, throughfall, stemflow, runoff, and infiltration. The precipitation patterns vary in frequency and amount. Different precipitation regimes and soil textures will produce different temporal and spatial patterns of exploitable water. The characteristics of a successful plant in one climatic and soil regime may not be present in a successful plant in another regime. Generalities may be hard to formulate without characterizing both the environmental and the plant characteristics. The plant characteristics are interrelated with each other and with the soil water regime so that more than one combination of characteristics may solve the plant's water balance problems on a site (Fig. 2). The amount of water the plant takes up relative to annual precipitation is limited by the annual precipitation through its effects on potential plant biomass, foliage area index, and associated plant properties. With increased photosynthesis, plant biomass can increase, more ground surface can be shaded, and water loss by soil evaporation can be reduced. The transpiration efficiency, calculated

Table 2. Examples of species or varietal differences in resistance due to avoidance rather than to tolerance. (After Levitt 1972)

Species in order of decreasing resistance	Avoidance factor correlated with resistance	Reference
(a) Avoidance due to water conservation		
1. Ponderosa pine, Douglas fir, Western arbovitae	Decreased rate of water loss	Parker (1951)
2. Grain sorghum, cotton, peanut	Lower transpiration rate	Slayter (1955)
3. Two oat varieties	More rapid stomatal closure	Stocker (1956)
4. *Pinus densiflora* versus *Chamaecyparis obtusa* and *Cryptomeria japonica*	Cuticular transpiration inversely related to resistance	Satoo (1956)
5. *Quercus ilex* versus *Q. pubescens*	Ten times as great a drop in transpiration on stomatal closure	Larcher (1960)
6. Sugarcane varieties	More rapid stomatal closure	Naidu and Bhagyalakshmi (1967)
7. *Pinus halepensis* versus *P. pirea*	Lower cuticular transpiration	Oppenheimer and Shomer-Ilan (1963)
8. Pine versus mulberry seedlings	Lower cuticular transpiration	Tazaki (1960)
9. Six clones of blue panic grass	Fewer stomata	Dobrenz et al. (1969)
10. Six wheat cultivars	Higher percentage water retention on desiccation at 81% r.h.	Salim et al. (1969)
(b) Avoidance due to greater water absorption		
1. Short-leaf versus loblolly pine	Higher transpiration rate	Schapmeyer (1939)
2. Grain sorghum, cotton, peanut	Greater water uptake	Slayter (1955)
3. *Pinus densiflora* versus *Chamaecyparis obtusa* and *Cryptomeria japonica*	Greater root development	Satoo (1956)
4. Varieties of corn	Larger root system	Misra (1956)
5. Pine species	More water removed from soil	Oppenheimer (1967)
6. Two-eared versus single-eared corn	Extracted 1–2% more water from soil	Barnes and Woolley (1969)
7. Spring wheats	Greater root development	Devera et al. (1969)
8. *Eucalyptus socialis* versus *E. incrassata*	Higher root-shoot ratio	Parsons (1969)

[a] Measurements of tolerance when made revealed no difference in Nos. 1, 6, and 10; differences in tolerance found in No. 4 correlated with drought resistance

as transpiration divided by precipitation, is directly related to the foliage area index. With increasing foliage area indices, soil evaporation decreases while transpiration and interception increase. Because interception losses are less than soil evaporation losses, the vegetation can transpire a greater fraction of the annual precipitation as the precipitation and foliage area index increase. The foliage area index does not increase indefinitely but reaches a maximum at which photosynthesis just sustains tissue maintenance and replacement costs. At these high foliage areas,

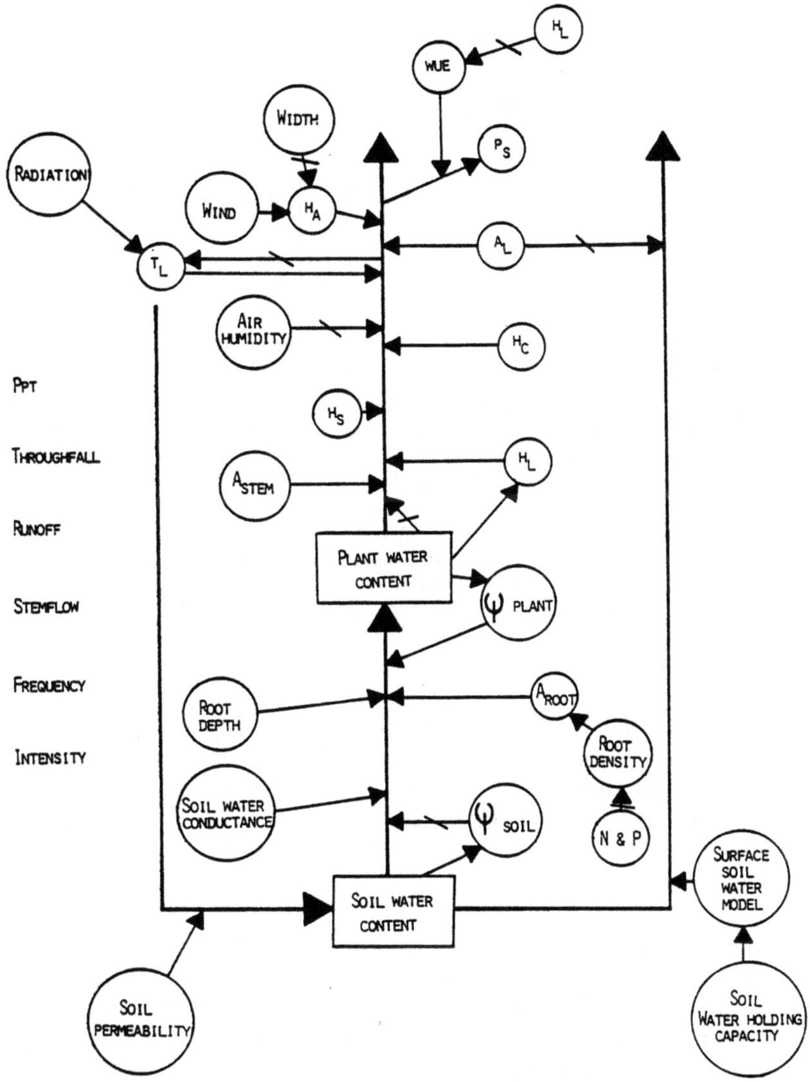

Fig. 1. Diagram of the flow of water through the soil–plant–atmosphere system and controlling plant and environmental variables. Variables are leaf, cuticular, air, and stem conductances (*H* plus first letter as subscript), leaf, stem, and root area (*A*), soil and plant water potential (ψ), leaf temperature (T_L), photosynthesis (P_S), water use efficiency (*WUE* g CO_2/g H_2O), leaf width, wind speed, root length density, nitrogen (*N*), and phosphorus (*P*). *Heavy lines* flows of water; *light lines* controls; *boxes* water storage; *circles* variables controlling the water flows

transpiration reaches a plateau even though precipitation increases, and in this range of precipitation water capture efficiency decreases.

The maximum foliage area which can be maintained is constrained by the amount of photosynthate fixed, which is related to the annual precipitation; the fraction of the precipitation which is transpired; the ratio of photosynthesis to

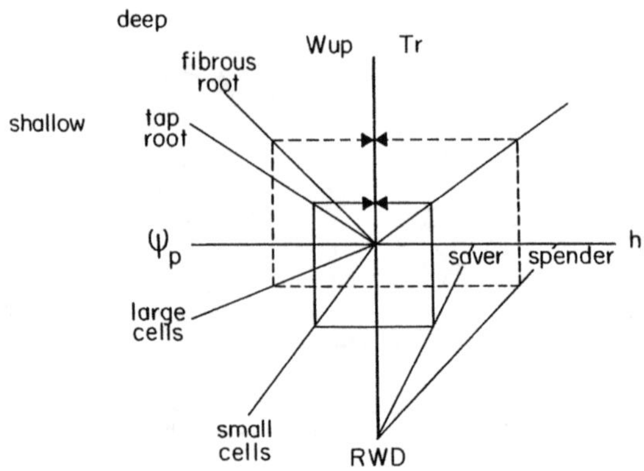

Fig. 2. Diagram of the interrelationships between leaf conductance (h_1) and transpiration (*Tr*), leaf conductance (h_1), and relative water deficit (*RWD*), relative water deficit (*RWD*), and plant water potential (ψ_p), and plant water potential (ψ_p), and water uptake (*Wup*), making simplifying assumptions about other controlling variables. The interrelations are illustrated for a hypothetical water saving plant, with small leaf cells and tap root system (*solid line*) and for a hypothetical water spending plant, with large leaf cells and fibrous root system (*dashed line*). When the plant is in steady state, water uptake equals transpiration and leaf conductance, relative water deficit, and plant water potential will have particular values depending on the interrelationships between values. If the relative water deficit is lower than the steady state value, transpiration exceeds water uptake, increasing the relative water deficit

transpiration; the respiratory and biomass costs of replacing leaves, stems, and roots; the turnover rates of leaves, stems, and roots; and the respiratory maintenance costs. In the steady state, photosynthate is completely used to maintain the biomass, where maintenance includes maintenance respiration and the replacement of shed parts. Thus, the steady state biomass (B) is approximately

$$B = (Ppt)(Tr/Ppt)(P_s/Tr)/(r_m + r_g\delta),$$

where Ppt is annual precipitation, Tr is transpiration, P_s is photosynthesis, r_m is maintenance respiration rate, r_g is growth respiration, and δ is the turnover rate. A relation between biomass and precipitation is given by Grier and Running (1977). With low precipitation (<400 mm), only low foliage areas can be maintained (<2.0); soil evaporation is high, giving a low water capture efficiency (Fig. 3). The site and the plant must be characterized as a system to understand the adaptive significance of any particular plant characteristic. This characterization has been reasonably well completed for several tundra chaparral and tundra species (Fig. 4).

Other plant characteristics should correlate with the precipitation gradient. At low precipitation (<400 mm yr^{-1}) and low foliage area index, plants compete with soil evaporation for water. Transpiration and transpiration efficiency increase with high leaf conductance and high leaf area to dry weight ratios. The latter increases soil shading. Because soil drought is lengthened by soil evaporation and high plant conductances, drought deciduous and drought escaping species are favored. According to leaf energy budget theory (Gates 1965), water is conserved by steeply inclined leaves, narrow leaves, and high leaf reflectances. Thus, in areas with low

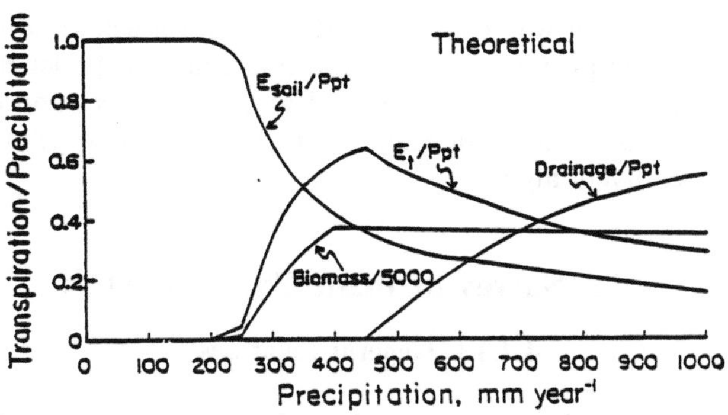

Fig. 3. Relations between transpiration efficiency and annual precipitation for Californian and Chilean shrubs at their steady-state foliage area indices for different precipitation amounts. Theoretical curves are calculated from the general relationships described in the text. E_{soil}/Ppt ratio of annual soil evaporation to annual precipitation; E_t/Ppt transpiration to precipitation; *Drainage/Ppt* water drained from the soil in relation to precipitation. Biomass is given in m^{-2} divided by 5,000. Data is from *Adenostoma fasciculatum, Arctostaphylos glauca, Ceanothus greggii,* and *Rhus ovata* at Echo Valley; and *Colliguaya odorifera, Lithraea caustica, Satureja gilliesii,* and *Trevoa trinervis* at Fundo Santa Laura. (After Miller 1981)

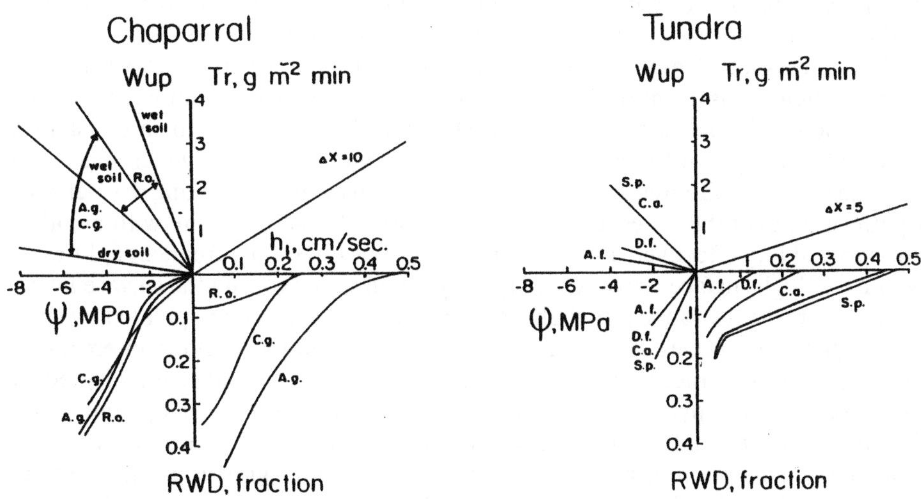

Fig. 4. Interrelationships between leaf conductance (h_1) and transpiration (*Tr*), leaf conductance (h_1), and relative water deficit (*RWD*), relative water deficit (*RWD*) and plant water potential (ψ), and plant water potential (ψ), and water uptake (*Wup*) for chaparral shrubs and wet meadow tundra graminoids using characteristic environmental conditions. (After Poole and P.C. Miller 1975; Stoner and P.C. Miller 1975)

precipitation, the transpiration efficiency is increased with leaves with high leaf area to dry weight ratios, steep inclinations, narrow widths, and high reflectances. At higher precipitation levels (> 550 mm yr^{-1}), water is lost by drainage, transpiration is increased with higher leaf conductances, and the length of the drought is short. Under these conditions, evergreen species are favored unless they are restricted by cold temperatures. Leaf width, inclination, and reflectance can be more variable without affecting water loss. At intermediate precipitation levels (400–550 mm yr^{-1}), the composition of the vegetation changes the length of the soil drought by including species with different leaf conductances and different leaf area indices. The transpiration efficiency is controlled by the vegetation composition and can be relatively high.

4.4.3 Survey of Plant Characteristics

4.4.3.1 General Relations

In general, few physiological or morphological measurements related to plant water balance have been made on either introduced species or early successional species. Artificial revegetation research has emphasized observations of vegetation and plant species trials sometimes combined with mulch, irrigation, and fertilizer treatments. One can conclude, tentatively, that with annual precipitation over about 500 mm, irrigation and mulching of mine spoils has had little effect, at least in regions with summer rain. Studies in Pennsylvania (Czapowskyj 1973; Schimp 1973), West Germany, and Montana (Farmer et al. 1974, 1976; Brown and Johnston 1976) support this conclusion. Mulching in areas subject to frost heaving of seedlings may improve seedling survival (Brown and Johnston 1979).

Mulching should improve the moisture conditions in the surface layer of soil. Frequent light rains promote the development of shallow-rooted annual grass species in the California annual grassland (Pitt and Heady 1978). In this system a heavy infiltrating rain followed by a dry fall promotes *Erodium* which has a tap root. Legumes are favored by heavy infrequent rains with intervening dry periods, a pattern unfavorable to establishment of grasses. A disturbed site with a soil of low permeability effectively may be wetted as if it received only frequent light rains because of high surface runoff during heavy, intense rains. Only the surface layers are wetted by heavy rains or by light rains and such sites can be populated by shallow-rooted grass species. However, as the vegetative canopy develops, interception increases. A larger fraction of light rains than of heavy rains will be intercepted by the developing canopy. Anderson et al. (1969) showed that the development of understory herbs was inhibited by interception of rain rather than interception of light. The interception of rain may inhibit the understory grasses as an overstory of shrubs or trees develops. The development of the vegetative cover also increases soil permeability and infiltration capacity. Thus, the effective precipitation regime shifts from frequent light rains to infrequent heavy rains as the vegetative cover develops. The plant species response can be expected to shift from shallow-rooted plants to deep-rooted plants. Mulching retains surface water and creates an environment analogous to that provided by frequent light rains, which favors shallow-rooted species.

4.4.3.2 Factors and Processes Affecting Water Loss

Most pioneer species on sandy material are xeromorphic (Darmer 1973), apparently in response to soil drought, but this xeromorphy may be caused by low nutrient conditions as well as by drought (Loveless 1962; Beadle 1968; Small 1972; Specht 1979). Xeromorphy may not necessarily be correlated with low rates of water loss (Oberbauer and P.C. Miller 1979, 1981).

Transpiration rates have been measured in a few studies of early successional or introduced species. Wieland and Bazzaz (1975) gave transpiration rates of early successional species in Illinois. These rates are well within the range of measured transpiration rates of later successional species (P.C. Miller et al. 1978). Brown (1973) measured very high transpiration rates of alpine species. The transpiration rates of two introduced grass species were higher than the native grass species. Brown concluded that the introduced species had no control on water loss, while the native species had control.

Water loss can be reduced by dropping transpiring surface or by decreasing leaf conductance. Orshan (1963) has emphasized leaf shedding in contrast to stomatal closure as a drought adaptation. Leaf shedding is costly in terms of carbon balance and nutrient loss.

The control of leaf conductance is complex. In several chaparral shrub species the maximum leaf conductance attained during the day decreased with decreasing soil water potential as the season progressed. The leaf conductance is composed of the stomatal conductance and the cuticular conductance. Cuticular conductance is commonly higher in plants from dry habitats. Several hypotheses concerning the control of leaf conductance have been proposed. Turner (1974), Running et al. (1975), P. C. Miller et al. (1976), and P. C. Miller and Poole (1979) have emphasized the relation of leaf conductances to leaf water potential or leaf water content. Schulze et al. (1972, 1975) emphasized the control of leaf conductance by the atmospheric vapor pressure. Penning de Vries (1972) related leaf conductance to the carbon dioxide concentration inside the leaf. All workers have recognized the relation of solar irradiance to leaf conductance. The physiological mechanisms and interactions between these controlling factors are not clear. At Echo Valley, in southern California, leaf conductances of a chaparral shrub, *Adenostoma fasciculatum* changed in a complex manner beyond the suggested controls. The changes were related to the time since sunrise and may involve abscisic acid accumulation (Roberts and Valamanesh, unpubl. data). Leaf conductances of other chaparral shrubs, *Arctostaphylos glauca, Ceanothus greggii*, and *Quercus dumosa* related more clearly to leaf water potential or atmospheric vapor pressure. The rate of transpiration also varies with density of water conducting tissues in the leaf.

Early successional species in a sense compete with soil evaporation; they should have high leaf conductances, shallow and fibrous roots. Late successional species may have lower conductances and deeper roots. The lower conductances will give higher water use efficiency, which will yield more carbohydrates for maintaining biomass. The higher biomass increases the possibility of shading other species and of capturing and holding the nutrients which are more easily mineralized.

4.4.3.3 Factors and Processes Affecting Water Uptake

Water uptake is affected by root length or surface area, soil hydraulic conductivity' and root permeability. Wieland and Bazzaz (1975) showed that the root depth distribution differed among early successional species. They related the root depth distribution to niche segregation and resource use. Parrish and Bazzaz (1976) emphasized the role of root depth distribution as niche segregation. Rapid root development is an important characteristic in species to be used for revegetation (R. Brown pers. comm.). Tadmor and Cohen (1968) and Cohen and Tadmor (1969) give root elongation rates of several mediterranean grasses and legumes of 1–3 cm day^{-1}. Rates for wheat and barley were 1–6 cm day^{-1}. Kummerow (unpubl. data) measured rates of 0.2–1.0 cm day^{-1} in several chaparral and matorral shrubs. Mycorrhizae are also important. Species without mycorrhizae or with endomycorrhizae could not grow on black coal spoils in Pennsylvania (Schramm 1966). Species with ectomycorrhizae could grow on these sites but were often killed by high surface temperatures. Jacoby (1973) pointed out that root system development was strongly influenced by mine spoil texture, with long, mostly vertical roots in the loose, coarse-textured spoils and shorter, more numerous roots in the fine-textured spoils. The total length of the first-order roots did not differ significantly. Harabin and Greszta (1973) showed that roots grow very close to the surface on toxic sites. Soil compaction reduces root growth but results vary with species (B.K. Taylor 1967; F.G. Taylor 1974; Pearson et al. 1970; Grimes et al. 1975, 1978). In the surface soil, soil drought will be accentuated. Surface drought will be affected by soil color, exposure to sun and wind, and high rock content (Curtis 1973b). Curtis suggested that moisture at depth does not appear to be limiting and that deep-rooted species may do better than shallow-rooted species. Birches (*Betula pendula* and *B. pubescens*) are commonly introduced species on mine spoils in northern Europe (Schlätzer 1973). The roots of young birch seedlings penetrate relatively deep into the soil, and their root length/shoot length ratio is high compared with that of black alder (*Alnus glutinosa*) (McVean 1956 cited in Schlätzer 1973). Early development of deep roots in seedlings may anchor the seedlings against soil movement caused by frost action (Savile 1974) and reduce mortality due to frost heaving. Fertilization reduced death of seedlings due to frost heaving (Zarger et al. 1973). H.J. Bauer (1973) noted that near Cologne, West Germany, *Calamagrostis epigeios* is an effective competitor because it has a dense root system extending to a depth of about 2 m. Trees are difficult to establish in this well exploited soil.

The utilization of water in the soil depends on the rate of water transport to the soil layer immediately adjacent to the root surface as well as the rate of the root moving into moist soil. Root growth, hydraulic conductivities, and water potentials of the soil and the plant interact in determining the water supply to the plant (Newman 1974; Caldwell 1976; Hsaio et al. 1976). Soil conductivity decreases rapidly as soil water content decreases. Where soil evaporation is high, plant uptake should be increased by increasing rates of root growth and root densities. Shallow-rooted plants with fibrous roots seem advantageous for surface water. Deep-rooted plants can absorb water at slower rates because soil evaporation does not remove water from these depths. These plants may have fibrous or woody roots. Nitrogen and phosphorus are often limiting on disturbed sites. The diffusion of these nutrients to roots is lower than is the movement of water (Bieleski 1973; Lambert and Penning de Vries 1975). Root densities required for the uptake of nitrogen and phosphorus should be adequate for the uptake of water. Low nutrient soils should promote greater investment in root development by the plant. Newman (1974) concluded that most plants have a higher root density than they need for water uptake and cites Devera et al. (1969) who found no consistent correlation between root density and drought tolerance in 15 varieties of wheat in Australia. He also pointed out that root pruning of wheat plants rarely reduced growth. Ion accumulations may occur at the root surface but only with high transpiration rates and low root lengths (Newman 1974). The ion accumulation may decrease the osmotic potential of the soil near the root and cause the plant to endure low water potentials throughout the 24 h day. Hsiao et al. (1976) placed more emphasis on root growth in water uptake. They presented an equation for water uptake, based on Acevedo (1975), combining root length density, soil hydraulic conductivity, root radius, and soil depth. Their equation implies that root elongation will be relatively unimportant when soil water content and hydraulic conductivity are high but will be an important factor when soil water content and hydraulic conductivity are low. They point out that root length is usually from branching rather than from continual elongation of the main root. The branching of roots leading to logarithmic growth of root length is seen in measurements of range grasses (Plummer 1943) but not in chaparral shrubs (Kummerow et al. 1978). Caldwell (1976) reviewed the question of rhizospheric resistance to water uptake. Consensus is that rhizospheric resistance is not important until soil water potential is below −1.5 MPa. Caldwell concludes that rhizospheric resistance may be important with soil potentials of −0.3 MPa at moderate rooting densities. Under such conditions root growth may be important for water extraction. At the higher soil water contents root growth is probably related to nutrient uptake more than water uptake. Water spenders with well-developed root systems may be the result of nutrient deficiences. However, deep rooting is advantageous and is frequently seen in halophytes because roots avoid the surface soil horizons where concentrations of salts are usually highest.

4.4.3.4 State of Water in the Plant

The state of water in the plant is indicated by relative water content or relative water deficit and by water potentials. Jarvis and Jarvis (1963) and Levitt (1972) emphasized the importance of both plant water content and plant water potential. Disruption of cellular processes can be caused by changes both in cell volume and in the spatial organization of subcellular structure (Vieira da Silva 1976; Kluge 1976). The relation between the water content and water potential varies by species and is not unique to a habitat (Poole and P. C. Miller 1975; Stoner and P. C. Miller 1975; Oberbauer and P. C. Miller 1979, 1981). Plant water contents are usually much higher than lethal water contents for the species (Höfler et al. 1941 cited in Levitt 1972), and no difference has been seen in early successional and introduced species compared with late successional native species. The critical water content changes with the season.

Several workers have pointed out the occurrence of small cells in xeric adapted plants. Cutler et al. (1977) developed the concept that with small cells the ratio of cell wall water to protoplasmic water is relatively high. With high ratios, relatively large quantities of water can be lost without loss of turgor. Hsiao et al. (1976) emphasized that cell wall water was held by matric attraction. A high ratio of cell wall to protoplasmic water leads to low matric potentials as water contents decrease. This increases the potential difference causing water uptake without requiring loss of turgor or increased osmotic concentration.

Water potentials have been measured on only a few introduced and early successional species. Wieland and Bazzaz (1975) showed a range in water potentials during the day in early successional species, although all species recovered to above −0.2 MPa at night. *Setaria faberii* commonly reached water potentials below −1.2 MPa and occasionally −1.9 MPa, while *Polygonum pensylvanicum* maintained water potentials above −0.3 MPa. The transpiration rates were lowest, although photosynthesis was highest, in the species with the lowest water potential. Water potentials were lower with shallower rooting. Johnston et al. (1975) indicated that water potentials near the mine spoil surface in an alpine site can be expected to decline from near 0 to about −2.0 to −3.0 MPa after a few days of clear weather. On undisturbed sites, minimum water potentials of about −1.0 MPa were noted for the same period of time. Leaf water potentials of plants growing on spoil material reach greater extremes during a single diurnal period than those on native undisturbed sites. On spoil materials, leaf water potentials of *Poa alpina* declined from −1.0 MPa in the morning to about −3.0 MPa by mid-afternoon; water potentials of the same species on an undisturbed site did not exceed −1.8 MPa. Van Kekerix et al. (1979) showed that mean leaf water potentials of *Alopecurus pratensis* and *Poa alpina* throughout the field season and for periods of more than 2 days without precipitation were lower on control than on treated plots. Mean leaf water potentials were highest in plants grown on plots with jute netting or peat moss plus jute netting. The *Poa* control group had at −1.96 MPa, the lowest mean leaf water potential. Coyne (pers. comm.) pointed out that a native grass species in the southern plains maintains water potentials lower than an introduced species. Courtin and Mayo (1975) showed that water potentials of *Dryas integrifolia* from the polar desert never rose above −1.5 MPa, even under well-watered, controlled environmental conditions. However, Oberbauer and P.C. Miller (1979, 1981) measured potentials near 0 bars in *Dryas octopetala* in a similar habitat but a different area. The mechanisms of the maintenance of low water potentials are unclear; it could be due to matric forces in the cell wall. Osmotic potentials of mesophytes and xerophytes commonly differ. The osmotic potential is related to drought tolerance.

In general, the water contents and water potentials of introduced and early successional species are within the range shown by native plants. It is unlikely that introduced and early successional species will fall outside the commonly encountered range of 0 to −5.0 MPa or even 0 to −13.0 MPa, values which have occasionally been measured; they are also unlikely to fall outside the range of 0% to 70% relative water deficit, which has been measured in vascular plants. The foregoing suggests that it will likely not be fruitful to concentrate on single measures of plant water characteristics in an attempt to distinguish between the water responses of plants occupying undisturbed or disturbed sites. The individual measures (tissue water deficit, water potential and its components, stomatal activity, transpiration, and root resistances) all constitute parts of an integrated system and should be interpreted with regard to their interdependencies. The pressure volume approach

(Tyree and Hammel 1972; Roberts and Knoerr 1977; Richter 1978; Wilson et al. 1979) offers an integrative means of relating water deficits to tissue water potential and its components. Characteristic species differences occur in the relationships measured by such analyses, and these differences are interpretable in the context of the environmental setting (Cheung et al. 1975; Roberts and Knoerr 1977; Hinckley et al. 1980).

4.4.3.5 Growth and Death in Relation to Plant Water Content

Seedlings tend to be more drought resistant than adult plants (Levitt 1972). New leaves, after attaining a degree of maturity, are more drought resistant than old leaves (Levitt 1972). Leaf shedding during drought usually involves dropping old leaves. Repp (1973) indicated that small leaves are advantageous because if shedding occurs the loss of photosynthetic surface is not as severe. The advantage of small leaves seems more important with low leaf area indices and in seedlings where dropping single large leaves can involve a significant fraction of the total leaf area. Drought induced death can be related to the loss of water and disruption of cell organization. The advantage in this case is with plants with small leaf cells. Drought induced death can also be related to the osmotic potential.

The death of seedlings may be caused by heat damage rather than by water stress. Chadwick (1973) indicated that the distribution of two species in the desert, which was believed to be brought about by the moisture regimes of the clay and sandy soil, was better explained by heat transmission and heat tolerance of the two soils. Repp (1973) pointed out that death can occur due to chemical shock. Chemical shock can occur by very rapid intake of the damaging agents, which happens when plants with a high transpiration rate grow on a moist soil containing injurious chemicals. In support of his generalization, he pointed out that halophytes are often damaged when rainy weather is followed by high evaporation due to sun, strong wind, and heat. Repp further pointed out that small cells seem preferable to large cells because small cells have a greater number of cell-physiological barriers against the damaging substances.

The mechanisms of how water stress leads to plant death are unclear. Vieira da Silva (1976) discussed the concept that water stress leads to decompartmentalization of enzymes which then hydrolyze other cell components. Levitt (1972) suggested that the buildup of amino acids and amides from the breakdown of proteins associated with water stress may be lethal.

4.4.4 Theoretical Considerations Relating Plant Characteristics and Successional Stage

4.4.4.1 Water Availability and Vegetative Recovery in the Semiarid Mediterranean Regions of Southern California

The patterns of water availability and water use during vegetative recovery from any disturbance are not well quantified for semiarid mediterranean regions,

Table 3. Plant characteristics related to water expected in early and late successional species in mediterranean climates

	Early successional	Late successional
Effective precipitation regime	Frequent light	Infrequent heavy
Depth of soil water	Shallow	Deep
Rooting patterns	Shallow	Deep
	Fibrous	Woody
Root length density	High	Low
Leaf longevity	Short	Long
Plant longevity	Annual	Perennial
Carbon/nutrient priority – annual	Seed	Seed
– perennial	Root	Root, leaf, stem
Growth period	Winter	Spring-summer
Flowering period	Spring	Anytime
Plant height	Close to surface	Short or tall
Leaf conductance	High	Low
Hydrostability	Low	High
Topographic influences via radiation, E_s, T_s	High	Low

but a theoretical scheme can be postulated based on available information on successional patterns and plant water characteristics (Bazzaz 1979; Table 3). The plant characteristics are closely linked with the spatial and temporal patterns of soil moisture. For a 1- to 2-year period after a fire, the soil surface is almost bare and is darkened by charcoal. Surface temperatures vary widely annually and diurnally. Temperatures will be slightly below 0 °C in winter and above 65 °C in summer (H.L. Bauer 1936; E.H. Miller 1947; P.C. Miller et al. 1977; Thrower and Bradbury 1977). The surface soil will be wetted by three processes, precipitation, dew, and distillation. The effective precipitation regime will consist of both relatively frequent light rains and a few heavy rains. On heavy soils, the heavy rains may not infiltrate the soils because of low permeability and high runoff. With high runoff, the regime of light rains is accentuated. On coarser soils, which are more commonly populated by shrubs, runoff will be low. Dew should condense on the majority of the nights since the relative humidity of the air usually reaches 100% and soil surface temperatures are cooler than air temperatures. Distillation and movement of water in the soil in the vapor phase requires high soil moisture within the soil and wide diurnal fluctuations, >20 °C, of surface soil temperature. These conditions are met in April, May, and June. Soil moisture will be held close to the surface in dry soils and will be deeper in coarser soils.

Soil evaporation will be high. Daily rates of 2 mm day^{-1} can be expected when the surface soil moisture is above 0.2 cm^3 cm^{-3}. Annual evaporation rates of 200–240 mm yr^{-1} can be expected (Shachori and Michaeli 1965; Patric unpubl. data). The high daily rates in late spring cause rapid drying. The surface 30 cm can dry from field capacity (0.20–0.25 cm^3 cm^{-3}) to air dry (about 0.05 cm^3 cm^{-3}) in 7 days in May (Poole and Miller unpubl. data). The rapid drying of the surface results in a relatively long summer drought in the surface soil, from about 1 May to 1 October or 5 months. During the winter, air temperatures are too cool to promote growth, but the soil is wetter. However, daytime temperatures at the soil surface are high enough to allow growth of plants of low stature (P.C. Miller 1981).

After a fire, soil nutrients in readily available form are most abundant at the surface (Christensen and Muller 1975).

The spatial and temporal patterns of soil moisture dictate the plant forms which are possible. Several plant forms are appropriate: annual grasses and forbs, succulents, seedlings of perennial shrubs, and resprouts of perennial shrubs. In order to "compete" effectively with soil evaporation, plant characteristics of high leaf conductance, shallow roots, and fibrous roots are appropriate. Water is exploited faster by root growth than by water movement along a water potential gradient (Hsaio et al. 1976). Water movement by a potential gradient should be faster to the surface than to roots because the potential difference from soil to air is much greater than the potential difference from soil to root. Consequently, the plant compensates by creating a shorter distance from soil water pool to root than from soil water pool to soil surface. This requirement to shorten the distance for water movement should lead to high root length densities and in high root surface areas in the surface soil layers – in other words, to fibrous roots.

The long drought period in the surface soil may lead to a predominance of drought-escaping plant forms which survive the drought in the seed stage especially among shallow-rooted forms. Seeds tolerate much lower water contents than adult plants and can become air dry, while adult plants must maintain water contents above equilibrium levels with the air (Levitt 1972). Thus, annuals are favored as a shallow-rooted form. In annuals, the accumulated carbon and nitrogen is diverted to seed production at the expense of the growth of leaves, roots, and stems. Growth must take place during winter, after the rains begin, and is possible at the soil surface but not above in the air (P.C. Miller et al. 1981). Consequently, the characteristic of low stature is reinforced. Any factor increasing the effectiveness of light rains relative to heavy rains, such as heavy soils and condensation of water from the marine air by nighttime cooling, should enhance the growth of annual grasses. Knowledge gained from agricultural crops indicates that water demands by the annual grasses should be especially high during the heading stage. The grass species should begin flowering in mid-April when the availability of surface water is ensured. The period of heading cannot be cued by surface drying because drying occurs periodically throughout the winter and is followed by rain. Therefore, daylength or some other cue related to calendar time or phenological development is more reliable. As a result of this timing mechanism, in many years the annuals pass through their complete life cycle while water is still available.

Thus, the exploitation of surface water is accomplished by shallow, fibrous-rooted annuals, which have low stature, high conductances to water loss, few adaptations to endure drought in seedlings and adult plants, and phenological development patterns which are relatively independent of the current environment. Such characteristics seem appropriate to search for in introduced species.

Succulents also are effective "competitors" with soil evaporation for surface water. The low osmotic potential of succulents should give higher rates of water uptake than occur in annuals. Shallow, fibrous roots are appropriate for succulents. Water is stored in the plant rather than in the soil and released slowly under plant control rather than under the control of the physical environment and rate of soil evaporation. The succulent form is adapted to drier conditions than are annual grasses.

Seedlings of perennial shrubs are also possible in mediterranean type climates, but many of their characteristics contrast with those of the annuals. Perennials exploit soil water below 30 cm which shortens the length of the drought which they experience. Deep soil moisture, which results from the infrequent heavy rains and greater permeability of coarser soils, is necessary for the establishment and survival of perennial shrubs. The perennial form avoids the surface drying and high soil evaporation. In the seedling stage, the priority of developing a deep root system should lead to low exploitation of the soil surface layers because the plant reserves are used for this purpose. Roots should grow downwards without lateral diversions. The rapid soil drying in the spring (P.C. Miller 1981) indicates that roots of seedlings should be below 40 cm by 1 May to avoid the dry soil surface layers. A rapid exploitation of deep water is not needed because evaporation loss of water is low from these depths. Water capture efficiency is less important in perennial shrubs than in annuals. Low leaf conductances are possible and advantageous if water is to be retained for use throughout the year. Because rapid water uptake is less important, water uptake can depend on water movement to the root rather than root growth into regions where soil moisture is available. Fibrous roots are less important below 40 cm than in the surface soil. In order to grow roots 40 cm long, rapid elongation rates are needed. For example, to grow 40 cm of root between 1 March and 10 April, a root growth rate of 1 cm day^{-1} is required. Root growth rates of *Adenostoma fasciculatum* were 1.04 cm day^{-1}, *Cercocarpus betuloides* 0.37 cm day^{-1}, and *Arctostaphylos glauca* 0.27 cm day^{-1} (Kummerow unpubl. data). *Adenostoma fasciculatum* appears best adapted to reproduce by seed on rapidly drying soils and is commonly found on south-facing slopes and lower elevations in chaparral. Root growth can begin in February (Kummerow et al. 1978) but will probably occur at slower rates in the field than measured in the laboratory because of colder temperatures.

In perennial shrubs, the carbon and nitrogen acquired goes to support root growth at the expense of leaf and stem growth and seed production. Seedlings must survive high soil surface temperatures during the summer. Shading seems desirable and can be provided by other plants, rocks, other non-living materials, and from the development of leaves and stems directly over the stem–soil contact. The north-facing slope has lower rates of surface drying and lower temperatures than the south-facing slope (H.L. Bauer 1936; E.H. Miller 1947; P.C. Miller and Poole unpubl. data). Seedlings should be drought tolerating for several years until a deep root system can be developed. Convectional cooling should be maximized by attaining heights over 10 cm and minimizing leaf size. Small leaf size also will reduce the shock of leaf drop by seedlings during the summer drought.

After a fire the highest nutrient concentrations are at the surface, with the result that deep rooting patterns may be accompanied by lower nitrogen and phosphorus availability. Perennial seedlings may experience limited nitrogen and phosphorus in spite of high surface concentrations. Leaf growth may not be possible during the winter when water is available because of low temperature, although photosynthesis is possible. The evergreen sclerophyll leaf apparently has been selected by the temperature and moisture regime and by the low nitrogen and phosphorus regimes in the deeper soil levels. As leaching of the nutrients from the surface to the deeper levels occurs, the nutrient regime should improve. Thus, shrub seedlings require deep roots to survive the drought, which results in poor exploitation of surface water, nitrogen, and phosphorus. Low leaf conductances are possible but fibrous roots are not. Drought tolerance is essential. Evergreen sclerophyllous leaves are important for drought tolerance and are reinforced by the low nutrient environment.

Perennial shrubs, which revegetate burned areas in mediterranean type ecosystems by sprouting from root crowns, seem to be in the most favored position. This form can include drought evaders because the root system is large relative to the transpiring surface after the fire. High water potentials are possible, and sprouts grow during the summer. The sprouts grow above the hot surface air layer and are cooled by convection and transpiration. Because of their preexisting root system resprouting shrubs do not have to invest as much carbohydrate as do shrub seedlings in growing a deep root system. Current photosynthate and nutrients taken up plus previously stored carbohydrates and nutrients can be used primarily for leaf and stem growth. Resprouters can develop surface feeding roots in early successional stages and exploit water and nutrients in the surface layers. Thus, resprouting perennial shrubs can be both evergreen and efficient exploiters of water and nutrients.

As the vegetation develops, the soil becomes increasingly shaded. Turbulent transfer of heat and water vapor from the soil surface is reduced. Surface temperatures during the day may increase as the leaf area index increases to 0.5 or 1.0, and then decreases with higher leaf area indices. Nighttime soil surface temperature under a canopy will be higher than without a canopy. Precipitation will be increasingly intercepted in the canopy. Because a larger fraction of light rains is intercepted and evaporated than of heavy rains, the effective precipitation regime will shift to less frequent soil surface wetting, and deep soil water will become more important. Dew and distillation will become less important as nighttime temperatures increase and soil surface temperature fluctuations decrease. Soil evaporation decreases as does annual mean soil temperature.

As the vegetation develops, these changes in the water regime mean that annual grasses and forbs become less favored by the decreasing amounts of surface water (Pitt and Heady 1978). Shrubs are favored and, after developing deep root systems, can develop surface feeding roots which will increase their efficiency for water capture and recycling of nitrogen and phosphorus. Annual grasses and forbs will be less able to capture nitrogen and phosphorus because the nutrients will increasingly be held in the shrub and recycled within the plants. Year-to-year variations in precipitation and nutrient availability will be more detrimental to the annuals than to the shrubs because the shrubs can recycle nutrients within the plant while annuals depend on external sources. Soil evaporation will be high in bare areas between shrubs. Shrubs should show their maximum water use with about 50% cover (Hibbert et al. 1974 cited in Cable 1975). Thus, the shrub form is favored and the annual form is disfavored by the changing water regime and the capture and retention of nutrients by the shrubs.

4.4.5 Conclusions

1. Other than data for agricultural plants, very few data are available to compare the characteristics, related to plant water balance, of plants found on modified sites with those of plants on natural sites. Past research has emphasized short-term answers to revegetating disturbed areas. The development of general principles has hardly been attempted. With respect to plant water characteristics, not enough measurements have been made to develop any generalities from an empirical basis. More data are needed on leaf conductance, water potentials, water contents, and root growth patterns combined with basic climatological and soils information.

2. It is unlikely that water-related characteristics of plants from modified sites will be outside the ranges measured for plants of natural habitats, e.g., 0 to -9.0 MPa water potential, 0–2.0 cm s^{-1} leaf conductance, and 70%–100% relative water content. Rather than concentrate on single measures of species differences, integrative approaches are required which account for the interdependency of the interacting parts of the plant water system. We need an integration of cellular and subcellular processes with whole plant life cycle problems. We have on the one hand species trials which are very empirical and descriptive and on the other hand detailed biochemical studies.

3. Water does not seem to be a major limiting factor for plants on disturbed sites in regions which receive summer rains and have an annual precipitation above about 500 mm; although some degree of drought hardiness is referred to and seems advantageous at all sites.

4. Deep, rapidly growing root systems are advantageous to avoid surface drying, high surface temperatures, and chemical toxicities.

5. High surface temperatures kill stem tissue at the surface contact.

6. Plant characteristics related to water balance that are advantageous for any site must be considered in the context of each other and of the climatic and soil water regime of the site. The effect of the prolonged drought of mediterranean areas on species replacement has not been considered.

7. Although a survey of plant characteristics related to water balance is needed, studies integrated by the concepts of the controls of water flow through the soil–plant–atmosphere system comparing a few successful native and introduced species should be more interesting scientifically. These studies would be carried out in two climatic regimes of less than and greater than 500 mm yr^{-1} precipitation. Such a study should involve descriptions of natural succession sequences of plant populations, basic physiologies of species, basic life histories of species, inhibitions by resource use, inhibitions of allelopathy and herbivory, and competitive experiments for species interaction effects.

Acknowledgements. I thank Drs. Douglas Johnson, Ray Brown, and Pat Coyne for their suggestions on literature on this topic. Ms. Patsy Miller, Ms. Lee Stuart, and Dr. Steve Roberts critically reviewed the manuscript and offered helpful suggestions.

Résumé

Pour subsister pendant la saison sèche, les plantes peuvent adopter plusieurs tactiques, en modulant leur système racinaire, leur système conducteur et leur système foliaire. En particulier, l'„efficacité de la transpiration", c'est-à-dire le quotient de la transpiration par les précipitations, est en relation directe avec l'indice foliaire. L'augmentation de cet indice coïncide avec une diminution de l'évaporation à la surface du sol, avec une forte augmentation de la transpiration et avec une faible augmentation de l'interception. L'indice foliaire ne s'accroît pas indéfiniment, et son maximum correspond au stade où la photosynthèse équilibre exactement les coûts de maintenance des tissus; la transpiration atteint alors aussi son maximum, ainsi que l'efficacité de la transpiration.

Les plantes pionnières ont des taux de transpiration compris à l'intérieur de la gamme générale, mais quelquefois très variables au cours de la journée; elles ont un feuillage caduc, des conductances souvent élevées, des racines fibreuses et nombreuses à croissance rapide (1 à 6 cm par jour); elles ont souvent des cellules de petite taille, qui peuvent rester turgescentes même quand elles perdent une assez grande quantité d'eau. Les plantules sont plutôt plus résistantes à la sécheresse que les plantes adultes, surtout dans le cas des plantes à petites feuilles; elles souffrent souvent plus de la chaleur que de la sécheresse.

En région méditerranéenne semi-aride, après un incendie, la température et l'humidité des horizons superficiels du sol subissent de fortes variations quotidiennes et annuelles; l'évaporation de l'eau du sol est intense (2 mm $jour^{-1}$ pour une humidité du sol égale à 0,2 cm^3 cm^{-3}) et peut atteindre 200 à 240 mm an^{-1}. Les horizons superficiels se dessèchent très vite. Les plantes qui s'installent sont surtout des annuelles, qui peuvent avoir des conductances élevées et une faible résistance à la sécheresse si elles sont capables d'accomplir tout leur cycle pendant les mois tièdes et humides du printemps (ou de l'automne). Les plantules d'espèces pérennes doivent alors produire des racines qui descendent en profondeur, et elles doivent ensuite résister à la sécheresse, grâce à une faible conductance.

Quand la végétation reconquiert le terrain, les transferts verticaux de chaleur et de vapeur d'eau diminuent; la température à la surface du sol peut atteindre un maximum pour des indices foliaires compris entre 0,5 et 1. La quantité d'eau disponible à la surface du sol diminue aussi, et les arbustes sont favorisés aux dépens des plantes annuelles (pour les arbustes, la situation optimale est celle où le recouvrement de la végétation est voisin de 50%).

References

Acevedo E (1975) The growth of maize (*Zea mays* L.) under field conditions as affected by its water relations. PhD thesis, Univ California, Davis

Anderson RC, Loucks OL, Swain AM (1969) Herbaceous response to canopy cover, light intensity, and throughfall precipitation in coniferous forests. Ecology 50:255–263

Barnes DL, Woolley DG (1969) Effect of moisture stress at different stages of growing. I. Comparison of a single-eared and two-eared corn hybrid. Agron J 61:788–790

Bauer HJ (1973) Ten years' studies of biocenological succession in the excavated mines of the Cologne lignite district. In: Hutnik RJ, Davis G (eds) Ecology and reclamation of devasted land. Proceedings of the international symposium on ecology and revegetation of drastically disturbed areas, University Park Pa, 1969, vol I. Gordon and Breach, New York Paris London, pp 271–283

Bauer HL (1936) Moisture relations in the chaparral of the Santa Monica Mountains, California. Ecol Monogr 6:409–454

Bazzaz FA (1979) The physiological ecology of plant succession. Annu Rev Ecol Syst 10:351–371

Beadle NCW (1968) Some aspects of the ecology and physiology of Australian xeromorphic plants. Aust J Sci 30:348–355

Bieleski RL (1973) Phosphate pools, phosphate transport, and phosphate availability. Annu Rev Plant Physiol 24:225–252

Brown RW (1973) Transpiration of native and introduced grasses on a high-elevation marsh site. In: Hutnik RJ, Davis G (eds) Ecology and reclamation of devastated land. Proceedings of the international symposium on ecology and revegetation of drastically disturbed areas, University Park Pa, 1969, vol I. Gordon and Breach, New York Paris London, pp 467–481

Brown RW (1977) Water relations of range plants. In: Sosebee RE (ed) Rangeland plant physiology. Soc Range Manage Range Sci Ser 4:97–140

Brown RW, Johnston RS (1976) Revegetation of an alpine mine disturbance: Beartooth Plateau, Montana. US For Serv Res Note INT-206, p 8

Brown RW, Johnstone RS (1979) Revegetation of disturbed alpine rangelands. In: Special management needs of alpine ecosystems. US Dep Agric For Serv, Washington DC, pp 76–94

Brown RW, Johnston RS, Cleve Van K (1978) Rehabilitation problems in alpine and arctic regions. In: Reclamation of drastically disturbed lands. Am Soc Agron, Madison Wis, pp 23–44

Cable DR (1975) Range management in the chaparral type and its ecological basis: The status of our knowledge. US For Serv Res Pap RM-155:30

Caldwell MM (1976) Root extension and water absorption. In: Lange OL, Kappen L, Schulze E-D (eds) Water and plant life. Springer, Berlin New York Heidelberg, pp 63–85

Chadwick MJ (1973) Discussion. In: Hutnik RJ, Davis G (eds) Ecology and reclamation of devastated land. Proceedings of the international symposium on ecology and revegetation of drastically disturbed areas, University Park, Pa, 1969, vol II. Gordon and Breach, New York Paris London, p 466

Cheung YNS, Tyree MT, Dainty J (1975) Water relations parameters on single leaves obtained in a pressure bomb and some ecological interpretations. Can J Bot 53:1342–1346

Christensen NL, Muller CH (1975) Effects of fire on factors controlling plant growth in *Adenostoma* chaparral. Ecol Monogr 45:29–55

Clements FE, Shelford VE (1939) Bio-ecology. Wiley, New York, p 698

Cohen Y, Tadmor NH (1969) Effects of temperature on the elongation of seedling roots of some grasses and legumes. Crop Sci 9:189–192

Collier BD, Cox GW, Johnson AW, Miller PC (1973) Dynamic ecology. Prentice-Hall, Englewood Cliffs, p 563

Connell JH, Slatyer RO (1977) Mechanisms of succession in natural communities and their role in community stability and organization. Am Nat 111:1119–1144

Courtin GM, Mayo JM (1975) Arctic and alpine plant water relations. In: Vernberg FJ (ed) Physiological adaptation to the environment. Intext Editions Publishers, New York, pp 201–224

Curtis WR (1973a) Moisture and density relations on graded strip-mine spoils. In: Hutnik RJ, Davis G (eds) Ecology and reclamation of devastated land. Proceedings of the international symposium on ecology and revegetation of drastically disturbed areas, University Park Pa, 1969, vol I. Gordon and Breach, New York Paris London, pp 135–144

Curtis WR (1973b) Effects of strip mining on the hydrology of small mountain watersheds in Appalachia. In: Hutnik RJ, Davis G (eds) Ecology and reclamation of devasted land. Proceedings of the international symposium on ecology and revegetation of drastically disturbed areas, University Park Pa, 1969, vol I. Gordon and Breach, New York Paris London, pp 145–157

Cutler JM, Rains DW, Loomis RS (1977) The importance of cell size in the water relations of plants. Physiol Plant 40:255–260

Czapowskyj MM (1973) Performance of red pine and Japanese larch planted on anthracite coal-breaker refuse. In: Hutnik RJ, Davis G (eds) Ecology and reclamation of devastated land. Proceedings of the international symposium on ecology and revegetation of drastically disturbed ares, University Park Pa, 1969, vol II. Gordon and Breach, New York Paris London, pp 237–245

Darmer G (1973) Grasses and herbs for revegetating phytotoxic material. In: Hutnik RJ, Davis G (eds) Ecology and reclamation of devastated land. Proceedings of the international symposium on ecology and revegetation of drastically disturbed areas, University Park Pa, 1969, vol II. Gordon and Breach, New York Paris London, pp 91–101

Dean KC, Havens R, Harper KT, Rosenbaum JB (1973) Vegetative stabilization of mill mineral wastes. In: Hutnik RJ, Davis G (eds) Ecology and reclamation of devastated land. Proceedings of the international symposium on ecology and revegetation of drastically disturbed areas, University Park Pa, 1969, vol II. Gordon and Breach, New York Paris London, pp 119–136

Deely DJ, Borden FY (1973) High surface temperatures on strip-mine spoils. In: Hutnik RJ, Davis G (eds) Ecology and reclamation of devastated land. Proceedings of the international symposium on ecology and revegetation of drastically disturbed areas, University Park Pa, 1969, vol I. Gordon and Breach, New York Paris London, pp 69–79

Devera NF, Marshall DR, Balaam LM (1969) Genetic variability in root development in relation to drought tolerance in spring wheats. Exp Agric 5:327–337

Dobrenz AK, Wright L, Humphrey AB, Massengale MA, Kneebone WR (1969) Stomate density and its relationship to water use efficiency of blue panicgrass (*Panicum antidotale* Ritz). Crop Sci 9:354–357

Farmer EE, Brown RW, Richardson BZ, Packer PE (1974) Revegetation research on the Decker coal mine in southeastern Montana. US For Serv Res Pap INT-162:12

Farmer EE, Richardson BZ, Brown RW (1976) Revegetation of acid mining wastes in central Idaho. US For Serv Res Pap INT-178:17

Gates DM (1965) Energy, plants, and ecology. Ecology 46:1–13

Goodman GT, Pitcairn CER, Gemmell RP (1973) Ecological factors affecting growth on sites contaminated with heavy metals. In: Hutnik RJ, Davis G (eds) Ecology and reclamation of devastated land. Proceedings of the international symposium on ecology and revegetation of drastically disturbed areas, University Park Pa, 1969, vol II. Gordon and Breach, New York Paris London, pp 149–173

Grier CC, Running SW (1977) Leaf area of mature northwestern coniferous forest: Relation to site water balance. Ecology 58:893–899

Grimes DW, Miller RJ, Wiley PL (1975) Cotton and corn root development in two field soils of different strength characteristics. Agron J 67:519–523

Grimes DW, Sheesley WR, Wiley PL (1978) Alfala root development and shoot regrowth in compact soil of wheel traffic patterns. Agron J 70:955–958

Harabin Z, Greszta J (1973) Abnormalities in the roots of trees growing on toxic dump material. In: Hutnik RJ, Davis G (eds) Ecology and reclamation of devastated land. Proceedings of the international symposium on ecology and revegetation of drastically disturbed areas, University Park Pa, 1969, vol I. Gordon and Breach, New York Paris London, pp 413–428

Hibbert AR, Davis EA, Scholl DG (1974) Chaparral conversion potential. Part I: Water yield response and effects on other resources. US For Serv Res Pap RM-126:36

Hinckley TM, Duhme F, Hinckley AR, Richter H (1980) Water relations of drought hardy shrubs: Osmotic potential and stomatal reactivity. Plant Cell Environ 3:131–140

Höfler K, Migsche H, Rottenberg W (1971) Über die Austrocknungsresistenz landwirtschaftlicher Kulturpflanzen. Forschungsdienst 12:50–61

Hsiao TC, Fereres E, Acevedo E, Henderson DW (1976) Direct and indirect water stress. F. Water stress and dynamics of growth and yield of crop plants. In: Lange OL, Kappen L, Schulze E-D (eds) Water and plant life. Springer, Berlin New York Heidelberg, pp 281–305

Isaac LA (1938) Factors affecting establishment of Douglas-fir seedlings. US Dep Agric Circ 486:45

Jacoby H (1973) Growth and nutrition of beech trees on sites of different soil texture in the lignite area of the Rhineland. In: Hutnik RJ, Davis G (eds) Ecology and reclamation of devastated land. Proceedings of the international symposium on ecology and revegetation of drastically disturbed areas, University Park Pa, 1969, vol I. Gordon and Breach, New York Paris London, pp 391–411

Jarvis PG, Jarvis MS (1963) The water relations of tree seedlings. IV. Some aspects of the tissue water relations and drought resistance. Physiol Plant 16:501–516

Johnston RS, Brown RW (1979) Hydrologic aspects related to the management of alpine areas. In: Special management needs of alpine ecosystems. US Dep Agric For Serv, Washington DC, pp 65–75

Johnston RS, Brown RW, Craven J (1975) Acid mine rehabilitation problems at high elevations. In: Proceedings of the watershed management symposium, Logan, Utah, August 11–13, 1975. Am Soc Civ Eng, New York, pp 66–79

Jones HE (1971) Comparative studies of plant growth and distribution in relation to waterlogging. An experimental study of the relationship between transpiration and the uptake of iron in *Erica cineria* L. and *E. tetralix* L. J Ecol 59:167–178

Jones R, Etherington JR (1971) Comparative studies of plant growth and distribution in relation to water-logging. IV. The growth of dune and dune slack plants. J Ecol 59:793–801

Kekerix van LK, Brown RW, Johnston RS (1979) Seedling water relations of two grass species on high-elevation acid mine spoils. US For

Kluge M (1976) Direct and indirect water stress. C. Carbon and nitrogen metabolism under water stress. In: Lange OL, Kappen L, Schulze E-D (eds) Water and plant life. Springer, Berlin New York Heidelberg, pp 243–252

Kummerow J, Krause D, Jow J (1978) Seasonal changes of fine root density in the southern Californian chaparral. Oecologia (Berlin) 37:201–212

Larcher W (1960) Transpiration and photosynthesis of detached leaves and shrubs of *Quercus pubescens* and *Q. ilex* during desiccation under standard conditions. Bull Res Counc Isr 8D:213–224

Lambert JR, Penning de Vries FWT (1975) Dynamics of water in the soil-plant atmosphere: A model named TROIKA. In: Hadas A, Swartzendruber D, Rijtema PE, Fuchs M, Yaron B (eds) Ecological studies 4. Springer, Berlin New York Heidelberg, pp 257–274

Lee R, Hutson WG, Hill SC (1975) Energy exchange and plant survival on disturbed land. In: Gates DM, Schmerl RB (eds) Perspectives of biophysical ecology. Ecological studies 12. Springer, Berlin New York Heidelberg, pp 239–247

Lemon E, Stewart DW, Shawcroft RW (1971) The sun's work in a cornfield. Science 174:371–378

Levitt J (1972) Responses of plants to environmental stresses. Academic Press, London New York

Loveless AR (1962) Further evidence to support a nutritional interpretation of sclerophylly. Ann Bot (London) 26:551–561

MacArthur RH (1970) Species packing and competitive equilibrium for many species. Theor Popul Biol 1:1–11

MacArthur RH (1972) Geographical ecology. Harper and Row, New York

McVean DN (1956) Ecology of *Alnus glutinosa* (L.) Gaertn. III. Seedling establishment. J Ecol 44:195–218

Miller EH Jr (1947) Growth and environmental conditions in southern California chaparral. Am Midl Nat 37:379–420

Miller PC (ed) (1981) Resource use by chaparral and matorral: A comparison of vegetation function in two mediterranean type ecosystems. Springer, Berlin Heidelberg New York

Miller PC, Poole DK (1979) Patterns of water use by shrubs in southern California. For Sci 25:84–98

Miller PC, Stoner WA, Hom J, Poole DK (1976) Potential influence of thermal effluents on the production and water use efficiency of mangrove species in south Florida. In: Esch GW, McFarland RW (eds) Thermal ecology, vol II. Proceedings of the thermal ecology conference. Energy Res Dev Agency, Washington DC, pp 39–45

Miller PC, Bradbury DE, Hajek E, LaMarche V, Thrower NJW (1977) Past and present environment. In: Mooney HA (ed) Convergent evolution in Chile and California mediterranean climate ecosystems. Dowden, Hutchinson and Ross, Stroudsburg Pa

Miller PC, Stoner WA, Richards SP (1978) MEDECS, a simulator for mediterranean ecosystems. Simulation 30:173–190

Miller PC, Hajek E, Poole DK, Roberts SW (1981) Microclimate and energy exchange. In: Miller PC (ed) Resource use by chaparral and matorral: A comparison of vegetation function in two mediterranean type ecosystems. Springer, Berlin Heidelberg New York

Misra D (1956) Study of drought resistance in certain crop plants. Ind J Agron 1:25–39

Naidu KM, Bhagyalakshmi KV (1967) Stomatal movement in relation to drought resistance in sugarcane. Curr Sci 36:555–556

Newman EI (1974) Root and soil water relation. In: Carson EW (ed) The plant root and its environment. Charlottesville Univ Press, Virginia, pp 363–440

Oberbauer S, Miller PC (1979) Plant water relations in montane and tussock tundra vegetation types in Alaska. Arct Alp Res 11:69–81

Oberbauer S, Miller PC (1981) Some aspects of plant water relations in Alaskan arctic tundra species. Arct Alp Res 13:205–218

Oppenheimer HR (1960) Adaptation to drought: Xerophytism. Plant-water relationships in arid and semi-arid conditions. Arid Zone Res 15:105–138

Oppenheimer HR (1967) Mechanism of drought resistance in conifers of the Mediterranean zone and the arid west of the U.S.A. I. Physiological and anatomical investigations. Final Report Project No. AID -FS7 Grant No. FG-IS-119

Oppenheimer JR, Shomer-Ilan A (1963) A contribution to the knowledge of drought resistance of Mediterranean pine trees. Mitt Florist Soziol Arbeitsgem 10:42–55

Orians GH (1975) Diversity, stability, and maturity in natural ecosystems. In: Dobbes van WH, Lowe-McConnell RH (eds) Unifying concepts in ecology. Junk, The Hague and Pudoc, Wageningen

Orshan G (1963) Seasonal dimorphism of desert and mediterranean chamaephytes and its significance as a factor in their water economy. In: Rutter AJ, Whitehead FH (eds) The water relations of plants. Blackwell, Oxford, pp 206–222

Parker J (1951) Moisture retention in leaves of conifers in the Northern Rocky Mountains. Bot Gaz 113:210–216

Parrish JAD, Bazzaz FA (1976) Underground niche separation in successional plants. Ecology 57:1281–1288

Parsons RF (1969) Physiological and ecological tolerances of *Eucalyptus incrassata* and *E. socialis* to edaphic factors. Ecology 50:386–390

Pearson RW, Ratliff LF, Taylor HM (1970) Effect of soil temperature, strength, and pH on cotton seedling root elongation. Agron J 62:243–246

Penning de Vries FWT (1972) A model for simulating transpiration of leaves with special attention to stomatal functioning. J Appl Ecol 9:57–77

Pickett STA, Bazzaz FA (1978) Organization of an assemblage of early successional species on a soil moisture gradient. Ecology 59:1248–1255

Pitt MD, Heady HF (1978) Responses of annual vegetation to temperatures and rainfall patterns in northern California. Ecology 59:336–350

Plummer AP (1943) The germination and early seedling development of twelve range grasses. J Am Soc Agron 35:19–34

Poole DK, Miller PC (1975) Water relations of selected species of chaparral and coastal sage communities. Ecology 56:1118–1128

Repp G (1973) Cytoecological investigations with regard to the mechanism of chemical resistance of plants. In: Hutnik RJ, Davis G (eds) Ecology and reclamation of devastated land. Proceedings of the international symposium on ecology and revegetation of drastically disturbed areas, University Park Pa, 1969, vol I. Gordon and Breach, New York Paris London, pp 445–466

Richter H (1978) A diagram for the description of water relations of plant cells and organs. J Exp Bot 29:1–7

Roberts SW, Knoerr KR (1977) Components of water potential estimated from xylem pressure measurements in five tree species. Oecologia (Berlin) 28:191–202

Running SW, Waring RH, Rydell RA (1975) Physiological control of water flux in conifers. A computer simulation model. Oecologia (Berlin) 18:1–16

Salim MH, Todd GW, Stutte CA (1969) Evaluation of techniques for measuring drought avoidance in cereal seedlings. Agron J 61:182–185

Satoo T (1956) Drought resistance of some conifers at the first summer after their emergence. Bull Tokyo Univ For 51:1–108

Savile DBO (1974) Arctic adaptations in plants. Can Dep Agrig Monogr 6:81

Schapmeyer CS (1939) Transpiration and physico-chemical properties of leaves as related to drought resistance in loblolly pine and shortleaf pine. Plant Physiol 14:447–462

Schimp PE (1973) Deep-mine waste reclamation experimentation in the bituminous regions of Pennsylvania. In: Hutnik RJ, Davis G (eds) Ecology and reclamation of devastated land. Proceedings of the international symposium on ecology and revegetation of drastically disturbed areas, University Park Pa, 1969, vol II. Gordon and Breach, New York Paris London, pp 457–467

Schlätzer G (1973) Some experience with various species in Danish reclamation work. In: Hutnik RJ, Davis G (eds) Ecology and reclamation of devastated land. Proceedings of the international symposium on ecology and revegetation of drastically disturbed areas, University Park Pa, 1969, vol I. Gordon and Breach, New York Paris London, pp 33–64

Schramm JR (1966) Plant colonization studies on black wastes from anthracite mining in Pennsylvania. Trans Am Philos Soc 56(1):1–194

Schulze E-D, Lange OL, Buschbom U, Kappen L, Evenari M (1972) Stomatal responses to changes in humidity in plants growing in the desert. Planta 108:259–270

Schulze E-D, Lange OL, Kappen L, Evenari M, Buschbom U (1975) The role of air humidity and leaf temperature in regulating stomatal resistance of *Prunus armeniaca* L. under desert conditions. II. The significance of leaf water status and internal carbon dioxide concentration. Oecologia (Berlin) 18:219–233

Shachori AY, Michaeli A (1965) Water yields of forest, maquis, and grass covers in semiarid regions. A literature review. In: Eckardt FD (ed) Methodology of plant ecophysiology. UNESCO, Paris, pp 467–477

Slatyer RO (1955) Studies of the water relations of crop plants grown under natural rainfall in northern Australia. Aust J Agr Res 6:365–377

Small E (1972) Photosynthetic rates in relation to nitrogen recycling as an adaptation to nutrient deficiency in peat bog plants. Can J Bot 50:2227–2233

Specht RL (ed) (1979) Heathlands and related shrublands. Part A: Descriptive studies. Elsevier, Amsterdam, p 498

Stocker O (1956) Die Dürreresistenz. In: Ruhland W (ed) Handbuch Pflanzenphysiologie vol 3. Springer, Berlin Göttingen Heidelberg, S 696–741

Stoner WA, Miller PC (1975) Water relations of plant species in the wet coastal tundra at Barrow, Alaska. Arct Alp Res 7:109–124

Tadmor NH, Cohen Y (1968) Root elongation in the preemergence stage of mediterranean grasses and legumes. Crop Sci 8:416–419

Taylor BK (1967) The nitrogen nutrition of the peach. I. Seasonal changes in nitrogenous constituents in mature trees. Aust J Biol Sci 20:379–387

Taylor FG Jr (1974) Phenodynamics of production in a mesic deciduous forest. In: Lieth H (ed) Phenology and seasonality modeling. Springer, Berlin Heidelberg New York, pp 237–254

Taylor SE (1975) Optimal leaf form. In: Gates DM, Schmerl RB (eds) Perspectives of biophysical ecology. Springer, Berlin Heidelberg New York, pp 73–86

Tazaki T (1960) Studies on the dehydration resistance of higher plants II. Theoretical considerations of dehydration resistance. Bot Mag (Tokyo) 73:205–211

Thrower NJW, Bradbury DE (eds) (1977) Chile-California mediterranean scrub atlas: A comparative analysis. Dowden, Hutchinson and Ross, Stroudsburg Pa, p 237

Tüxen R (1960) Wesenszüge der Biozonose. Gesetz des Zusammenlebens von Pflanzen und Tieren. Biosoziologie: Ber Int Symp, Stolzenau, 1960. Junk, Den Haag

Turner NC (1974) Stomatal behavior and water status of maize, sorghum, and tobacco under field conditions. II. At low soil water potential. Plant Physiol 53:360–365

Tyree MT, Hammel HT (1972) The measurement of the turgor pressure and the water relations of plants by the pressure-bomb technique. J Exp Bot 23:267–282

Vieira da Silva J (1976) Direct and indirect water stress. A. Water stress, ultrastructure, and enzymatic activity. In: Lange OL, Kappen L, Schulze E-D (eds) Water and plant life. Springer, Berlin Heidelberg New York, pp 207–224

Wieland NK, Bazzaz FA (1975) Physiological ecology of three codominant successional annuals. Ecology 56:681–688

Wilson JR, Fisher MJ, Schulze E-D, Dolby GR, Ludlow MM (1979) Comparison between pressure-volume and dewpoint hygrometry techniques for determining the water relations characteristics of grass and legume leaves. Oecologia (Berlin) 41:77–88

Zarger TG, Bengtson GW, Allen JC, Mays DA (1973) Use of fertilizers to speed pine establishment on reclaimed coal-mine spoil in northeastern Alabama: II. Field experiments. In: Hutnik RJ, Davis G (eds) Ecology and reclamation of devastated land. Proceedings of the international symposium on ecology and revegetation of drastically disturbed areas, University Park Pa, 1969, vol I. Gordon and Breach, New York Paris London, pp 227–236

Section 5 Population Characteristics

5.1 Reproductive Strategies and Disturbance by Man

P.H. Gouyon, R. Lumaret, G. Valdeyron, and Ph. Vernet

5.1.1 Introduction

Man has developed a technology which allows him to modify his environment to a large extent. However, he is not fully in control of every modification which he induces. The most common of these modifications is the change of a stable or "natural" environment into a disturbed or modified one, through means such as fire or grazing. Here we examine some of the consequences of habitat modification on the characteristics of plant populations.

Although it is difficult to characterize a disturbed environment precisely in relation to a plant population, a definition of a stable environment is relatively easy to develop. A perfectly stable environment is an environment where the individuals of different generations find exactly the same conditions throughout a long period of time. On the contrary, in a disturbed environment the individuals of different generations do not encounter the same conditions.

It might appear that these concepts are not very helpful, since in a strict sense a stable environment probably does not exist. A purely stable environment is an abstraction and thus only relative stability will be considered here. There is a continuum between stable and unstable habitats but initially we attempt to differentiate between the two extreme situations and to compare populations living on them. For this analysis, we consider results concerning two perception levels and two wild species. The genotypic level will be studied using *Dactylis glomerata* L., and the phenotypic level using *Thymus vulgaris* L.

5.1.2 Chemical and Sexual Polymorphism in Thyme

In *Thymus vulgaris* L., two elements of variability have been studied.

5.1.2.1 Sexual Polymorphism

T. vulgaris is a gynodioecious species with two kinds of plants occurring in natural populations:

a) male-fertile individuals (noted mF), which are functionally hermaphrodite; they produce both ovules and pollen.

b) male-sterile individuals (noted mS), which are functionally female, producing only ovules. They therefore must be fertilized by mF plants.

These two kinds of plants coexist in every population studied but their respective proportions vary widely, from 5% to 95% mS.

The genetic basis for this polymorphism is still not known. It has been investigated by Assouad and Dommée (1974) who have shown that a large number of nuclear and probably cytoplasmic genes are involved. In spite of this, the phenotypic differentiation between the two sexual forms is very sharp, and it is generally easy to recognize a mF individual (bearing four stamens) from a mS one (bearing no stamens and generally showing a reduction in corolla size).

The biology and ecology of these sexual forms has been investigated. Domée et al. (1978) have shown that populations with a high proportion of mF could be found in rocky environments whereas mS were more frequent on deeper soils. Studies on the pollination mechanisms have shown that in the progeny of mF plants, the proportion of selfed individuals could be very different from one plant to another and could reach 70% (Valdeyron et al. 1977). This selfing rate has been shown to be influenced by the immediate surroundings of the plant (Brabant et al. 1980). Competition experiments have shown that mS plants are generally better competitors than mF ones (Bonnemaison et al. 1979) and that individuals obtained by selfing are less vigorous than those originating from outcrossing (Assouad et al. 1978).

5.1.2.2 Chemical Polymorphism

Granger et al. (1963) and Passet (1971) have shown the existence of an important polymorphism in thyme in regard to the leaf terpene production. These terpenes can be detected by vapor phase chromatography. The genetic basis for this polymorphism was elucidated by Vernet (1977). Six different terpenes can be found with any given individual producing mainly one of them, the others being present at very low levels. Hence, there are six types of individuals (i.e., six chemotypes). These chemotypes will be designated as follows:

G (geraniol) L (linalol)	non cyclic compounds
A (alpha-terpineol) U (thuyanol)	cyclic non phenolic compounds
C (carvacrol) T (thymol)	phenolic compounds.

They are classed here in the order of their synthesis. Although they probably do not derive directly from one another, they all belong to the same biosynthetic chain. They are genetically regulated by a series of 5 loci. Each of these loci can be occupied by a gene or its recessive allele. The relations between these alleles and loci are indicated in Fig. 1.

Each locus governs a branching of the synthetic chain and the epistatic order corresponds to the order of synthesis for every type except for the (L) one. This latter type has a rather special developmental program. Although the other chemotypes produce the same substance during their whole life, young L individuals produce thymol or carvacrol according to the genes which are present at the Ph locus.

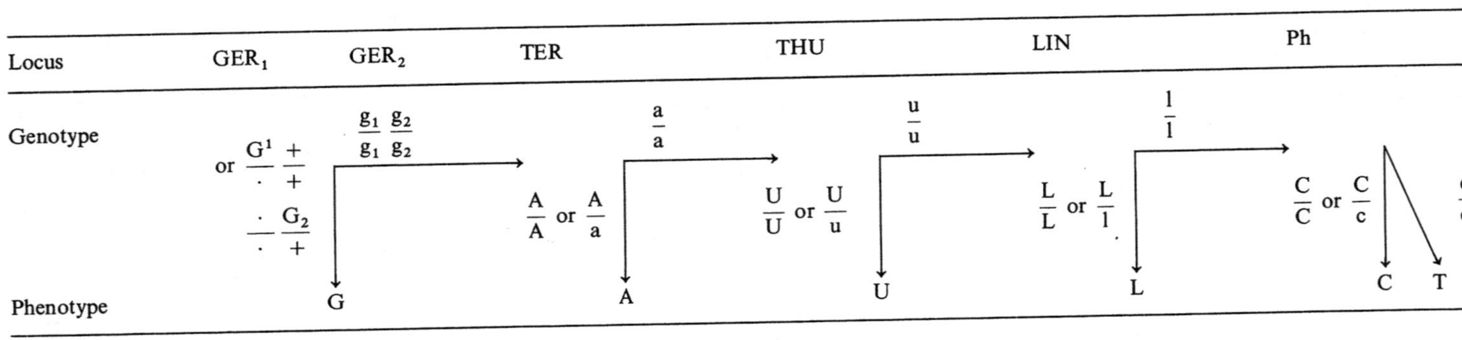

g_1 or G_1 : + : g_2 et G_2.

Fig. 1. Genetic control of chemotype in *Thymus vulgaris* L

It takes about three months for the L leaves to appear and after this period the plant produces only leaves containing L.

The exact function of these terpenes is not known. They can play a role in defense against herbivores or they could be a form of excretion; their role could now differ from that when they first arose in thyme evolution. Their ecological distribution has been studied on different scales (Gouyon 1975; Vernet et al. 1977; Gouyon et al. 1979) but particularly in a study area measuring about 10×8 km^2. This area, located arount St Martin de Londres, near Montpellier (France), is composed of a basin surrounded by a calcareous plateau. The basin is more fertile than the plateau and, even if the whole area was cultivated in the past, only the basin is still cultivated. Thyme, which is nearly always present under the evergreen oak forest (*Quercus ilex*) on the plateau, is only found in the abandoned fields of the basin. The distribution of the chemical forms in this area is such that only phenolic forms are found on the plateau while the non-phenolic chemotypes are predominantly found in the abandoned fields of the basin. The transition between the types can be very sharp, sometimes over distances of less than ten meters. This transition occurs all around the basin indicating that this distribution is not caused by an effect of random drift. Within each group (phenolic and non-phenolic), it is, at present, impossible to eliminate the hypothesis of the occurrence of drift or foundation effects (Kimura and Maruyama 1971).

Thus, thyme provides two different traits both showing habitat specificity and presumed adaptive significance. Male sterility would not be maintained if it were not associated with a superiority in viability or seed production (Lloyd 1975) at least at the level of their progeny (Valdeyron et al. 1973). Also the strict distribution of the chemical types in relation to the edaphic environment could not be maintained without a very strong selection. The relation between population structure concerning these traits and the degree of environmental disturbance will be discussed below.

5.1.2.3 Environment and Population Genetic Structure

In the area of St Martin de Londres, the distribution of sexual and chemical forms show the following patterns: Most of the populations, where the proportion of mS is low, have a low index of chemotypic diversity (Table 1). This relationship

Table 1. Contingency table showing the number of plots according to the proportion of females and chemotypic diversity

Diversity index (entropy)	Proportion of females	
	< 50%	> 50%
E 0.35	34	59
E 0.35	1	16

$E = \Sigma P_i \log P_i$ $P(\chi^2 = 6.2) < 0.02$

Table 2. Contingency table showing the number of plots according to the proportion of females and prevalent chemical type (>50%)

Chemical type	Proportion of females	
	<50%	>50%
Phenolic	20	23
Nonphenolic	13	44

$P(\chi^2 = 6.2) < 0.02$

Table 3. Contingency table showing the number of plots according to the population type and the presence or absence of chemotypic diversity

Diversity index	Population type	
	Phenolic	Non phenolic
Monomorphic ($E = 0$)	54	37
Polymorphic ($E \propto 0$)	80	156

$P(\chi^2 = 17.5) < 0.001$

is equivalent to the relationship between mS abundance and non-phenolic types given in Table 2. These two facts are not independent; Table 3 shows that many populations where the predominant chemotype is phenolic are monomorphic while few monomorphic populations have been found among the predominantly non-phenolic ones. This result can be related to the genetic controls indicated in Fig. 1. (T) individuals are homozygous at every locus while (G) individuals can be heterozygous at every locus. Under these conditions, selection could easily homogenize populations on the plateaux (where phenolic types are found) while in the basin the genes determining the phenolic form, protected by their recessive nature, could be maintained and appear under suitable circumstances.

A mutual-information analysis on the results collected in 106 stations was performed using the technique of Godron (Godron 1968; Daget et al. 1972). The proportion of mS was introduced under three classes considered as three different species in the mathematical treatment. These classes were:
- class 1 less than 50% mS (low proportions)
- class 2 from 50% to 70% mS (medium proportions)
- class 3 more than 70% mS (high proportions)

In this analysis the information given by the presence of a species on the environmental factors and the reverse, is weighted by the quality of the sampling of the factors; the final index is then $I = MI/FE$ (where MI is the mutual information and FE is the entropy factor). Two levels of significance were chosen.

1. Strong interrelations: $I > 4\%$
2. Medium interrelations: $I > 2\%$

Table 4. Observations significantly correlated with the proportion of females in thyme populations

Strength of relationship (MI/FE)[a]	Proportion of females		
	Low 0%–50%	Medium 50%–70%	High 70%
High (>4%)	Bedrock: hard limestone Xeric environments Phenolic types Most dominant species: *Quercus Ilex* and *Thymus vulgaris* Frequency of dominant chemotype >0.70 (low polymorphism)		
Medium (2%–4%)	Shallow soil Trees T and C chemotypes Low proportion of fine soil	Relative paucity of T or TC populations Most dominant species: *Genista scorpius* Frequency of dominant chemotype: 0.50–0.70 (polymorphic population) Soil free from stones	Most dominant species: *Aphylantes monpeliensis* *Rosmarinus officinalis* *Quercus lanuginosa*

[a] MI/FE: mutual information index

Table 4 gives the results of this analysis. It shows that low proportions of mS are associated with the less modified environments, while medium proportions correspond to more open and disturbed habitats.

It also shows that strong relations with the environment are only found for low proportions of mS. From this, it is possible to deduce that, for the plant, as noted earlier, it is easier to define a stable environment than an unstable one. In stable Mediterranean environments, thyme is mainly found as a phenolic form (the one which is derived by the most homozygous genotype) and in the mF sexual form (which allows selfing and, thus, favors homozygosity).

When the environment is disturbed, for whatever reason, diverse genotypes are favored and the mS individuals selected (1) because their progenies, derived from outcrossing, have higher genetic potentialities (Valdeyron et al. 1973) and (2) because mS individuals themselves are more heterozygous than mF (Gouyon and Vernet 1982). Even so it is difficult to explain how this selection for diversity can adjust the proportion of mS individuals at values as high as 90%, as can be found in some populations. This situation is likely maintained by non-equilibrium conditions. Maynard-Smith (1978, p. 6) noted that "my own insight in the field may have been obscured by an obsession, which I share with most population biologists with equilibrium situations." It is true that, studying disturbed situations while looking for an equilibrium situation is probably not a correct approach. A system, even if it does possess an equilibrium condition does not necessarily attain it by a monotonic transformation; oscillations can occur. Such oscillations have been described in predator-prey interactions; they can also occur when nucleocytoplasmic interactions are involved. Indeed, since we are studying functional gynodioecy (when hermaphrodites actually produce seeds), it should be remembered that most of the cases which have been studied in detail (perhaps all of them) have shown that male sterility has been determined by a joint action of cytoplasmic and nuclear factors. The maintenance of a cytoplasmic polymorphism in these conditions raises multiple questions. If a purely phenotypic selection (i.e., involving only selective differences between the two sexes whatever their cyto-genotype) is assumed, it is impossible to find polymorphism at the level of the cytoplasm (Charlesworth and Ganders 1979).

A model presented by Delannay et al. (1981) shows that, if genes act directly on fitness (i.e., individuals of the same sexual form have different fitnesses if their genotype is different), there could be a joint equilibrium. Starting from a situation near to it, the equilibrium is reached by a monotonic transformation. On the contrary, if the frequencies are drastically modified, important oscillations can occur. The reason is that in a given cytoplasmic environment, the genes controlling sex evolve toward an equilibrium frequency. This evolution changes the genetic context of the cytoplasm. The cytoplasmic frequencies then evolve as a function of the genic frequencies which are not the equilibrium frequencies. Computer simulations show that these oscillations can (but need not) lead to a polymorphic equilibrium. In the two cases, it is possible to find very high values (80%) at the beginning of the oscillations even if the equilibrium value is low (less than 40%).

In a disturbed environment, thyme populations could be far from an equilibrium. In these conditions, it is probably not useful to speculate about what an observed frequency can mean. It is however important to notice that, when disturbance occurs, diversity seems generally to increase and the reproductive system evolves toward an increase of gene exchange.

5.1.3 Enzymatic Polymorphism in Orchard Grass

Dactylis glomerata L. is predominantly an auto-tetraploid species, except for a few diploid subspecies, which are widespread in Eurasia (Borrill 1977). This species is predominantly allogamous and pollen is generally carried by the wind. Pollen migration distance does not usually exceed 100 m (Naghedi-Ahmadi 1977) and in most cases seeds do not seem to migrate any great distance.

Among the tetraploid forms of *D. glomerata* two infraspecific units, which are generally considered subspecies, are found; *D. glomerata* ssp. *glomerata* and *D. glomerata* ssp. *hispanica* both of which grow in Europe, Asia, and North Africa. The former is found in northern and moist regions while the latter occurs in drier and warmer regions.

Seven polymorphic loci have been studied, belonging to five enzymatic systems (Lumaret and Valdeyron 1978; Lumaret 1981). These loci are:

- GOT 1 (7 alleles) and GOT 2 (3 alleles)
- Ac Ph 1 (7 alleles)
- Pox 1 (7 alleles) and Pox 2 (8 alleles)
- M.D.H. 1 (5 alleles)

- ADH 1 and ADH 2 (8 alleles) which probably originated from one another by duplication.

(GOT = Glutamate oxaloacetate transaminase, Ac Ph = Acid phosphatase, Pox = peroxydase, M.D.H. = malate dehydrogenase, ADH = alcoholdehydrogenase).

The proportions of these alleles are different in the two subspecies but they can also change according to the population within a subspecies. For some of these loci, the allele distribution can be shown to have a habitat relationship. The results are particularly clear with the GOT 1 locus. In this locus, the alleles 1 (migration index 0.1) and 2 (migration index 0.72) are precisely linked with the environment; from northern to southern Europe, the frequency of allele 1 decreases regularly from 0.9 to 0 with variations related to altitude changes (Lumaret in prep.). The frequency of allele 2 changes in the opposite way. Two other alleles (allele 3, migration index 0.38 and allele 4, migration index 0.10) are only found in Mediterranean regions. In fact, this large scale pattern is the result of adaptations occurring over very short distances. Two transects showed similar tendencies. The first was made in Great Britain by Ashenden et al. (1975) and the other in France, near Montpellier, by Lumaret (1981). Allele 1 appears to be significantly more frequent in the wet part of the transect while allele 2 shows the opposite tendency. In the transect near Montpellier, where alleles 3 and 4 were present, it was not possible to find any correlation between the environmental conditions and the distribution of these alleles. The fact that alleles 1 and 2 show the same trend in three situations (and on two scales) indicates that these alleles are directly related to a presumed adaptative trait.

The area studied here is a plain (area about 50,000 ha) about 100 km north of Marseille called La Crau. It consists of two parts:

- a rocky portion (representing about 3/5 of the area) with silicious or calcareous rock but little available mineral nutrient. This area is grazed by sheep.
- an irrigated area with calcareous silt (45% lime). After removal of stones, and irrigation, numerous plant species appeared spontaneously.

In the rocky part of the plain, plants are of the *hispanica* type, which probably arrived long ago. In the irrigated part, on the contrary, one can find only the *glomerata* type. These latter plants probably arrived by migration, carried by irrigation water from the Durance River. They likely originated in the Southern Alps. Irrigation and mowing created an environment allowing the establishment of the immigrants in place of the native *hispanica* type in this part of the plain.

At the La Crau plain, the disturbances caused by man are known; irrigation, ploughing, and hay production. The irrigated part of the plain is more characteristic of the habitat of the subspecies *glomerata* than of the subspecies *hispanica* which is native in the dry pastures of this area.

In the irrigated pasture, the *glomerata* subspecies completely replaced the *hispanica* subspecies. The genotypic composition of the populations has been studied in an area where the irrigated and non-irrigated pastures meet (Lumaret and Valdeyron 1980) (Table 5). Ssp. *glomerata* usually flowers at the end of April, whereas ssp. *hispanica* flowers at the beginning of June. Hybridization should then not be possible. However, hay is harvested at the beginning of May in the irrigated area. This favors a second period of flowering which occurs at the beginning of

Table 5. Allelic frequencies at four electrophoretic loci in different populations of *Dactylis glomerata*

Locus	GOT 1 Alleles					AcPH 1 Alleles			PX 1 Alleles					MDH 1 Alleles	
Population	1 (1.00)	2 (0.72)	3 (0.38)	4 (0.10)	7 (1.26)	1 (1.00)	2 (0.88)	3 (0.95)	1 (1.00)	2 (0.93)	3 (0.90)	4 (0.84)	0 (nul)	1 (1.00)	2 (0.88)
D.g. glomerata (ssp) (mean values)	77	23	–	–	–	67	27	6	70	18	5	6	2	99	1
CRAU A[a] (irrigaded)	59 (±9)	23 (±8)	10	6	2	56 (±9)	38 (±9)	6	67 (±8)	18 (±7)	7	2	6	99	1
CRAU B (dry)	4	38 (±8)	49 (±7)	8	–	88 (±5)	7	5	39 (±8)	52 (±8)	4	4	1	100	–
D.g. hispanica (ssp) (mean value)	13	33	40	12	–	85	13	2	43	46	4	3	4	100	–

[a] In CRAU A and B populations, 27 and 41 individuals were respectively sampled representing respectively 108 and 164 homologous genes per locus. Migration distances are given in parentheses relatively to the migration distance of allele 1

June. Table 5 shows that, in the irrigated area, important genotypic changes are occurring. Whereas the environmental conditions of this area favors the *glomerata* type, these plants, which originated from ecologically different areas are not as well adapted to the climatic conditions of the region as is the native *hispanica* type. With disturbance, the reproductive system permits the new immigrant population to receive alleles coming from the neighboring native population. This is apparent at the GOT 1 locus where the frequencies of the genes are neither of a typical *glomerata* population nor of a typical *hispanica* one. These genes probably do not originate alone in the population. At other loci where no precise adaptative significance could be detected, this phenomenon does not appear. It seems here that natural selection is able to sort out very efficiently the genes which are of adaptative value in the gene pool of the population and that only these are maintained.

This sort of "use," by a population of disturbed areas, of older and less disturbed neighboring populations, is probably a widespread phenomenon in the evolution of cultivated species. Two striking examples can be found in the literature. First, pearl millet hybridizes, at each generation, with its wild equivalent (Pernes et al. 1980). Secondly, maize, when sympatric, hybridizes with teosinte (Beadle 1980). In these two cases, it seems that these genetic exchanges are beneficial for the cultivated species. These species, probably more than others, have a reduced gene pool through selection. The wild ancestor then plays the role of a gene bank. The same biological process happens between the two orchard grass populations; genes are transferred from the old population to the new one without any intervention of man. In pearl millet and maize, the farmer simply replaces the natural selection occurring in *Dactylis*. He sorts the plants on morphological criteria in order to maintain the specific characteristics of the crop. These farmers from so-called under-developed regions seem to show an intuition superior to the trained breeder. A western farmer, faced with a field where his crop hybridizes with a weed, would try to suppress the weed in order to homogenize his production (such as happens with *Beta vulgaris* L. in France). Faced with the same situation, the African farmer respects the weed and calls it "the millet's father;" he then chooses the best ears for his seed (Pernes pers. commun.).

This is exactly comparable to the system described by Beadle (1980) for maize where the Mexicans call the teosinte "madre del mais" and say that it is "good for the corn" when it grows around, or even inside, the corn fields.

5.1.4 Conclusions

In the examples shown here, the natural populations are able to adapt to disturbance because they possess a certain amount of diversity. This diversity increases when the population is disturbed. This favors open reproductive systems and, in such conditions, mechanisms increasing the proportion of cross fertilization can then be found. The proportion of females increases in hermaphroditic thyme populations when disturbance occurs. Cross pollination allows *Dactylis* populations to increase their adaptation by including genes from the neighboring native subspecies in their gene pool. Some cultivated plants are able to keep their diversity by crossing with their wild equivalent.

The value of maintaining genetic diversity in cultivated plants has been recently rediscovered. An important question still remains however, how to keep this variability? Whereas the African and Mexican farmers favor each year crosses between wild and cultivated plants, each of these populations evolving separately under the action of their own evolutive pressures, advanced civilizations collect material together in "gene banks." In this latter situation, populations are prevented from evolving and are less likely to be useful. It might be better to improve our understanding of the mechanisms acting in natural populations and to allow our crops to adapt themselves rapidly to disturbance.

Résumé

Les effets de la perturbation par l'homme dans les populations sont décrits en utilisant deux espèces, *Thymus vulgaris* L. (thym) et *Dactylis glomerata* L. (dactyle) et différents types de marqueurs: marqueurs morphologiques et fonctionnels chez *T. vulgaris* et marqueurs enzymatiques chez *D. glomerata* devant un environnement perturbé. *T. vulgaris* réagit en augmentant ses potentialités d'allogamie par un accroissement de la proportion de plantes mâley stériles (qui sont incapables de s'autoféconder car elles ne produisent pas de pollen) et de sa diversité génétique au niveau des produits synthétisés dans ses feuilles. Une population de *D. glomerata* récemment introduite par migration après une perturbation dans une plaine de la région méditerranéenne augmente sa diversité par des croisements avec des populations voisines qui n'ont pas subi de perturbation. Une réaction normale à la perturbation chez les systèmes génétiques naturels, peut-être la plus commune, semble donc être une augmentation de la diversité dans différentes directions, en particulier par une modification du système de reproduction.

Certaines espèces cultivées comme le mil ou le maïs sont toujours capables d'échanger des gènes avec leur parent sauvage. On connaît des systèmes culturaux primitifs capables de profiter de ces situations. Dans le plupart des cas cependant, l'homme ne contrôle pas consciemment la manière dont la diversité disponible est utilisée dans les programmes d'amélioration. Ces derniers pourraient probablement être perfectionnés grâce à l'étude des réponses aux perturbations des populations naturelles.

References

Ashenden TW, Stewart WS, Williams W (1975) Growth responses of sand dune populations of *Dactylis glomerata* L. to different levels of water stress. J Ecol 63:97–107

Assouad MW, Dommée B (1974) Recherches sur la genetique ecologique de *Thymus vulgaris* L. Etude expérimentale du polymorphism sexuel (M.W.A.). Determinisme genetique et répartition écologique des formes sexuelles (B.D.) CR Acad Agric Fr 9:57–62

Assouad MW, Dommée B, Lumaret R, Valdeyron G (1978) Reproductive capacities in the sexual forms of the gynodioecious species *Thymus vulgaris* L. Bot J the Linn Soc 77:29–39

Beadle GW (1980) The ancestry of corn. Sci Am 242:96–103

Bonnemaison F, Dommée B, Jacquard P (1979) Etude experimentale de la concurrence entre formes sexuelles chez le thym, *Thymus vulgaris*. Oecol Plant 14:85–101

Borrill M (1977) Evolution and genetic resources in cocksfoot. Rep Welsh Plant Breed Stn 1977. Univ College Wales, Aberystwyth

Brabant Ph, Gouyon PH, Lefort G, Valdeyron G, Vernet Ph (1980) Pollination studies in *Thymus vulgaris* L. Oecol Plant 15:37–45

Charlesworth D, Ganders FR (1979) The population genetics of gynodioecy with cytoplasmic-genic male-sterility. Heredity 43:213–218

Daget Ph, Godron M, Guillerm JL (1972) Profils ecologiques et informations mutuelles. CR 13e Syn Assoc Int Phytosociol „Grundfragen und Methoden in der Pflanzensoziologie“ W. Junk, Den Haag, pp 121–149

Delannay X, Gouyon PH, Valdeyron G (1981) Mathematical study of the evolution of gynodioecy with cytoplasmic inheritance under the effect of a dominant restorer gene. Genetics 99:169–181

Dommée B, Assouad MW, Valdeyron G (1978) Natural selection and gynodioecy in *Thymus vulgaris* L. Bot J Linn Soc 77:17–28

Godron M (1968) Quelques applications de la notion de fréquence en écologie végétale. Oecol Plant 3:185–212

Gouyon PH (1975) Note sur la carte provisoire de la répartition des différentes formes chimiques de *Thymus vulgaris* L. Oecol Plant 10:187–194

Gouyon PH, Vernet Ph (1982) The consequences of gynodioecy in natural populations of *Thymus vulgaris* L. Theor Appl Genet 61:315–320

Gouyon PH, Valdeyron G, Vernet Ph (1979) Selection naturelle et niche écologique chez les végétaux supérieurs. Bull Soc Bot Fr 126:87–95

Granger R, Passet J, Verdier R (1963) Diversite des essences de *Thymus vulgaris* L. Fr Ses Parfums 7:225

Kimura M, Maruyama T (1971) Pattern of neutral polymorphism in a geographically structured population. Genet Res 18:125–131

Lloyd DG (1975) The maintenance of gynodioecy and androdioecy in angiosperms. Genetica 45:1–14

Lumaret R, Valdeyron G (1978) Les glutamate oxaloacetate transaminases du dactyle *(Dactylis glomerata* L.): génétique formelle d'un locus. CR Acad Sci Ser D, 287:705–708

Lumaret R, Valdeyron G (1980) Les dactyles de la crau: mise en évidence de relations entre les différents écotypes par le polymorphisme enzymatique. CR Acad Agric 66:229–238

Lumaret R (1981) Structure génétique d'un complexe polyploïde: *Dactylis glomérata* L. Thèse de doctorat d'état mention Sciences USTL Montpellier France

Maynard-Smith J (1978) The evolution of sex. Cambridge Univ Press, Cambridge p 222

Naghedi-Ahmadi I (1977) Zur Frage der Pollenflugweite bei *Dactylis glomerata.* Z Pflanzenzucht 78:163–169

Passet J (1971) *Thymus vulgaris:* chémotoxonomie et biogenese monoterpénique. Thèse doct, Pharmancie, Montpellier

Pernes J, Nguyen Van E, Beninga M, Belliard J (1980) Analyse des relations genetiques entre formes cultivées et spontanées chez le Mil à chandelle II. Amelior Plant 30:229–269

Valdeyron G, Dommee B, Valdeyron A (1973) Gynodioecy: another computer simulation model. Am Nat 107:454–459

Valdeyron G, Dommee B, Vernet Ph (1977) Self fertilization in male-fertile plants of a gynodioecious species: *Thymus vulgaris* L. Heredity 39:243–249

Vernet Ph (1977) Les variations de composition de de l'essence de *Thymus vulgaris* L. mode de transmission héréditaire de trois terpènes. CR Acad Sci Ser D, 284:1289–1292

Vernet Ph, Guillerm JL, Gouyon PH (1977) Le polymorphisme chimique de *Thymus vulgaris* L. I. Répartition des formes chimiques en relation avec certains facteurs écologiques. II. Carte à l'échelle 1/25000e des formes chimiques dans la région de St Martin de Londres (Herault-France). Oecol Plant 12:159–194

5.2 Demographic Strategies and Originating Environment

P. Jacquard and G. Heim

5.2.1 Introduction

Within the study of population dynamics, demography is generally developed in relation to adaptive strategies. But among plants, not only do individuals, populations and/or species comprise levels of study, but so do organs (Jacquard 1980). The analysis of small-scale patterns in the distribution of species allows the discrimination among plants of three primary adaptive strategies, according to Grime (1979): competitors, stress-tolerators, ruderals. The first two categories are somewhat over-lapping. The models proposed to describe the dynamics of plant populations on the basis of demography of individuals do not account for plant plasticity. The specific characteristics linked with the development of green plants have led to the elaboration of a demography of populations of organs (Harper 1977). The study of strategies in relation to specific environments allows the establishment of the relations between the co-existence of plant species and niche differentiation or co-adaptation. As well, it enables the evolutionary significance of demographic parameters of strategies to be clarified.

In the present chapter, the comparative study of plant strategies is developed at several levels:

1. species (a great number of studies have been undertaken at this level);
2. populations of the same species: in this approach, attention is directed at the questions of "how" and "in function of what" the strategy of a species is changing among its populations (Barbault 1976);
3. individuals of the same populations. If the preceding level allows the identification of mechanisms, it is at this level that physiological aspects can be clarified.

The ecological characteristics of 4 species of grasses, *Arrhenatherum elatius* (L.) Beauv. ex J. et C. Presl., *Dactylis glomerata* L. (Cocksfoot), *Bromus erectus* Huds. and *Brachypodium phoenicoides* R. et S., are analyzed in relation to the specific environment of several populations.

5.2.2 Description of the Originating Environments

The populations studied originated from two contrasting types of environments:

1. One is modified by man and has been managed for hay production with irrigation. It is a grassland at Crau which has been used for a long time for forage production. The population of *A. elatius* which was studied was represented by

clones maintained by the Plant Breeding Station of INRA, in Montpellier. The study population of *D. glomerata* originated as 30 units of tufted tillers randomly distributed in an INRA plot called Pré du Mistral. The tufts were cloned and multiplied.

2. The other study habitats had been more or less abandoned by man, although there was infrequent grazing:

a) An herbaceous layer of a "garrigue" vegetation was used as a source for *A. elatius*. For this taxon, called Puech, the sampling was conducted by transplanting a great number of individuals which were subsequently vegetatively multiplied.

b) A low-land site (population from St. Mathieu de Tréviers) was used as a source for *D. glomerata* which was sampled as 30 shoots from vegetation dominated by *Brachypodium pheonicoides*, from a semi-mesic site. This population is noted A in the text whereas B is the Crau population.

The population of cocksfoot originating from Crau, belongs to the typical *glomerata* form, but the two other populations are the *hispanica* form.

For each population, the procedure of sampling and of further vegetative propagation resulted in the use of genotypically different individuals (genets).

5.2.3 Between and Within-Population Variations of Strategies in Arrhenatherum elatius

A study has been conducted to evaluate the performance, in competition, of populations of *A. elatius* originating from contrasting environments. The effect of several watering regimes was measured, notably, on the aggressive traits of the two populations (Azocar 1978).

In a first experiment, including replacement series (Van Den Bergh and Braakhekke 1978), the dynamics of the two populations was analyzed for three watering regimes: natural, optimal irrigation, and alternation of the two conditions. The variables measured were biomass (dry matter) or the number of individuals, in pure stands and in associations. In a second design, binary associations of four genotypes for each population were compared in wet and dry conditions using a diallel analysis (Gallais 1970; Jacquard et al. 1978), according to a model that allows characterization of the compared biological entities (genotypes, populations, species) using relative parameters based on pure stand and association performances. For each experiment, the Relative Yield (or Abundance) Totals (RYT from de Wit 1970) have been calculated. The values of the RYT permit the comparison of yield within pairs of populations or individuals, with limiting resources.

From the resulting data the following conclusions can be made:

1. When the two populations are in experimental competition, the exclusion of the populations depends on the water regime. Under the regime without irrigation, there is a severe elimination of individuals as a consequence of the reduction of biomass, whatever the composition of the stand. Such conditions represent an under-exploitation of the niche ($RYT < 1.00$) (Table 1). This stress results in either the elimination of the two populations or the decline of only the Puech population, as

Table 1. Relative yield totals (RYT) of two populations of *Arrhenaterum elatius* with the no-irrigation regime. (Means for 3 replications)

Dates of observations	On the basis of dry biomass	On the basis of the number of surviving individuals
21/04/76	1.00±0.13	0.96±0.07
21/06/76	0.55±0.30	0.35±0.10
15/06/77	0.98±0.18	0.96±0.05
15/08/77	1.01±0.58	1.05±0.32
15/10/77	0.63±0.27	0.76±0.63

Table 2. Relative yield totals (RYT) of two populations of *Arrhenaterum elatius* with the "ad libitum" regime. (Means for 3 replications)

Dates of observations	On the basis of dry biomass	On the basis of the number of surviving individuals
21/04/76	1.06±0.05	0.96±0.01
21/06/76	1.14±0.09	0.99±0.08
14/09/76	1.25±0.22	1.04±0.05
15/06/77	0.96±0.07	0.96±0.18
15/08/77	1.06±0.11	0.96±0.18
15/10/77	0.99±0.04	0.96±0.18

the result of a limitation of growth and whatever is the starting proportion. With conditions of optimal moisture, the two populations co-exist, with a domination of the Crau population which exhibits a better competitive capacity. The RYT for seven observations are reported in Table 2. They are always approximately equal to 1. Under these conditions there is a trend to equilibrium which is shown in terms of demography (number of survivals). In alternating conditions (unstable environment), the Puech population declines as a result of competition. The trajectories of the relative replacement rates, which express the temporal variations of the stands, show that for all the mixtures it is always the Crau population that wins.

2. Independent of the moisture treatment, the population of the modified environment (Crau) is always at an advantage when in association. The best performance of this population is obtained when its proportion in the mixture is in the minority (Fig. 1). The trajectories of the relative replacement rates, by reference to the number of surviving individuals, for the several associations show that, starting with the 5th observation, the values of ϱ were stabilized for each mixture and mortality reached a minimal value. The stabilization of the values of ϱ indicate a mixture in equilibrium. Despite the fact that the Crau population always wins (values of $\varrho > 1$), the Puech population is not eliminated. That is to say a stabilized mixture is obtained (Fig. 2).

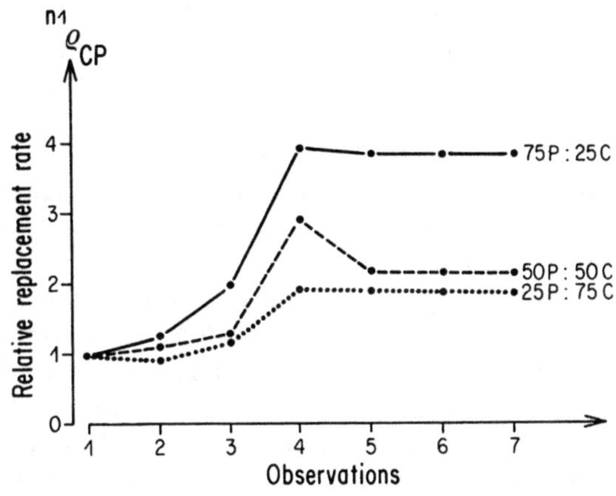

Fig. 1. Relative replacement rate ($^{n1}\varrho$ CP) based on the number of surviving individuals. The values of ϱ are computed by reference to the frequency of the individuals at the start of the experiment (initial observations); *C* Crau population; *P* Puech population; *75* Proportion in the mixture. (Azocar 1978)

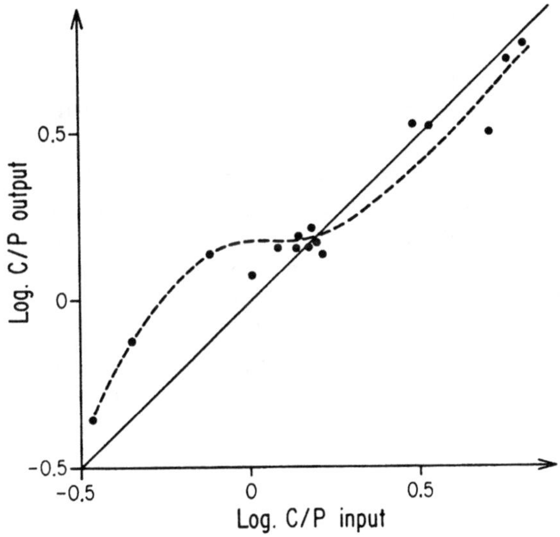

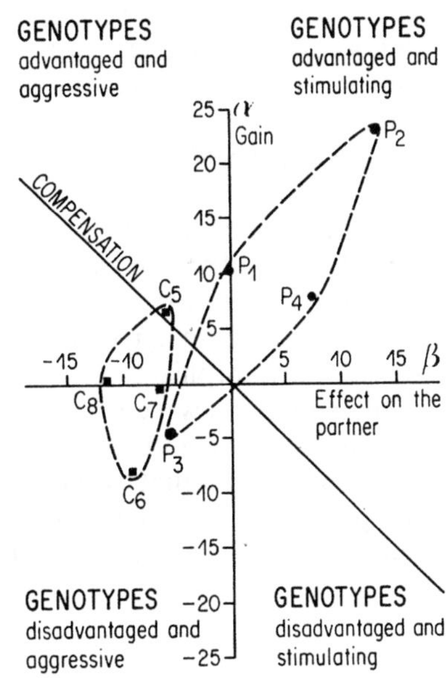

Fig. 2. Diagram of the relationship between the input and output number of two populations of *Arrhenatherum elatius* under competition; *C* Crau population; *P* Puech population. (Azocar 1978)

Fig. 3. Relationships between the direct effect (α) and the associated effect (β) for four genotypes of the population of *Arrhenatherum elatius* from garrigue (*P* Puech) and four genotypes of the population of *A. elatius* from an irrigated meadow (*C* Crau). The effects are centered on the general mean for the eight genotypes and measured in terms of the ratio of the number of panicles to the number of tillers. The results for two environments (irrigated and not irrigated) have been pooled. (Azocar 1978)

3. The individuals of the Crau population exhibit a higher level of aggressiveness than do those of the Puech population. There is no evidence of interaction between the genotypic parameters (resistance or susceptibility to neighbors, aggressiveness, association, and domination capacities) and the environment (Fig. 3). Dominance is the result of a positive feed-back between:

a) the mechanisms which cause the dominating plant to achieve a better demographic performance in comparison with its neighbors (response to neighboring or direct α effect of the diallel analysis),

b) the negative effects that prolific plants can exert on the fitness of neighbors with lower performances (aggressivity or associated β effect of the diallel analysis). According to Grime (1978), the direct effects change in relation to the positive action of the habitat on *strategies*. This variation is not evidenced here; in contrast, the variability between genotypes is large, mainly within the population originating from the environment with a lesser control by man. In contrast to direct effects, the associated effects should not change strongly with respect to habitat or strategies and should mainly pertain to several types of stress induced by plants such as the stress due to shading, or to a decrease of the levels of nutrients and water into the soil. For the two populations studied, the comparative variability of these associated effects is very different; very low for the populations from the human-controlled environment, large for the other population. The diallel analysis design, of the competitive abilities of the two populations of *Arrhenatherum elatius*, discriminates between inter and intrapopulation variability. Phenotypic and genetic variability can be segregated, providing the estimation of the necessary parameters is conducted in several environments, for example, irrigated and non-irrigated, as here.

4. In conclusion, for the distribution of the populations of *A. elatius*, as the Puech and Crau types, water seems not to be the critical selective factor. Instead the type of management of the habitat determines the more critical selective pressure. The modification by man of the management system, shifting from grazing to mowing, seems to lead to an advantage of the genotypes adapted to competition. These results are in agreement with those obtained by Mahmoud et al. (1975) who studied polymorphism in *Arrhenatherum elatius* in an area where the populations of this species show a continuum of variation. In the range of morphological types, erect and prostate individuals represent the extremes. The field and laboratory experiments suggest that the polymorphism results from a balance between the selective pressures imposed by competition for light in the tall vegetation and the defoliation found in the grazing lands.

5.2.4 Between and Within-Population Variations of Strategies in D. glomerata

This study compared two populations (A, abandoned by man, B, modified by man), grown under optimal irrigation, with respect to tiller, dry matter production, partitioning and reproductive effort.

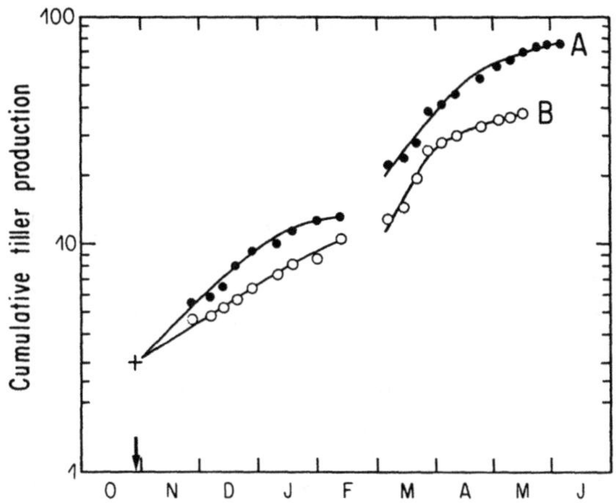

Fig. 4. Cumulative tiller production for plants of two populations of *Dactylis glomerata* originating from *A* abandoned by man; *B* modified by man (hay-making combined with irrigation). *Arrow* indicates the date of planting

Each population was represented by 20 genotypes, which were subsequently cloned. For each genotype, six individuals, comprising one fully developed tiller, were planted in autumn, 1978. Water regime was maintained near the optimum, using an automatic watering system, controlled by evaporation. Nutrients were added regularly.

Due to poor regrowth, or a fungal disease, some genotypes had to be discarded from the experiment, especially in population B. Therefore the results are for 19 genotypes in population A, and 14 genotypes in population B.

Tillering was studied by recording at weekly intervals number of living and dead tillers, for one individual per genotype. Dry matter (aerial parts only) of the six individuals was harvested twice (at flowering and three weeks later), and dry matter partitioning was recorded for one individual per genotype (Bremond 1979).

The results with respect to tillering are shown in Fig. 4. The cumulated number of tillers, produced up to flowering was much less in plants of the man-modified population (B), roughly 50% of the number of tillers produced by population A. For the whole period relative tillering rate was, on the average, 1.2% in population B and 1.5% in population A. However, this difference is not constant throughout the period, and the reverse situation (relative tillering rate higher in population B) was noticed in winter, when tillering rate was slowing down in population A. Maximum values occurred in early spring, and were of the same order of magnitude (3% per day) in both populations.

The number of panicles was also much less in population B (Table 3) as was the proportion of reproductive tillers (34.2% compared with 40.2% in population A).

The proportion of reproductive tillers varies considerably with genotype (Fig. 5). In population A this proportion is correlated with the number of tillers produced; the higher the number of tillers, the higher the proportion of reproductive tillers. Such a correlation does not appear in population B.

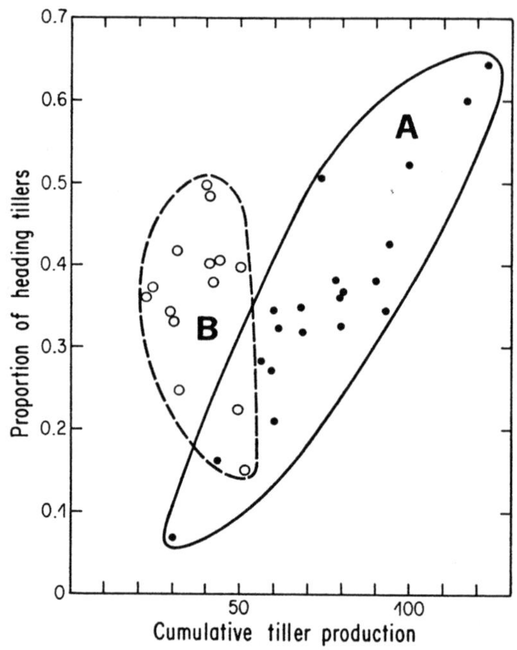

Fig. 5. Relationship between cumulative tiller production and proportion of heading tillers for two populations of *Dactylis glomerata: A* abandoned; *B* modified by man. Data correspond to one individual per genotype

Table 3. Number of tillers in two populations of *Dactylis glomerata*. A, abandoned by man; B, modified by man (hay-making systems combined with irrigation)

	A	B
Number of living tillers at first harvest (flowering)	48.1±3.52	30.4±2.24
Number of panicles at flowering	31.9±4.19	13.4±1.25
Number of dead tillers (cumulated since planting date)	29.9±2.46	7.2±1.09
Cumulative tiller production[a]	78.0±4.65	37.6±2.45

[a] Cumulative tiller production was obtained by adding to the number of tillers present at harvest the dead tillers registrated weekly between planting and harvest. The data are the means followed by standard-error. The means correspond to 19 genotypes in the case of population A and to 14 genotypes in the case of population B (1 individual per genotype)

Contrary to the number of tillers produced, which is lower in plants from the man-modified population (B), total dry matter production is slightly higher in B at the first harvest (flowering) and distinctly higher after regrowth (Table 4). This difference in regrowth might be due to differences in the proportion of living vegetative tillers which is much higher in the man-modified population (B) [45% against only 20.9% in population A (Table 3)].

Dry matter production is highly dependent on genotype (Fig. 6). At the first harvest (flowering), variation between genotypes (expressed as relative standard

Table 4. Dry matter harvested from two populations of *Dactylis glomerata* (A, abandoned by man; B, modified by man (hay-making combined with irrigation)[a]

	A	B
Dry matter at flowering (g)	71.3 ± 10.08	99.9 ± 4.04
Coefficient of variation (%)	63.2	25.7
Dry matter three weeks after first harvest (g)	13.3 ± 1.39	25.3 ± 2.12
Coefficient of variation (%)	47.8	32.5

[a] The data presented are the means (followed by standard-error) corresponding to dry matter harvested in aerial parts of 6 individuals per genotype (19 genotypes in population A, 14 genotypes in population B)

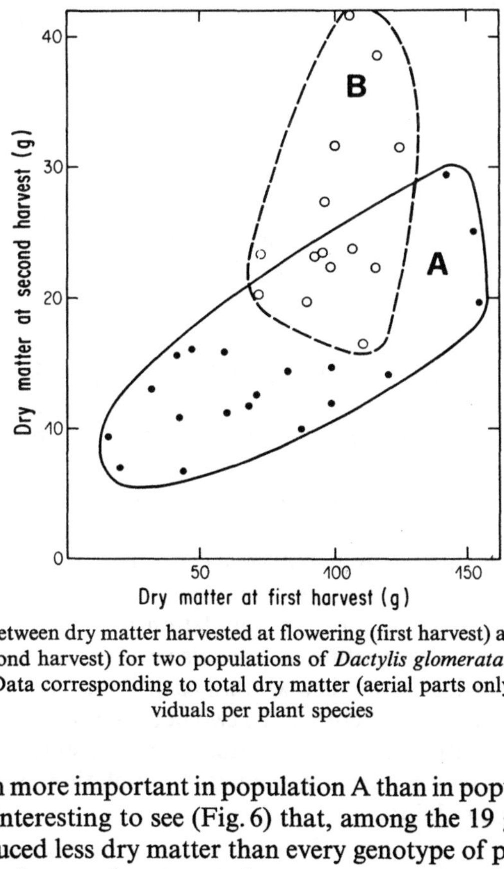

Fig. 6. Relationship between dry matter harvested at flowering (first harvest) and dry matter harvested three weeks later (second harvest) for two populations of *Dactylis glomerata: A* abandoned by man; *B* modified by man. Data corresponding to total dry matter (aerial parts only) harvested on six individuals per plant species

deviation) is much more important in population A than in population B (Table 4). It is particularly interesting to see (Fig. 6) that, among the 19 genotypes of population A, 11 produced less dry matter than every genotype of population B, whilst 3 of them produced more dry matter than every genotype of population B.

Dry matter harvested at the second harvest is correlated with dry matter at first harvest, but only in the case of population A (Fig. 6). Due to a decrease in variability of population A and an increase in variability of population B, variations within each population are less marked.

The ratio between dry matter production and number of tillers show big differences between populations; for a given level of dry matter production the number

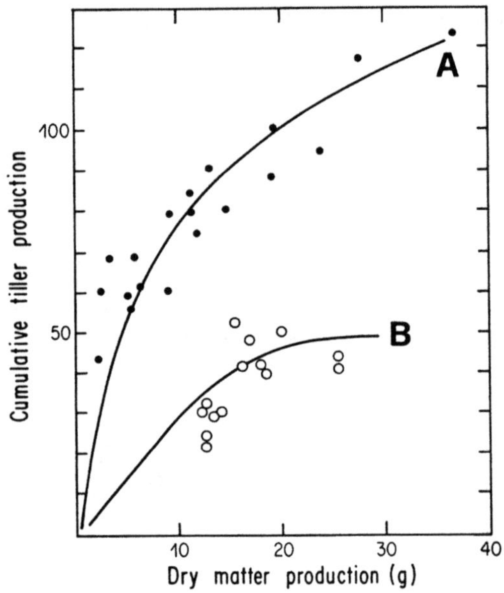

Fig. 7. Relationship between dry matter production (dry matter harvested from aerial parts at flowering) and cumulative tiller production for two populations of *Dactylis glomerata: A* abandoned by man; *B* modified by man. Data correspond to one individual per genotype

of tillers is three times higher in population A, than in population B (Fig. 7). This ratio varies also within each population, principally in the case of population A. This population has big differences in the "vigor" of the tillers of the different genotypes.

Dry matter partitioning at flowering (Fig. 8) shows that in both populations (excluding roots) 1/3 of dry matter is stored in leaves and 2/3 in panicles and stems. The two populations differ mainly with respect to partitioning of dry matter between panicle and stem; the panicle/stem ratio in the man-modified population is 0.42, compared with a ratio of 0.80 in population A.

Differences in dry matter partitioning are also evident from Fig. 9, which shows additional and important variability whithin each population.

Some clear-cut differences between the two populations, concerning mainly tillering, dry-matter partitioning and regrowth, have been shown. Differences of the same type are found between genotypes of each population, which display a large within-population variability.

5.2.5 Conclusions

The evolution of the populations of a species during the progress of domestication illustrates the modifying influence of Man. For example, Sano and Morishima (unpublished, in Oka 1977) observed the pattern of variation in reproductive

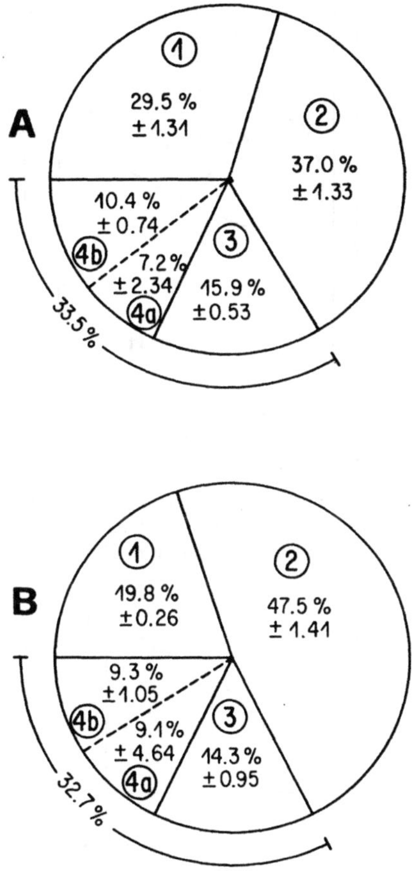

Fig. 8 A, B. Partitioning of dry matter harvested from aerial parts at flowering between panicles (*1*), stem (*2*), leaf sheaths (*3*), green leaf blades (*4 a*), and standing leaf blades (*4 b*) for two populations of *Dactylis glomerata:* **A** abandoned by man; **B** modified by man. Data correspond to averages of percentages corresponding to one individual per genotype

activity, as shown by allocation of resources, in a sample of 30 strains of *Oryza perennis*. They classified characters into two groups: vegetative and reproductive, the first linked to vigor, the second to the effort for seed production. Within each category, the characters were positively correlated and among categories the correlations were negative. In the first group were included the culm, panicle length, tillering angle, the regenerating capacity of excised stem segments, etc. In the second group seed weight to total plant weight ratio, panicle number to tiller number ratio, etc., were considered.

They found among-lines variation in r– and K–strategies (Gadgil and Solbrig 1972), the r characteristics being reinforced with domestication with the transition from a perennial to an annual form domesticated for seed production. Conversely, the K characteristics were reinforced during a selection oriented toward leaf production. For this reason, forage breeders have often been confronted with a de-

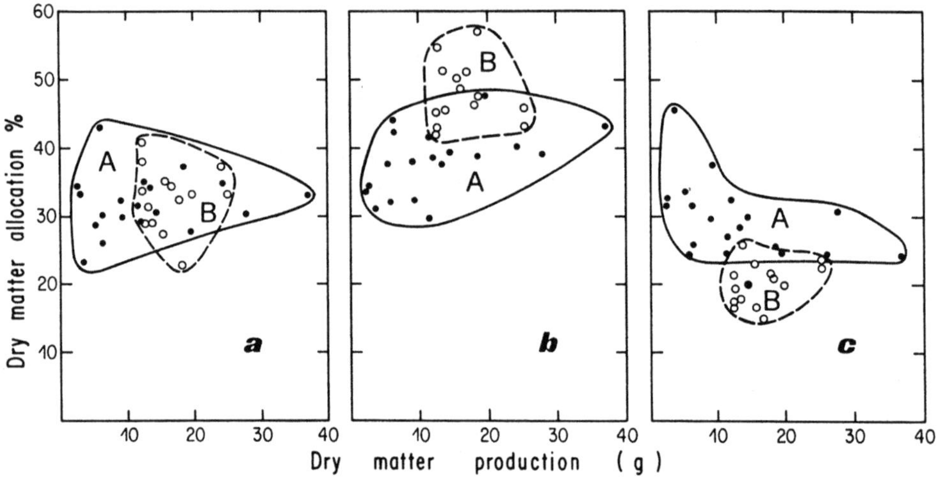

Fig. 9 a–c. Partitioning of the dry matter harvested from aerial parts at flowering between leaves (**a**), stem (**b**), and panicles (**c**) for 19 genotypes of *Dactylis glomerata* from a population abandoned by man (*A*), and 14 genotypes from a population modified by man (*B*)

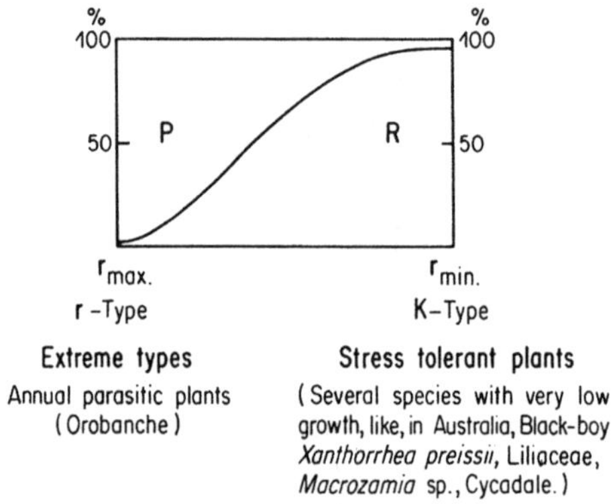

Fig. 10. Variation in the relative values of the two sub-sets of one population: the sub-set of potential individuals (*P*) and the sub-set of realized individuals (*R*). (Jacquard 1980)

crease of reproductive effort – sometimes substantial – in some promising families during domestication.

The conventional model for r– and K– strategies is being replaced by a view of a continuum of r– strategies in which the variation affects the ratio between realized individuals and potential individuals considering the global population to be constituted by the existing individuals and the viable seeds (the "bank") at any given time (Fig. 10). Each species, each population of a species, and each individual in a population is located on a portion of this continuum. At the level of the individual of an iteroparous species, an individual is not necessarily defined by a fixed position but has a fluctuating value of r (tactic), according to its interactions with the immediate environment. Utilizing this concept in its extreme, the several modules of the same individual would express contrasting tactics (flowering or clonal mul-

Table 5. Number of organs/individuals and allocation of resources for several genotypes of two populations of *Arrhenatherum elatius* with contrasting strategies. (Azocar 1978)

Populations		Number of tillers[a] (T)	Number of panicles[a] (T)	P/T	Allocation of resources[b]
Puech	P_1	105	115	0.52	0.50
	P_2	97	149	0.78	0.31
	P_3	157	120	0.49	0.53
	P_4	83	87	0.55	0.39
	P	111	118	0.59	0.43
Crau	C_1	79	66	0.39	0.16
	C_2	94	171	0.82	0.59
	C_3	104	47	0.22	0.53
	C_4	80	43	0.26	0.19
	C	89	82	0.42	0.37

[a] In % of the general mean (genotypes and populations combined)
[b] Ratio of the dry weight of the reproductive part to the dry weight of the total above ground biomass

tiplication), as a consequence of their hierarchical position in morphogenesis, or from year to year. This demographic aspect of plant strategy is illustrated in Table 5 which shows inter- and intrapopulation variability.

The modification of the environment by man can induce evolution in alternative directions; increasing the number of realized individuals or the number of potential individuals, corresponding to an increase in perenniality or annuality.

In a common environment, species belonging to the same layer of vegetation and growth form and sharing comparable "ecological niches" and neighborhood relations, can evolve very contrasting strategies (Table 6). Among the perennial grasses originating from the garrigue, large variations occur (Abouzakhem 1975), when the species are grown without competition in an experimental field, from species which are predominantly seed dispersers to those which have a predominance of clonal growth. Biomass production of spaced individuals varies by a factor of seven. The percentage of that biomass invested in the sexual reproduction reaches 60% in *Lolium perenne*, a species represented by a population with colonization through seeds, against 15% in *Phalaris tuberosa*, of which the growth is to a large extent accomplished by vegetative regeneration. The numbers of seeds per individual varies from more than 20,000 to little less than 400 (dormancy explains the lack of germination in *Oryzopsis paradoxa*).

In the context of the model presented above, *Festuca rubra* is a species with r_{max} and *Brachypodium phoenicoïdes* a species with r_{min}. There is a low relationship between the seed "rain" and the number of established seedlings/individual. The maximum value of this demographic parameter is about ten (for *Arrhenatherum elatius*, the probability of a seed producing a seedling is of 4×10^{-3} and only of 8×10^{-6} for *Festuca rubra*). These values are of the same order as the estimates for some British grasslands where 5×10^{-2} seedlings/individual are found for *Dactylis glomerata*.

Table 6. Adaptive strategies for colonization of several grasses of the "garrigue"

Species	Mechanism of colonisation G=seed C=tiller CC=predominance of tillering	Weight of a spaced plant (dm, g)	Weight of the inflorescences of an individual (dm, g)	Percentage of resources allocated to sexual reproduction	Number of seeds with ability to germinate (in greenhouse) per individual	Number of established seedlings/individual
Bromus erectus	G	141	24	17	4,835	9.1
Brachypodium phoenicoides	CC	138	29	21	374	0
Oryzopsis paradoxa	G	130	39	30	0	0
Phalaris tuberosa	G	100	15	15	1,224	0.7
Festuca arundinacea (Control Cultivar)	G	89	31	35	7,015	2.2
Festuca rubra	CC	64	39	61	23,801	0.2
Arrhenatherum elatius	G	43	14	33	2,672	10.6
Dactylis glomerata forme hispanica	G	43	7	16	988	0
Lolium perenne	G	42	26	62	13,966	1.9
Festuca glauca	G	25	7	28	2,109	0.8

In conclusion, in assessing the impact of Man on the evolution of population characteristics of plants, comparisons should be made of performances of populations of the same species as well as of individuals within a population. Further an analysis should be made among populations of species of the same growth form within a vegetation.

References

Abouzakhem A (1975) Etude de la croissance de quelques espèces de graminées pérennes de la garrigue méditerranéenne française (production fourragère et facteurs écologiques). Thése doct Ing, Univ Paul Sabatier, Toulouse, p 166

Azocar AI (1978) Biométrie des relations entre deux populations *d'Arrhenatherum elatius* (L.) Beauv. ex J. et C. Presl. soumises à deux milieux contrastés. These 3e Cycle, USTL, Montpellier, p 102

Barbault R (1976) La notion de stratégie démographique en écologie. Bull Ecol 7:373–390

Bergh JP van den, Braakhekke WG (1978) Coexistence of plant species by niche differentiation. In: Freysen AHJ, Woldendorp JW (eds) Structure and functioning of plant populations. Elsevier, North-Holland, Amsterdam New York, pp 125–138

Bremond B (1979) Etude de la croissance de dactyles de trois provenances sous un régime hydrique contrôlé. Rapp DEA Ecol, USTL, Montpellier, p 36

Gadgil M, Solbrig OT (1972) The concept of r- and K-selection: Evidence from wild flowers and some theoretical considerations. Am Nat 106:14–31

Gallais A (1970) Modèle pour l'étude des relations d'association binaire. Biométr Praxim 11:51–80

Grime JP (1978) Interpretation of small-scale patterns in the distribution of plant species in space and time. In: Freysen AHJ, Woldendorp JW (eds) Structure and functioning of plant populations. Elsevier, North-Holland, Amsterdam New York, pp 101–124

Grime JP (1979) Plant strategies and vegetation processes. Wiley, Chichester, p 222

Harper JL (1977) Population biology of plants. Academic Press, London New York, p 892
Jacquard P (1980) Stratégies adaptatives chez les végétaux: aspects démographiques et niveaux d'étude (organe, individu, population, espéce). In: Barbault R, Blandin P, Meyer JA (eds) Recherches d'écologie théorique: Les stratégies adaptatives. Maloine, Paris, pp 159–191
Jacquard P, Rotili P, Zannone L (1978) Les interactions génotype x milieu biologique: analyse diallèle des aptitudes à l'association entre populations de trèfle violet. Ann Amélior Plant 38:309–325
Mahmoud A, Grime JP, Fourness SB (1975) Polymorphism in *Arrhenatherrum elatius* L. Beauv. ex. J. et C. Presl. New Phytol 75:269–276
Oka HI (1977) The ancestors of cultivated rice and their evolution. In: Iyama S, Morishima H (eds) The ancestors of cultivated rice and their evolution. Natl Inst Genet, Mishima, pp 1–9
Wit de CT (1970) On the modelling of competitive phenomena. Proc Adv Study Inst Dynamics Numb Pop, Oosterbeck, pp 269–281

5.3 Genetic Characteristics of Populations

S. Jain

5.3.1 Introduction

Man's influence on plant populations is often discussed in terms of habitat destruction (leading to species extinction, greater stress, habitat subdivision), pollution (greater stress, novel environments), increased dispersal (resulting in greater intermixing of gene pools, plant hybridization and even origin of new taxa), and in relation to the species replacements in agriculture and forestry. All of these changes in ecosystems involve evolutionary processes; population genetics and ecology provide tools for studying these processes (both short and long term) and possibly for making some predictions about the fate of populations. Plant species invading or living in temporary habitats of roadsides, fields, sites of secondary succession, etc., are often treated collectively as colonizers, weeds, fugitives, or r-strategists. Literature on their population ecology provides a wealth of information on the life history components of survivorship and reproduction (e.g., see Harper 1977). For example, several recent discussions of secondary succession (Noble and Slatyer 1980; Peet and Christiansen 1980) have emphasized the need for understanding the population dynamics of species in terms of their dispersal, longevity of seed banks, and the life history characteristics of growth, reproduction and competitive success (see Bazzaz's paper in this symposium). Likewise, population genetic studies of variation, natural selection, migration, random drift and overall comparisons of the adaptive versatility of open versus closed recombination systems, have often involved colonizing annuals (viz. members of *Avena*, *Trifolium*, *Poa*, *Bromus*, and *Xanthium*).

In a recent review of some evolutionary characteristics of colonizers, Brown and Marshall (1981) defined colonizing plant species as "those which can establish populations in an area or habitat different from their origin," and noted the following common features "(1) fixed heterozygosity through polyploidy, (2) propagation by self-fertilization or asexual means, which allows single propagules to establish colonies, and adapted genotypes to multiply, (3) genetically depauperate populations, when colonies stem from one or a few seeds following long distance dispersal, (4) multilocus associations providing a reduced number of successful genotypes, (5) substantial population differentiation due to founder effects, and to variable environments, and (6) greater phenotypic plasticity, which permits genetically uniform populations to adapt to a wide range of environments." Then, they proceeded to test these deductive ideas with evidence from a critical survey of literature and their own studies. Likewise, in this paper, I shall briefly review a few population studies on colonizing species in order to analyze evolutionary issues such as these: *Which parameters describe the genetic structure and demography of plant pop-*

ulations living in disturbed or "successional" habitats? Do successful colonizers represent certain specific genetic or ecological strategies? Can we develop predictive theories of their adaptive changes and their relatives?

Baker and Stebbins' (1965) volume provided a good treatment of colonizing species as well as a synthesis of population genetic and ecological points of view. Earlier writings (e.g., Salisbury 1942; Dansereau 1957; Cain 1944) treated colonizers under categories of species that (1) invade disturbed habitats, (2) are expanding range of distribution, (3) are aggressive but not necessarily abundant, or (4) are getting to islands first. Often these categories are not easily separated unless the history and patterns of colonization are clearly described in terms of habitat parameters and appropriate time scales. We shall review population studies in grassland annuals and perennials as well as briefly for a number of potential crop-weed hybrid complexes, all occurring in habitats which are highly disturbed by grazing or urbanization activities. This review will emphasize the population genetic aspects of colonizing plant species in an attempt to describe their evolutionary strategies.

5.3.2 Population Studies in Avena spp.

Two wild oat species (*Avena fatua* L. and *A. barbata* Brot.) have become successful colonizers of the grasslands in California within the past 200 years. Their comparative ecology and genetics as well as certain features of microevolution were reviewed by Jain (1975). Population studies in *Avena barbata* have led to several findings, namely: (1) a pattern of regional differentiation in central California in which a large region in the Central Valley showed virtually no genetic polymorphism (Marshall and Jain 1968), (2) a highly patchy distribution of polymorphism within the Bay and North Coast regions (Rai 1972), and (3) a highly localized cline along xeric-mesic gradients (Hamrick and Allard 1972). In general, the variation patterns for morphological marker loci (lemma color, leaf sheath hairiness) were highly concordant with those derived from several allozyme loci (Clegg and Allard 1972). A summary of these studies (Table 1) emphasizes the microevolutionary processes that have been postulated but not yet explicitly demonstrated. Rai (1972), for instance, analyzed four geographic regions in relation to some of the temperature and precipitation variables of climate. Hamrick and Holden (1979) invoked selection in their analysis of clinal variation through a detailed study of linkage disequilibrium and correlations between the gametic frequencies and arbitrary microhabitat scores. Random drift, essentially based on founder effects, requires a study of population numbers, stability of patches, recolonization pattern, and estimates of gene flow rates. The experimental phase has now begun for testing the various hypotheses on the relative importance of selection, migration and drift.

Artificially founded colonies of *Avena barbata* were utilized in two experiments in order to examine the role of natural selection in the evolution of regional differentiation in central California (Jain and Rai 1980). One experiment involved a total of 41 colonies founded in three different areas and scored over a 10–year period; these colonies, started with known genotypes, showed that although both Valley and Bay region genotypes establish successful colonies, their relative survivorship

Table 1. A summary of variation studies in *Avena barbata*

Scale of variation	Nature of evidence	Evolutionary hypotheses	Reference
1. Regional (Valley, Bay, North Coast and South Coast)	Morphological loci: genotypic frequencies show distinct regional differentiation	Climatic variables as selective factors; role of polymorphism in Bay region unknown	Jain (1969), Rai (1972)
	Allozyme loci: genotypic frequencies and linkage disequilibria	Epistatic selection	Clegg and Allard (1972)
2. Interpopulation (within Bay and North Coast regions)	Differentiation on a local scale (100–1,000 m); genotypic frequencies show highly mosaic distributions	Habitat factors, e.g. slope, aspect, mesicness, not obviously correlated with selective forces; drift and low dispersal are important locally	Rai (1972)
3. Sampling sites within local populations			
a) Calistoga cline	Allozyme frequencies associated with xeric and mesic patches at a selected site	Selection and low rates of gene flow	Hamrick and Holden (1979)
b) Absence of clines but highly patchy variation	Morphological marker loci and quantitative variation commonly show no clinal variation	Selection-drift balance	Rai (1972)
c) Local patches over 1–10 m² scale	Small effective neighborhoods and patchy gene frequency distributions	Gene flow restricted and local colonization-extinction	Jain and Rai (1981)

and fecundity suggest weak selective forces consistent with those predicted from the observed patterns in natural populations. The second experiment involved two localities, representing two climatic regimes of temperature, in each of which ten colonies were started from identical seed sources. These colonies also showed relatively higher fitness of the genotypes matching with those sampled from the Valley and Coastal regions, respectively, again in line with their regional pattern of distribution. Artifically founded colonies appear to be useful in gathering evidence for potential rapid evolutionary changes in colonizers.

The variation patterns in *A. barbata* on the regional versus highly localized scales apparently represent a wide range of outcomes of the interacting processes of microevolution. In a recent study (Jain et al. 1981), the so-called polymorphic region in the San Francisco Bay and adjacent coastal areas were further surveyed in terms of large stands in the intermontane coastal grasslands and the isolated roadside colonies along a coastal highway. Most of these isolated stands showed lowered or no genetic variation in terms of per cent polymorphic loci (PLP) and an index of allelic diversity (H′) (Table 2). Significant effects of size bottlenecks on allelic frequencies, level of polymorphism, multilocus associations, and greater interpopulation differentiation (D'_{ST}) were attributed to random drift during the colonizing process. Data on genetic polymorphism, scored at two morphological loci,

Table 2. Summary of genetic variation parameters estimated from allozyme polymorphisms in *A. barbata* populations

Habitat type	No. of sites	PLP[a]		H[a]		$D_{ST/H}$[b]
		$\bar{X}$	Range	$\bar{X}$	Range	
Grassland	22	29.8	11–48	1.85	0.82–2.14	16
Roadside isolates	14	12.5	0–26	0.73	0–1.40	37

[a] PLP = percent polymorphic loci, and H, is the mean within-subpopulation diversity index defined as $H_j = 1 - \sum p_{ij}^2$, where p_{ij} are allelic frequencies at i^{th} locus and in j^{th} population, measure the relative amount of variation within populations

[b] $D_{ST/H}$ is a measure of between-subpopulation differentiation based on genetic distances (see Brown 1979)

showed geographically isolated roadside colonies to be significantly less polymorphic than the large central populations in continuous stands. The role of random drift (founder effect) was evident in the genetic structure of such roadside colonies which were, however, not monomorphic. Multilocus associations also resulted from the so-called Hill-Robertson effect in which certain gene combinations become frequent due to random drift in small populations. Such isolates with varying amounts of elapsed time since the founder events can lead to new ecological races, or simply nonadaptive geographical differentiation among isolates.

More recently (Rai and Jain 1982) the pollen and seed dispersal patterns were analyzed in both natural and experimental populations of *Avena barbata*. Localized estimates of gene flow rates and plant densities gave estimates of neighborhood size in the range of 40 to 400 plants; the estimates of mean rates and distances of gene flow seemed to vary widely due to wind direction, rodent activity, microsite heterogeneity, etc. The relative sizes of neighborhoods in several populations were correlated with the patchy distribution of different genotypes (scored for lemma color and leaf sheath hairiness) within short distances, but patch sizes had a wide range among different sites. Highly localized gene flow patterns seemed to account for the observed pattern of highly patchy variation. However, many experimental and statistical variables are involved in the evolution of clines versus patchiness (Fig. 1). Note that long term observations on gene frequency changes and a demographic study of colonies are needed to begin any analysis of numerous processes. Further research on both natural and founded colonies will be needed in order to understand the genetics and demography of colonizing success of *Avena* spp. Moreover, variation in *Avena* species and ecotypes should now be studied with the tools of physiological ecology such that some of the genetic and demographic components of selection are understood in terms of relative carbon gain, resource allocation, nutrient and water use efficiency, and responses to stress.

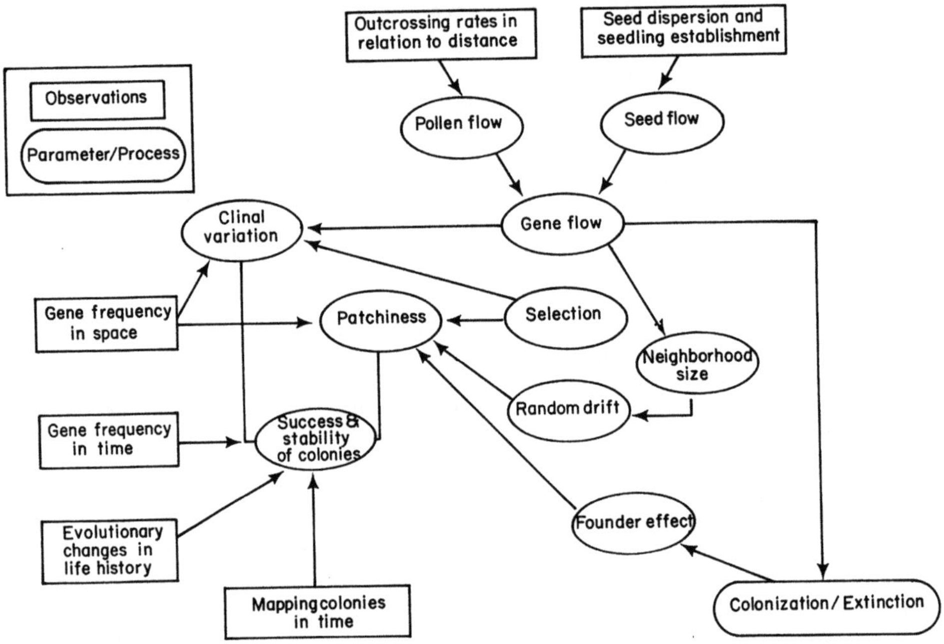

Fig. 1. Chart showing the items (in *rectangles*) derived from observations of pollen and seed movement and of gene frequencies, life history components, etc. in marked colonies (or neighborhoods) and the evolutionary parameters or processes (in *ellipses*) that are to be described based on observations. The point to note is that studies of a rather large number of parameters require long-term observations to provide adequate "degrees of freedom" for both estimation and hypothesis-testing statistics

5.3.3 Rose Clover, a Case History of Recent Colonization

Rose clover (*Trifolium hirtum* All.), a native annual forage legume of the Mediterranean region and introduced into California during 1944–1946 through extensive plantings, was recently observed to be colonizing certain roadside areas in several counties (e.g., Placer, Nevada, El Dorado). Some of the oldest plantings in Shasta, Madera and Glenn counties were extensive on the range but apparently not successful in their expansion into the adjacent roadside areas. These observations raised the following issues. How much genetic differentiation has occurred among the populations of different regions? In what ways are successful roadside colonies different from the neighboring range populations? Can we ascertain common colonizing characteristics from the comparisons of demographic and genetic variables in range versus roadside populations?

Thirteen populations sampled from among the oldest plantings and 12 roadside populations established through natural colonization were compared for their genetic and demographic features (Jain and Martins 1979). Roadside colonies showed a greater amount of reproductive effort in terms of a larger number of heads per plant and larger calyx, lower rate of seed carryover, lower seedling survivorship, and earlier flowering (Tables 3,4). The calyx was more hirsute in roadside collections and remained attached with the seed, a feature accounting for increased germination probabilities on the soil surface or in litter along the roadsides. Outcrossing rates were slightly higher in roadside colonies in which genetic poly-

Table 3. Comparisons for some genetic and demographic variables of roadside and range populations of *Trifolium hirtum*

Site location	Outcrossing rate		Genetic diversity[a]		Plant density[b]		Seed carryover		Seedling survivorship	
	Mean	SE	Mean	SE	Mean	SE	Mean	SE	Mean	SE
	(percent)				(No. of plants m^{-2})		(percent)		(percent)	
Range	3.8	0.6	1.17	0.031	9.82	0.501	56.0	4.6	60.3	4.3
Roadside	5.1	0.8	1.36	0.077	6.36	0.405	41.0	2.6	46.7	3.4
Mann-Whitney test of significance	ns[c]		ns[c]		$P<0.05$		$P<0.05$		$P<0.05$	

[a] Genetic diversity was measured by $H' = -\sum p_{ij} \ln p_{ij}$, where p_{ij} are allelic frequencies at i^{th} locus in j^{th} population

[b] Average over four years. Variance in plant density over years was large in both range and roadside populations (C.V.'s were 72.2 and 78.6, respectively)

[c] ns = not significant ($P \geqq 0.05$)

Table 4. Comparisons for three quantitative traits ($\bar{X}$ = mean and CV_w = coefficient of genetic variation among families) of range and roadside populations of *Trifolium hirtum*

Location	Seed size index				Flowering time code[a]				No. of heads/plant			
	$\bar{X} \pm SE$		$CV_w \pm SE$		$\bar{X} \pm SE$		$CV_w \pm SE$		$\bar{X} \pm SE$		$CV_w \pm SE$	
	(wt of 100 seed, g)											
Range	0.255	0.021	17.6	1.51	2.91	0.076	22.4	1.30	19.8	0.81	31.0	1.58
Roadside	0.243	0.009	18.6	1.07	2.66	0.074	19.8	0.94	24.2	0.80	27.3	1.58
Mann-Whitney test of significance	ns[b]		ns[b]		$P<0.05$		ns[b]		$P<0.05$		ns[b]	

[a] Flowering time, i.e., the number of days to the time of first bud opening, was scored on a scale of 1 = early, and 5 = late

[b] ns = not significant ($P \geqq 0.05$)

morphisms at three marker loci represented as high levels of genetic variability as in the range populations. The colonizing success of rose clover seemed to be determined largely by a few, rapid morphological changes and by the retention of some outbreeding characteristics and genetic variability.

Jain and Martins (1979) argued that certain dogma about the role of founder effects during colonization and the reduction in genetic variation (cf. Nei et al. 1975) need to be tested. For example, the clover example and one in *Lupinus* reported by Harding and Mannkinen (1972) did not show the colonization in predominantly selfing species to involve an increase in monomorphism (unlike *Avena bar-*

Table 5. Summary of some life history comparisons among three perennial grass species

Variable	Comparisons among species (rank orders)
Amount of genetic variation	De > An > Ho
Seed output/year	An > Ho > De
Reproductive effort: roadside	Ho > An > De
range	An > Ho > De
Vegetative Propagation	De > An > Ho
Longevity	De > An > Ho
Proportion of seed carried in the soil	An > Ho > De
Seedling survivorship	Ho > An > De
Colonizing ability	Varies in different situations (e.g. *Ho* succeeds in open, disturbed areas; *De* succeeds in relatively undisturbed areas and replaces *Ho* or *An*)

De = *Deschampsia holciformis*
An = *Anthoxanthum odoratum*
Ho = *Holcus lanatus*

bata discussed earlier) or the postulated role of increased autogamy in fixing certain unique gene combinations. The question whether inbreeding or outbreeding, higher or lower recombination rates, and plasticity or genetic polymorphisms contribute to colonizing success, will be discussed in a later section.

5.3.4 Population Dynamics of Species in a Coastal Grassland Ecosystem

Populations of six grass species living together in a coastal grassland were studied in order to describe their life histories, relative competitive ability, and colonizing success (Jain, unpubl. data). Several experimental approaches were used, including marked plots, transplanting to provide cohorts of single species versus mixed two-species stands, genetic analyses of variation within and among populations, and monitoring of species composition changes in natural stands. Longevity, competitive ability or colonizing success were not directly correlated with the patterns of early seedling survivorship. A highly plastic response in reproductive amounts and timing was observed (Table 5). Genetic variation, assayed by allozyme surveys as well as quantitative genetic study of between and within half-sib families showed two annuals, *Bromus mollis* and *Avena barbata*, to be highly variable. The three perennial grasses rank-ordered as follows for genetic variability: *Deschampsia holciformis* > *Anthoxanthum odoratum* > *Holcus lanatus*. This was unexpected since their longevity (and asexuality) ranked in the opposite order. Dispersal rates and colonizing ability were significantly higher in *Holcus* than *Anthoxanthum* and *Deschampsia;* reproductive effort through seed was highly plastic but a comparison of the roadside (colonizing) vs. rangeland (noncolonizing) stands showed *Anthoxanthum* and *Holcus* to be more of r-strategists than *Deschampsia*. Marked *Anthoxanthum* stands showed aggressive colonization (approx. 50% increase per year) and high tolerance to intraspecies competition. In general, the

Table 6. Examples of adaptive differentiation in weedy plant species. (After Oka and Morishima 1980)

Species	Life form	Habitat variables	Major differences[a]
Andropogon scoparius	P	2- vs. 40-year-old fields	Reproductive effort, heading time, seed size
Anthoxanthum odoratum	P	Fertilizer application and liming	Growth rate, mildew resistance
Anthoxanthum odoratum	P	Marginal vs. central sites	Seed size, negatively skewed age distribution
Anthoxanthum odoratum	P	Roadside vs. pasture	Morphological variation, heading time
Alopecurus pratensis	A	Lowland vs. upland field	Seed size, seed dormancy, photoperiod sensitivity
Poa annua	A-P	Grade of disturbance	Heading time, morphological variation, mortality
Amphicarpum purshii	A	Grade of disturbance	Resource allocation pattern
Trifolium hirtum	A	Roadside vs. pasture	Reproductive effort, flowering time, mortality, outcrossing rate, polymorphism

P = Perennial; A = Annual

[a] Under uniform garden comparisons, populations from roadside, marginal, disturbed or lowland sites generally showed early reproduction, smaller seed size, larger reproductive effort, less seed dormancy, lower photoperiod sensitivity, etc. (cf. Baker and Stebbins 1965 for similar observations)

shortlived perennials, *Holcus lanatus* and *Anthoxanthum odoratum*, seemed to be favored by higher dispersability and early reproduction in disturbed or stressed areas; however, in many roadside openings and permanent grassland sites, *Deschampsia holciformis* appeared to succeed over a three-year period.

Oka and Morishima (1980) reviewed a number of examples of significant evolutionary changes under disturbed or grazed environments; the examples (Table 6) include a wide range of characters related primarily to the so-called r-strategy to be discussed below.

5.3.5 Variation and Colonization Success of Crop-Weed Hybrids

The genetics and ecology of the so-called crop-weed complexes are of interest to a plant evolutionist. Harlan (1975) summarized the dynamic and diverse nature of crop-weed interactions in relation to habitat disturbances, gene flow, adaptive mimicry, and various historical processes (also see Heiser 1973). A recent and highly dynamic example of crop-weed hybridization giving origin to a successful colonizer is weedy rye in Northern California. Suneson et al. (1969) reported on the extent of phenotypic variability in rather extensive collections from California's northeastern counties (Lassen, Modoc, Plumas, and Siskiyou). Based on historical and genetic facts, they concluded that weedy rye "probably developed from an introduced interspecific hybrid subjected to recurrent introgression from the many cultivars of rye." The ancestor might be Michel's grass which was introduced in

1938 into the Fall River area and looked like the hybrid *S. (Secale) cereale* L. x *S. montanum* Guss. which showed significantly high sterility due to the presence of chromosomal translocations. Perenniality, fragile rachis, winter habit, and longer awns appear to have come from the wild parent, *S. montanum*.

Recently, eight populations of weedy rye from the Cascade Range counties in northern California were scored for characters describing the wild and cultivated gene complexes (Jain 1977; Jain et al. unpubl. data). The relative frequencies of individuals with nonfragile rachis, short awn, blue aleurone, and bronze lemma were used as a measure of gene flow by introgression from cultivars into the original hybrids giving rise to high levels of intrapopulation and geographical variation in weedy rye. The characters representing the wild and cultivated forms do not appear to be associated into distinct gene complexes, and the weedy rye populations have a varied mixture of them. In some areas there might be a continuing genetic shift in weedy rye toward the cultivated forms.

Another example of recently evolved roadside colonizer is weedy sunflower, *Helianthus bolanderi*, in California. Olivieri and Jain (1977) reported on the morphological and allozyme variation in *Helianthus bolanderi-exilis* complex and although no clearcut evidence for introgression was found, variation in the weedy sunflowers was large and showed almost no unique gene combinations. Nevo et al. (1979), on the other hand, reported numerous unique alleles and local races in *Hordeum spontaneum*, a wild barley species, with probable gene exchanges between the wild and cultivated species.

Finally, population studies of variation in a large array of colonizers are summarized in Table 7 (after Brown and Marshall 1981). In general, both PLP (% polymorphic loci), n_α (number of alleles per locus) and mean heterozygosity per locus as measures of genetic variation showed a majority of colonizers to be less variable than noncolonizers (see Brown 1979 and Hamrick et al. 1979 for detailed reviews). It is interesting to note, however, that shifts in mating system occurred in both directions, viz. toward increased selfing as well as outbreeding so that increased variation and genetic recombination should not be ruled out as an evolutionary features of colonizing species.

5.3.6 Alternative Strategies of Colonizing Success

Some theoretical aspects of dispersal, initial establishment, values of r and K in the logistic growth curve, and other fitness components were discussed by MacArthur and Wilson (1967) in relation to the probability of founding colonies and relative duration of colonies. Other theoretical developments include the work of Schaffer and Gadgil (1975), Levin (1975), Charlesworth (1980), among others. A comparative study of six grassland annuals showed that different colonizing strategies are involved in each case (Foin and Jain 1977). For example, rose clover has low rates of seed dispersal but occasional long distance dispersal with the aid of grazing animal in pastures. Man-made plantings allow new colonies to be founded which are capable of surviving over long periods of time through high seed output per plant and long-term seed storage in the soil. However, wild oats and bromes

Table 7. Colonizing species[a] populations depauperate in allozyme variation, (After Brown and Marshall 1981)

Species	No. of Loci	No. of popu-lations	PLP[b]	n_a[b]	Diversity (H′)	Het[b]
Xanthium strumarium L. (within 4 subspecies)	13	12	0.04	1.06	0.006	0.001
X. spinosum L.	8	3	0	1.0	0	–
Emex spinosa Campd.	15	5	0	1.0	0	–
Avena barbata, Brot (Valley Region, I)	5	9	0.02	1.02	0.001	0
Lycopersicon pimpinellifolium (marginal)	11	16	0.02	1.29	0.078	0.016
Typha latifolia L.	~20	74	0	1.0	0	–
T. domingensis Pers.	~20	52	0	1.0	0	–
Chenopodium fremontii S. Watson	6	40	0.05	1.05	0.018	0.002
C. atrovirens Rydberg	12	5	0.02	1.02	0.003	0
C. desiccatum A. Nelson	12	6	0.01	1.01	0.003	0
C. pratericola Rydberg	12	5	0.05	1.05	0.017	0
C. hians Standley	12	4	4.08	1.08	0.019	0
C. incognitum Wahl	12	4	0.08	1.08	0.019	0
C. leptophyllum Nutt. ex Moq.	12	7	0.08	1.08	0.030	0

[a] All examples are either known or presumed to be self-pollinated

[b] PLP = Proportion of loci polymorphic per population, n_a = number of alleles per locus, H′ = genetic diversity index defined as $-\sum\sum p_i \ln p_{ij}$, where p_{ij} are allelic frequencies, Het = mean heterozygosity per individual for all scored loci

differ in their dispersal and reproductive patterns from rose clover. Seed dormancy is relatively unimportant in the grasses but plastic response to environments allows their germination to occur under a wide range of conditions. Seedling survival is also higher in the grasses and in some situations, these species quickly build up large stands. As noted above, three coastal grassland perennials (*Holcus lanatus*, *Anthoxanthum odoratum*, and *Deschampsia holciformis*) also provided a lack of expected correlations among the variables of genetic variation, longevity and colonizing features.

Certain comparisons of colonizing features are suggested by recent discussions of the evolution of sex. Asexual reproduction is claimed to be favored through an intrinsically more efficient reproduction and is evolutionarily favored in stable environments. Accordingly, sexuality is favored when colonization of new environments requires dispersal and strong selection for unique gene combinations. Williams (1975) reviewed several population studies to illustrate this model. Maynard Smith (1978) developed a model of sib competition in relation to the roles of sex and genetic variation in the patchy survival of a colonist. A model for inbreeding versus outbreeding in *Leavenworthia* suggested the evolution of optimal strategy as a compromise between the costs of pollination mechanisms and advantages of genetic variation generated by outbreeding (Solbrig 1980). A review of evolutionary ideas on inbreeding in plants, on the other hand, brought out a number of alternative hypotheses none of which could be unconditionally supported by the existing examples (Jain 1976).

We may ask whether colonizing species evolve optimal strategies. Perhaps we need to define the term "strategy." Strategies evolve as "character combinations" which Grime (1980) succinctly defined as "groupings of similar or analogous genetic characteristics which recur widely among species or populations and cause them to exhibit similarities in ecology." Under heterogeneous environments, adaptive response by a population could be described in terms of phenotypic variances and covariances for a set of relevant traits. Another way of defining optimal strategies involves the study of polymorphisms, quantitative genetic variation, plasticity, and homeostasis. Strategies, or what might be better termed tactics, rely upon genetic and phenotypic capacities to adapt to varying environments which may or may not be reflected in the demographic analyses of the survivorship and reproductive success. Accordingly, discussions of adaptive strategies vary in emphasis on genetic, physiological or demographic analyses.

5.3.7 Evolutionary Genetics of Adaptive Responses

As outlined by Levins (1968), Harper (1977), and others, there are several outcomes of variation in populations living in heterogeneous environments: (1) genotypes may produce the same phenotype (canalization), or (2) produce different phenotypes (phenotypic plasticity); (3) distinct developmental classes (phases) appear in response to some threshold factor, or (4) natural selection favors different coexisting genotypes (genetic polymorphism). Merely describing the amount of total phenotypic variation or genotypic variation does not allow one to explain the adaptive role of such variation. It is intuitively easy to see how a population of genotypes might respond to varying environments through polymorphism-based gene frequency changes (Hedrick et al. 1976), through plasticity of certain genotypes (Bradshaw 1965) or based on homeostasis of heterozygous genotypes (Lerner 1954). A heritability study of plasticity in *Bromus mollis* involving the use of parent-progeny regression estimates showed that families of highly plastic individuals grown under similar environments were also plastic (Jain 1978). Evolution of plasticity, like any other trait under selection, could then proceed by changes in the genotypic proportions in successive generations. Developmental and physiological studies of phenotypic plasticity are needed as part of evolutionary studies under carefully monitored and heterogeneous environmental conditions.

Often comparisons of life histories have shown an increased reproductive effort in the colonizing species. The so-called r- and K-strategies are primarily defined in terms of greater reproductive allocation in the shortlived colonizing species, but the uncritical use of these terms and inadequate evidence have generated confusion. Regardless of an exact terminology, we need to know about the genetics and evolutionary potential of reproductive rates. In order to illustrate a quantitative genetic model of such an evolutionary change, an experiment was undertaken in two populations of *Medicago polymorpha*, a common annual weed of the Mediterranean region (Jain and Stebbins unpublished). Four components of total seed output per plant, namely, number of branches, number of pods/branch, number of seeds per pod, and seed weight, were studied for the heritable component of variation and the correlations among various components. Selection to increase seed output through selection for seed number, and in another parallel study through number of seeds per pod, showed that response could not be predicted from the phenotypic variances and only partially from the heritability estimates. Selection to modify various fitness traits in many cultivated plant species also illustrate numerous genetic constraints of evolution, such as linkage, gene interactions and pleiotropy. In a recent simulation study of life history evolution, Ritland and Jain (unpublished) discovered rather complex interactions among the means and variances of life history traits, the pattern of environmental fluctuations and initial conditions. Thus, even though genetic and phenotypic variation for individual life history characteristics might be readily available within and among populations, evolutionary changes may not be predictable for their direction, rate, or strategic outcomes. Evolutionary ecologists should, however, explicitly incorporate genetic analyses involving selection response and estimation of genetic analyses involving selection response and estimation of genetic variances and character correlations in their strategy-evolution models. Both allelic and multilocus (and therefore, multitrait) components of variation depend on the variables of genetic recombination.

5.3.8 Recombination Properties of Genetic Systems

Genetic systems are often discussed in relation to the regulation of recombination through such variables as sexual versus asexual reproduction, cross- versus self-fertilization, levels of ploidy, and other karyotypic variations. Their evolution toward certain optimal levels of recombination is then discussed in relation to the following largely unproven statements: (1) "open" genetic systems with higher rates of outbreeding, genetic recombination, etc. allow populations to carry a large amount of genetic variation and heterozygosity which are adaptively useful; (2) since genetic variation also imposes "genetic load" in the sense of not allowing the most fit genotype to be fixed, conflict between immediate fitness and long-term flexibility require genetic systems to maintain certain optimal rates of outbreeding, sex ratio, etc.; (3) most species represent compromises in terms of several variables in genetic systems. Evolution of genetic systems has been explicitly studied in relation to the models of heterostyly, sex ratios, male sterility, etc. Whether sexuality (which is uncritically equated to genetic recombination) evolved in relation to the colonization of highly temporally varying environments or spatial heterogeneity, or as a hitchhiking effect between linked favorable and neutral genes, can only be resolved by increased information on the occurrence and behavior of genes affecting variables of genetic system (modifiers?), selection pressures, and appropriate demographic data (Maynard Smith 1978; Lloyd 1980).

Due to the parallel development of interest in the evolution of breeding systems and in the life history strategies in plants, a search for correlations between colonizing ability and uniparental reproduction (selfing, apomixis) has been a recurring theme in evolutionary ecology. There are at least two different lines of arguments put forth in writings on this subject. Genetic recombination and the opportunities for adaptive genetic changes by natural selection are emphasized on the one hand, and the relative costs and benefits of various life history and breeding system tactics in terms of the reproductive efficiency are emphasized on the other.

For example, several writers have argued that after an initial success in colonization, inbreeding or agamospermy allows the rapid fixation of one or a few successful genotypes (Stebbins 1957, 1958), while Grant (1967) stated that weeds are associated with diverse and in most cases flexible genetic systems ... "successful weeds develop where a heterotic advantage may be fully utilized in conjunction with some means of effective dissemination" (also see Williams 1975; Maynard-Smith 1978). Weeds and colonizers of various sorts are often pooled together in these discussions. In the cost-benefit model one argues that under uniparental reproduction higher reproductive rates evolve or they are simply a preadaptive factor somehow associated with selfing or apomictic genotypes.

In order to test whether inbreeders are better colonizers, the Flora of the British Isles was surveyed and tabulated for the distributions of colonizers and noncolonizers in relation to breeding systems and longevity (Price and Jain 1981). No overwhelming evidence for association was found between longevity and colonizing ability although predominant selfing was found to be significantly more common among the colonizers. However, such evidence for associations among various genetic and ecological factors accounting for high colonizing ability must be interpreted with caution when the role of longevity, ploidy, interfamily and interpopulation genetic variation, uncertainties about the habitat classification, etc., are taken into account.

For example, in an elegant monograph of *Gaura*, Raven and Gregory (1972) noted that several weedy species are self-incompatible and include both diploids and tetraploids, but invoked Baker's rule to explain the widespread distribution of *G. parviflora*. Several other monographic surveys have likewise suggested Baker's rule (e.g., *Crepis*, Babcock 1947; *Clarkia*, Lewis and Lewis 1955; *Epilobium*, Raven and Raven 1976). Pandey (1980) examined in detail the evolution of different incompatibility systems and noted that an inherently complementary (multilocus) incompatibility can make colonization highly feasible through temporary shift to self-compatibility and reversal to self-incompatibility. As Pandey (1980) himself noted, this sort of exception proves Baker's rule, but it could make inferences from statistical surveys hazardous, and thus biology of individual examples must be adequately incorporated in such surveys.

5.3.9 Interspecies Interactions in Community Dynamics

Bazzaz (this symposium) has thoroughly reviewed several aspects of succession in relation to community structure and dynamics, viz. changes in species composition, niche evolution and physiology of competition. Here, we shall briefly explore the role of genetic variation in the outcome of such interspecies interactions.

Although Pimentel's (1964) classical work on genetic feedback demonstrated the role of genetic variation, many ecological writings on community structure have commented on the paucity of evidence for any "coadaptive properties" of species living in a community. Experimental tests as well as indirect measures of coadaptedness point to the need for a closer examination of character displacement, niche partitioning, and the performance of mixtures. There is a wealth of literature in plants, for instance, dealing with such variables as density and relative frequency of competitors, pattern of neighborhood, mortal versus plastic response and conditions for co-existence. That polymorphisms can regulate the outcome of interspecies competition was elegantly shown by Lerner and Dempster (1962). They used inbred strains of *Trifolium confusum* and *T. castaneum* to show that the so-called indeterminate outcome, reported by Park (1954, et seq.), resulted from using random mixtures of genotypes within each species. In a recent study (Yazdi-Samadi et al. 1978), three different series of population samples of two *Avena* species co-occurring in California were grown for an analysis of the role of genetic variation in interspecies competition: I, samples from mixed *fatua-barbata* sites in nature, grown in mixed stands; II, samples from pure sites and grown in pure stands, and III, the same sites as in II but grown in mixed stands. Four macroenvironments and four densities were used giving a total of sixteen entries for each genetic/competitive unit in order to measure both mean and variance of survival and reproductive rates as fitness characters. Sites used in each series included low versus high levels of genetic polymorphism within each species. In general, high polymorphism favored *A. fatua* in competition with monomorphic *A. barbata*, and high polymorphism in *A. barbata* allowed it to compete better with monomorphic *A. fatua*. This observation fits well into the pattern of reduced polymorphism in natural mixed stands. Mean performance of individuals in a polymorphic mixed stand was not consistently higher than the monomorphic combinations or pure stands but the greater relative stability over environments seemed to favor polymorphisms in one or both of the competitors. A relatively less regular pattern of density or competitive response in series III was interpreted as evidence for the lack of coadaptedness between samples drawn from pure sites. These laboratory studies provided several promising clues for more critical field studies.

Parallel to the work in *Avena*, Wu and Jain (1979) studied the effects of density on competition between individuals in pure and mixed populations of *B. mollis* and *B. rubens*. In both species, increasing density induced greater mortality and a striking plastic reduction in the size and reproductive potential of the individuals. Further, individuals of *B. rubens* showed a relatively greater mortality and less plastic response to densities than those of *B. mollis* in both pure and mixed stand. Two different types of plasticity were considered: one in response to changing density (d-plasticity); and the other in response to changing environmental conditions (e-plasticity). High plasticity in one of them need not imply that the other one is high too. *B. rubens* showed higher e-plasticity, but lower d-plasticity than *B. mollis*.

The relationships between r, K, and competitive ability were also examined by Wu and Jain (1978). Two types of K-strategy were distinguished: one involving greater non-reproductive effort with longer life span, or lowered mortality (Type-I) and the other with density-induced adjustments in body size along with survival in higher numbers (Type-II). Different populations of these two *Bromus* species showed different values of r and K (Type-II) and different competitive abilities. It was found that higher r was usually accompanied by lower K (Type-II), while higher K (Type-II) was accompanied by lower competitive ability, which in turn is correlated with higher d-plasticity. In general, coexistence was predicted on the basis of estimates derived from such interspecific competition experiments.

5.3.10 Conclusions

In this paper I reviewed certain genetic and demographic properties of three groups of colonizing plant species, namely, grassland annuals, grassland perennials, and crop-weed complexes living in ruderal habitats. Several recent reviews on the life history strategies of colonizers have put forward some generalizations as a part of what one author called "belief-webs." In this essay, admittedly, the emphasis was on describing the population genetic parameters and on the potential role of genetic systems in the evolution of adaptive strategies. For example, wild oats, rose clover and bromes are all predominantly inbreeding annuals and one would have simply ascribed their colonizing success to the consequences of selfing alone. However, one has to examine closely how or why inbreeding might theoretically result in colonizing success and then verify certain predictions about their population structure or demographic characteristics (e.g., see Stebbins 1957, 1958; Jain 1976). Our comparative studies of three grassland annuals and perennials are not evenly detailed for different species but we have found their colonizing features to vary significantly in terms of the relative roles of various adaptive traits, e.g. seed dormancy, reproductive effort, etc. Moreover, genetic variation as studied in the past using marker genes describes certain evolutionary processes formally in population genetic terms, but is of rather little help in understanding the evolutionary changes or potential for such changes in the life history characteristics or physiological adaptedness. The labeling of colonizing strategies or predictability of success for different species in different environments based on our present understanding is still a risky venture.

Can population genetics and population ecology provide meaningful inputs to the ecosystem level research? Many of the International Biological Program Projects emphasized biomass and energy-nutrient variables rather than population-genetic phenomena. Foin and Jain (1977) argued that an understanding of community-level process, at least in terrestrial plant communities, will largely require analyses of life history tactics in natural communities, ecotypic variation, genetic regulation, and species interactions. Mooney et al. (1977) also noted that "the evolutionary approach to ecosystem study is aimed at the identification of the adaptive strategies of the various members of the ecosystem and of the selective forces that account for these strategies." Accordingly, in their outstanding review of convergent evolution in Mediterranean ecosystems in Chile and California, several features of photosynthetic adaptations (e.g., timing, form, and amount of carbon apportionment to the roots, stem, leaves, and reproductive organs) were discussed as components of strategies. New advances in population biology could well depend largely on a close tie-in between such physiological traits and the ones reviewed earlier (viz. life tables, population size, and regulation, measures of genetic and phenotypic variation, dispersability, competing ability). The optimism for such developments in the near future is well-justified, since the origin, genetics, physiology, and ecology of adaptations have become the research emphasis at all four levels: individual, population, species, and ecosystem.

The r- and K-strategies have been described using a wide range of characteristics (see Grime 1980 for an extensive list; also Bazzaz 1979; Stearns 1976). Cook (1980) arrived at a similar dichotomy in terms of species with dispersal in space (colonizers?) versus dispersal in time (high dormancy and long-lived seed banks in soil). Since colonization is the key feature of secondary succession, it is no wonder that recent reviews of succession (especially, with shift in emphasis toward population processes) have also arrived at a similar classification of strategies (Noble and Slatyer 1980; Peet and Christiansen 1980). Even though descriptions of r- and K-strategy descriptions in terms of longevity, reproductive effort and opportunism appear rather robust, I suspect that new critical syntheses of available information will reveal the following:

1. Colonizing species of various man-made habitats and of various kinds of disturbed habitats in natural communities will have to be carefully treated in certain "logical" groups based on community type, nature of disturbance, etc.

2. Life history comparisons among these groups will show that r- and K-concepts will need to be carefully redefined in specific contexts, e.g., within groups of sexual versus asexual, and short-lived versus long-lived iteroparous species, life forms and phylogenetic lineages.

3. An attempt to standardize various terms and empirical "measures" of heterogeneity, disturbance, grain, patchiness, etc. in spatial and temporal environments is essential. How else can we utilize theories of evolution and adaptation under heterogeneous environments (Levins 1968; Roughgarden 1979) with the ecologists' model of patchiness, disturbance, stress, and cycles (Snaydon 1980; Grime 1980).

4. Finally, parameters describing population structure, population size, units of changes in life histories, and responses to selection need to be explicitly defined and certain minimal criteria for acceptable ecogenetic evidence have to be prescribed.

Résumé

1. Depuis plus de quarante ans, des écologues essaient de préciser les caractères des plantes colonisatrices. De nombreux exemples ont été récemment étudiés dans cette pespective.

2. *Avena fatua* L. et *Avena barbata* Brot. Pour ces deux especes, plusieurs phénomènes, visibles à des échelles différentes, peuvent expliquer les variations observées (Tableau 1). Cinquante et une colonies artificielles d'*Avena barbata* ont été implantées dans une large gamme de milieux; elles ont suffisamment bien réussi pour que les pressions de sélections puissent être considérées comme faibles.

Quatorze populations isolées d'*Avena barbata* possédaient en moyenne nettement moins de polymorphisme intra-populations que vingt deux populations échantillonnées dans des pâturages de la région de la baie de San Francisco.

Une étude récente a porté sur les relations entre la dimension du voisinage (qui va de 40 à 400 individus) et le mode de distribution spatial des populations: les voisinages de petite taille semblent liés aux mosaïques très contrastées.

3. *Trifolium hirtum* All. Cette espèce colonise des bords de routes, par des populations qui ont un plus grand nombre d'inflorescences, un calice plus large et plus hirsute, un taux de transport de graines plus faible, une moindre résistance des plantules, une floraison plus précoce et un moindre polymorphisme.

4. Six Graminées. Pour ces six Graminées, des paramètres démographiques et génétiques (nombre de graines, variabilité génétique, effort de reproduction, propagation végétative, longévité, résistance des plantules, etc.) n'ont pas donné régulièrement les classements attendus d'après les théories récentes.

5. Hybrides entre cultivars et formes sauvages. Hétérozygotie et polymorphisme. Pour le Seigle et pour *Helianthus bolanderi-exilis*, les caractères des formes sauvages et des formes cultivées ne semblent pas associés en complexes distincts.

Le tableau 7 montre que le polymorphisme et le taux d'hétérozygotie de 14 espèces colonisatrices sont assez faibles.

6 et 7. Des tactiques variées. Plusieurs exemples de taxons sont en contradiction avec la théorie des stratégies „r" et „K", et il est préférable d'admettre que les plantes pratiquent des tactiques variées, et de savoir que la connaissance des variations phénotypique et génotypique totales ne permet pas d'expliquer le rôle adaptatif de ces variations (cf. les travaux récents sur *Medicago polymorpha*).

8. Systèmes génétiques et recombinaisons. Le degré d'„ouverture" du système génétique n'est pas constant chez les espèces colonisatrices, et des opinions contradictoires ont été avancées sur ce point.

9. Interactions interspécifiques et diversité génétique. La co-adaptation d'espèces est difficilement démontrable. Pour être probantes, les expériences doivent combiner le degré de polymorphisme, le type de milieu, la stabilité du milieu, les densités absolues et relatives, et l'interprétation doit tenir compte de plusieurs types de plasticité et de plusieurs types d'espèces „K".

10. Conclusion: „Caveant emptores".

1) Parmi des espèces colonisatrices des milieux perturbés, il faut d'abord distinguer des groupes logiques, en fonction des types de communautés, des types de perturbations, etc.
2) A l'intérieur de ces groupes, les concepts traditionnels doivent être redéfinis.
3) Il est essentiel de tenir compte de l'hétérogénéité spatiale (grain, taches, etc.) et des perturbations qui rompent l'homogénéité du temps.

References

Babcock EB (1947) The taxonomy, phylogeny, distribution, and evolution of *Crepis*. Univ California Press, Berkeley, p 197

Baker HG, Stebbins GL (eds) (1965) The genetics of colonizing species. Academic Press, London New York, p 458

Bazzaz FA (1979) The physiological ecology of plant succession. Annu Rev Ecol Syst 10:351–372

Bradshaw AD (1965) Evolutionary significance of phenotypic plasticity in plants. Adv Genet 13:115–155

Brown AHD (1979) Enzyme polymorphisms in plant populations. Theor Popul Biol 15:1–42

Brown AHD, Marshall DR (1981) The evolutionary genetics of colonizing plants. Paper presented at the II Int Congr Syst Evol Biol, Vancouver
Cain SA (1944) Foundations of plant geography. Harper, New York, p 556
Charlesworth B (1980) Evolution in age-structured population. Cambridge Univ Press, Cambridge
Clegg MT, Allard RW (1972) Patterns of genetic differentiation in the slender wild oat species *Avena barbata*. Proc Natl Acad Sci USA 69:1820–1824
Cook R (1980) The biology of seeds in the soil. In: Solbrig OT (ed) Demography and evolution in plant populations. Univ California Press, Berkeley, pp 107–130
Dansereau P (1957) Biogeography: An ecological perspective. Ronald Press, New York, p 394
Foin TC, Jain SK (1977) Ecosystems analysis and population biology: Lessons for the development of community biology. BioScience 27:532–538
Grant WF (1967) Cytogenetic factors associated with the evolution of weeds. Taxon 16:283–293
Grime JP (1980) Competition and the struggle for existence. In: Anderson RM, Turner BD, Taylor LR (eds) Population dynamics. Blackwell, Oxford, pp 123–140
Hamrick JL, Allard RW (1972) Microgeographic variation in allozyme frequencies in *Avena barbata*. Proc Natl Acad Sci USA 69:2100–2104
Hamrick JL, Holden LR (1979) The influence of microhabitat heterogeneity on gene frequency distribution and gametic phase disequilibrium in *Avena*. Evolution 33:521–533
Hamrick JL, Linhart YB, Mitton JB (1979) Relationships between life history characteristics and electrophoretically detectable genetic variation in plants. Annu Rev Ecol Syst 10:173–200
Harding J, Mannkinen CB (1972) Genetics of *Lupinus*. IV. Colonization and genetic variability in *Lupinus succulentus*. Theor Appl Genet 42:267–271
Harlan JR (1975) Crops and man. Am Soc Agron, Madison
Harper JL (1977) Population biology of plants. Academic Press, London New York
Hedrick PW, Ginevan ME, Ewing EP (1976) Genetic polymorphism in heterogeneous environments. Annu Rev Ecol Syst 7:1–32
Heiser CB (1973) Introgression re-examined. Bot Rev 39:347–366
Jain SK (1969) Comparative ecogenetics of two *Avena* species occurring in Central California. Evol Biol 3:73–118
Jain SK (1975) Patterns of survival and microevolution in plant populations. In: Karlin S, Nevo E (eds) Population genetics and ecology. Academic Press, London New York, pp 49–89
Jain SK (1976) The evolution of inbreeding in plants. Annu Rev Ecol Syst 7:469–495
Jain SK (1977) Genetic diversity of weedy rye in California. Crop Sci 17:480–482
Jain SK (1978) Inheritance of phenotypic plasticity in soft chess (*Bromus mollis* L.), Gramineae. Experientia 34:835–836
Jain SK, Martins PS (1979) Ecological genetics of the colonizing ability of rose clover (*Trifolium hirtum* All.). Am J Bot 66:361–366
Jain SK, Rai KN (1980) Population biology of *Avena*. VIII. Colonization experiment as a test of the role of natural selection in population divergence. Am J Bot 67:1342–1346
Jain SK, Rai KN, Singh RS (1981) Population biology of *Avena* XI. Variation in peripheral isolates of *A. barbata*. Genetica 56:213–215
Lerner IM (1954) Genetic homeostasis. Wiley, New York, p 134
Lerner IM, Dempster ER (1962) Indeterminism in interspecific competition. Proc Natl Acad Sci USA 48:821–826
Levin SA (1975) Dispersal and population interactions. Am Nat 108:207–228
Levin SA (1976) Population dynamics in heterogeneous environments. Annu Rev Ecol Syst 7:287–310
Levins R (1968) Evolution in changing environments. Princeton Univ Press, Princeton, p 120
Lewis H, Lewis M (1955) The genus *Clarkia*. Univ Calif Publ Bot 20:1–241
Lloyd DG (1980) Demographic factors and mating patterns in angiosperms. In: Solbrig OT (ed) Demography and evolution in plant populations. Univ California Press, Berkeley, pp 67–88
MacArthur RH, Wilson EO (1967) The theory of island biogeography. Princeton Univ Press, Princeton, p 199
Marshall DR, Jain SK (1968) Phenotypic plasticity of *Avena fatua* and *A. barbata*. Am Nat 102:457–467
Martins PS, Jain SK (1979) The role of genetic variation in the colonizing ability of rose clover (*Trifolium hirtum* All.). Am Nat 113:591–595
Maynard Smith J (1978) The evolution of sex. Cambridge Univ Press, New York, p 222

Mooney HA, Kummerow J, Johnson AW, Parsons DJ, Keeley S, Hoffmann A, Hays RI, Giliberto J, Chu C (1977) The producers – their resources and adaptive responses. In: Mooney HA (ed) Convergent evolution in Chile and California: Mediterranean climate ecosystems. Dowden, Hutchinson and Ross, Stroudsburg, pp 85–143

Nei M, Maruyama T, Charaborty R (1975) The bottleneck effect and genetic variability in populations. Evolution 29:1–10

Nevo E, Zohary D, Brown AHD, Haber M (1979) Genetic diversity and environmental associations of wild barley, *Hordeum spontaneum* in Israel. Evolution 33:815–833

Noble R, Slatyer RO (1980) The use of vital attributes to predict successional changes in plant communities subject to recurrent disturbances. Vegetatio 43:5–21

Oka HI, Morishima H (1980) Ecological genetics and the evolution of weeds. In: Holzner W, Numata U (eds) Biology and ecology of weeds. Junk, The Hague, pp 110–138

Olivieri AM, Jain SK (1977) Variation in *Helianthus exilis – H. bolanderi* complex: A reexamination. Madroño 24:177–188

Pandey KK (1980) Evolution of incompatibility systems in plants: Origin of "independent" and "complementary" control of incompatibility in Angiosperms. New Phytol 84:381–400

Park T (1954) Experimental studies of interspecies competition. II. Temperature, humidity, and competition in two species of *Tribolium*. Physiol Zool 27:177–238

Peet RK, Christiansen NL (1980) Succession: a population process. Vegetatio 43:131–140

Pimentel D (1964) Population ecology and the genetic feedback mechanism. In: "Genetics today." Proc XI Int Congr Genet, The Hague, pp 483–488

Price S, Jain SK (1981) Are inbreeders better colonizers? Oecologia 49:283–286

Rai KN (1972) Ecogenetic studies on the patterns of differentiation in natural populations of slender wild oat, *Avena barbata* Brot. PhD Dissert, Univ California, Davis

Rai KN, Jain SK (1982) Population biology of *Avena*. IX Gene flow and neighborhood size in relation to microgeographic variation in *A. barbata*. Oecologia 53:399–405

Raven PH, Gregory DP (1972) A revision of the genus *Gaura* (Onagraceae). Mem Torrey Bot Club 23:1–96

Raven PH, Raven TE (1976) The genus *Epilobium* in Australasia. N Z DSIR Bull 216:321

Roughgarden J (1979) Theory of population genetics and evolutionary ecology: an introduction. MacMillan, New York, p 634

Salisbury EJ (1942) The reproductive capacity of plants. Bell, London, p 244

Schaffer WM, Gadgil MV (1975) Selection for optimal life histories in plants. In: Cody ML, Diamond JM (eds) Ecology and evolution of communities. Harvard Univ Press, Cambridge, pp 142–157

Snaydon RW (1980) Plant demography in agricultural systems. In: Solbrig OT (ed) Demography and evolution in plant populations. Univ California Press, Berkeley, pp 131–160

Solbrig OT (1980) Genetic structure of plant populations. In: Solbrig OT (ed) Demography and evolution in plant populations. Univ California Press, Berkeley, pp 49–66

Stearns SC (1976) Life-history tactics: a history of the ideas. Q Rev Biol 51:3–47

Stebbins GL (1957) Self-fertilization and population variability in the higher plants. Am Nat 91:337–354

Stebbins GL (1958) Longevity, habitat, and release of genetic variability in the higher plants. Cold Spring Harbor Symp Quant Biol 23:365–378

Suneson CA, Rachie KO, Khush GS (1969) A dynamic population of weedy rye. Crop Sci 9:121–124

Williams GC (1975) Sex and evolution. Princeton Univ Press, Princeton, p 200

Wu KK, Jain SK (1978) Genetic and plastic responses of geographic differentiation of *Bromus rubens* populations. Can J Bot 56:873–879

Wu KK, Jain SK (1979) Population regulation in *Bromus rubens* and *B. mollis*. Life cycle components and competition. Oecologia 39:337–357

Yazdi-Samadi B, Wu KK, Jain SK (1978) Population biology of *Avena*. VI. The role of genetic polymorphisms in the outcome of interspecific competition in *Avena*. Genetica 48:151–159

5.4 Characteristics of Populations in Relation to Disturbance in Natural and Man-Modified Ecosystems

F. A. Bazzaz

5.4.1 Introduction

Catastrophic, large-scale disturbances generate much of the observed community dynamics in nature (see reviews by Pickett and Thompson 1978; White 1979). Fire has been a major factor in the organization of plant communities and in the evolution of their species strategies. In the boreal forest of North America (Tande 1979) and other northern forests (Heinselman 1973) patches of differing species composition, age structure, etc. are created by fires. In the southern Wisconsin forest γ-diversity is generated and maintained by fires (Loucks 1970). Fire plays a similar role in the Mediterranean-type vegetation in the the Americas, Europe, Africa, and Australia (see reviews in Mooney et al. 1981). Much of the structure of the vegetation is determined by fire frequency and intensity in the Garrigue of southern France (Trabaud 1980). Wind throw, sometimes together with fire, seems to generate much of the pattern in the New England forest vegetation (Henry and Swan 1974; Oliver and Stephens 1977) and in the tropics (Gomez-Pompa 1971; Whitmore 1975). Other disturbance agents, e.g., landslides, earthquakes (Garwood et al. 1979), major climatic shifts, herbivores, predators, pathogens also play major roles in ecosystem structure and function.

Man has been a factor in the organization of communities for a long time. However, his impact on ecosystems has been accelerating rapidly through:

1. Extensive clearing of natural vegetation for agricultural and other purposes.
2. Selective harvest of desirable species and the introduction of alien ones.
3. Abandonment of unproductive agricultural land.
4. Mining and reclamation of mined lands.
5. Drainage of wetland for agricultural or fuel usage.
6. Introduction of biocides and other chemicals in the environment.
7. Creation of war-impacted ecosystems by bombing, defoliation and movement of men and material.

Ecologists have long recognized the similarities between natural and some man-made disturbances. Fire is one agent of disturbance whose consequences may be the same whether they are started by man or by lightning (Spurr and Barnes 1973). There are strong similarities between selective removal from forests of certain tree species by lumbering and killing of plants by selective pathogens or defoliators. Thus, the distinctions between natural and man-made disturbances are less important and what matters is not what caused disturbance but what are the nature and consequences of disturbance and how do species populations respond to them over ecological and evolutionary times. It may be possible to begin to develop a general

Table 1. Some correlated disturbance, population, and life history characteristics in plant communities

A. Disturbance characteristics with relevance to plant response
Size
Intensity
Frequency and regularity, and predictability of occurrence
Duration
Seasonal time of occurrence
Level of environmental heterogeneity within the disturbed area
Nature of the biotic neighborhood
B. Population characteristics responsive to disturbance
Density and dispersion
Growth rate
Survivorship and age and size structure
Levels of gene flow in the population
Degree of relatedness among the members of the population
Organization of variation within the population
Strength of competitive interactions
Niche breadth and niche overlap
Strength of interactions with other trophic levels
C. Plant life history characteristics responsive to disturbance
Spatial and temporal dispersibility
Seed germination
Seedling establishment and growth
Reproductive strategies
Breeding system
Fecundity
Reproductive allocation and packaging

theory of disturbance of ecosystems. I define disturbance as sudden change in the resource base of a unit of the landscape that is expressed as a readily detectable change in population response.

In this paper I first begin by identifying disturbance characteristics that are relevant to population response, population characteristics that are sensitive to disturbance, and life-history attributes that are influenced by disturbance. These are given in Table 1. Next, I discuss some interactions between these three categories of attributes that seem to be more important than others or about which we know more. Clearly the interactions between these various parameters may be much more complex than indicated and certainly involve more than one trophic level. I cite only a few relevant references to document my discussion, rather than reviewing the literature on disturbance. The emphasis is placed on coupling certain disturbance characteristics with certain population and life history traits of plants.

5.4.2 The Nature of Disturbance

Disturbance plays a significant role in the organization and function of all ecosystems – natural as well as man-modified. Even in equilibrium communities, classically called climaxes, some level of disturbance is always operative. There is now much evidence that disturbance interferes with the exclusion of species by competition and therefore is necessary for the maintenance of species diversity in many plant communities (Levin and Paine 1974; Grubb 1977; Connell 1978) and that it plays a significant role as a selective agent in these communities (Pickett 1976). Disturbance produces spatial pattern as well (Greig-Smith 1979).

When large scale disturbances occur, succession is initiated and an assortment in dominance of species populations takes place through time. Different species guilds replace each other depending on their colonizing and competitive ability (Connell and Slatyer 1977) and other life-history characteristics (Noble and Slatyer 1980). The site will eventually approach a new equilibrium community dominated by a lesser number of strongly competitive species. Thus maximum species diversity is usually attained prior to this point and, in forest succession, is achieved when the shade-tolerant species have not yet excluded the shade-intolerant ones from the community (Loucks 1970; Horn 1974). Autogenic processes (e.g., Watt 1947) also play a role in community dynamics, but their contribution relative to that of external disturbance varies between communities.

5.4.3 Disturbance Characteristics with Relevance to Population Response

5.4.3.1 Size

Size ranges from very small gaps created by the death of an individual in an herbaceous community, to the breakage of a tree limb to the fall of a single tree creating a canopy gap, to the fall of several trees, to very large disturbances created by fires, windstorms, landslides, earthquakes or on a different time scale, the clearing of much of the North American prairies for agriculture and removal of whole vegetation types by glaciation. Animal activities also create disturbances of different sizes and distributions ranging from hoof marks or ant hills to intensive and extensive disturbance of vegetation over large geographic areas by large herds of mammals. Disturbance of vegetation by defoliating insects may also be small and localized or may be spread over large areas. Very small gaps in forest canopy may be filled by adjacent trees via crown extension, but closing of very large gaps depends on dispersal, establishment and growth of the colonizers and their replacement with more equilibrium-type species populations. In small gaps, shape and orientation are important as well.

The native human inhabitants of many vegetation types of the world created small clearings, usually by slash and burn methods in forests, and used these clearings to grow crops. The clearings were abandoned soon afterwards when their productivity declined after the subsidence of the resource pulse caused by clearing and burning. Natural succession quickly closed the clearings due to the proximity of colonization source, and because the sites had lost little nutrients. These man-made disturbances may be no larger than those caused by some naturally occurring disturbances such as those creating canopy gaps in tropical and temperate forests.

However, in recent times disturbance caused by man has been increasing in size and in intensity. The removal of the North American prairie or the central European forest vegetation have no parallels in nature except those created by large climatic shifts over a geologic time scale such as the destruction of the post-Miocene vegetation by the Pleistocene glaciation.

5.4.3.2 Frequency of Occurrence

Like size, the frequency of occurrence of disturbance varies widely in natural systems, and in many situations small scale disturbances occur more frequently than large scale disturbances. Fire in the eastern part of the North American prairies occurred frequently and was caused by lightning and by human activities. Major burns occur every 100–200 years in southern Wisconsin deciduous forests (Loucks 1970), and more often in boreal forest (Johnson and Rowe 1975; Tande 1979). Humid tropical forests burn only rarely. Some of these disturbances occur in irregular fashion and present a high level of unpredictability of resource change to the species involved; others occur more regularly. Disturbance occurs at more-or-less regular intervals in areas that lie in the path of tropical storms (Webb 1958) or in areas where a threshold of combustible material is reached over a certain time span.

In the California chaparral the growth of the shrubs and the increase in their cover results in suppression of the herbs by competition and interference. After a certain amount of litter accumulates, fire usually sweeps these areas removing the litter and killing many of the shrubs. Competition and chemical interference are greatly reduced, and the germination and establishment of the herbs are vastly stimulated. The shrubs begin to grow again from seeds stimulated by fire as well as from sprouts and the fire cycle is initiated again (Muller et al. 1968). The patterns of growth and accumulation of combustible fuel tend to regulate the frequency of fire such that it occurs once every 30–40 years. Very likely in other fire-prone ecosystems, the life history characteristics become selected to fit and even contribute to the regularity in timing of occurrence of disturbance.

The change in the frequency of disturbance by introduction of new species or by man's activities will lead to change in structure of the vegetation. Clearly it would favor species whose life history characteristics fit the new situation. For example, Forman and Boerner (1981) show that the change in fire frequency from an average of 20 years to about 65 years in the pine barrens of New Jersey, U.S.A. resulted in a shift from dominance of the fire-adapted *Pinus rigida* to non-fire-adapted species and the replacement of cedar swamps by hardwood swamps. Adaptations to fire include release of seed from serotinous cones, fire-stimulated seed germination and prolific sprouting even in genera that usually do not sprout, e.g., *Pinus rigida*. Fire frequency may also influence growth morphology and the pattern of photosynthate allocation to maximize seedling survivorship, e.g., in *Pinus palustris*. Here again the rate of accumulation of combustible litter influences the frequency and regularity of disturbance and perhaps its intensity as well.

5.4.3.3 Intensity

Disturbances caused by defoliation in the eastern deciduous forests of North America may differ in intensity, as well as in size and frequency (Stephens 1981). These three characteristics may interact in influencing the response of the plant population to disturbance. Trees may be defoliated severely or slightly, frequently or rarely, in successive years or intermittently. These events have different consequences for the structure of species populations and subsequently for their interactions and the resulting community characteristics. A single defoliation has little effect but repeated and especially consecutive defoliations are followed by increased mortality. Mortality is higher when canopy defoliation is severe. Defoliation intensity influences different species differently. Mortality is higher in *Quercus* than in other groups (e.g., *Acer*). Within *Quercus* mortality is greater in *Q. alba* and *Q. prinus* than in *Q. rubra*, *Q. velutina*, and *Q. coccinea* (Stephens 1981). The frequency and intensity of grazing in the Serengeti Plains control the primary productivity of the ecosystem. The level of grazing at which productivity is maximum may be close to the normal level of wildebeest grazing of the system (McNaughton 1979).

5.4.3.4 Time of Disturbance

Time of disturbance may be predictable to a certain degree within a year. Fires occur more frequently during the dry season. Wind throw may also be seasonally predictable but earthquakes and landslides may not be. Plowing and flooding may be predictable in seasonal environments, but their occurrences within the season may not be. Time of disturbance selects species whose germination and growth requirements are such that they can take advantage of the resource suddenly made available by disturbance. In the eastern United States soil disturbance by plowing early in autumn results in dominance by winter annuals in the following growing season and the suppression of the spring and summer annuals. Disturbance in the spring leads to dominance by the latter group. Within a season, disturbance time selects for certain species within the spring-summer guild. Late disturbance is disadvantageous to *Polygonum pensylvanicum* but favorable to *Setaria faberii* and *Ipomoea hederacea* (Fig. 1). The temperature requirements of germination control much of this response (Bazzaz 1979). Different European herbs in disturbed sites emerge at different times of the year (Ellenberg 1963), and doubtless different population groups will dominate as a result of timing of disturbance.

5.4.3.5 Level of Environmental Heterogeneity

Heterogeneity within a disturbed site depends on its size, mode of formation and the intensity of disturbance. Small gaps are most likely less heterogeneous than large gaps. Severe disturbances may create less heterogeneous gaps than do mild disturbances. Nevertheless, disturbed sites are never homogeneous and the levels of

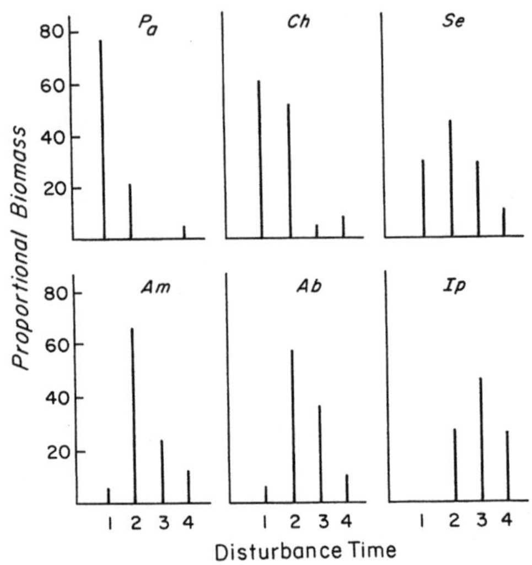

Fig. 1. Relationship between time of disturbance during the growing season and the proportional contribution of major species to total biomass of an annual community. *Polygonum pensylvanicum* (*Pa*); *Chenopodium album* (*Ca*); *Setaria faberii* (*Sa*); *Amaranthus retroflexus* (*Am*); *Abutilon theophrasti* (*Ab*); *Ipomoea hederacea* (*Ip*). (Perozzi and Bazzaz 1978)

heterogeneity in a site are perceived differently by different species occupying the site. This makes quantification of heterogeneity quite difficult. Single tree gaps, common in tropical as well as temperate forests, may be quite heterogeneous (e.g., Oldeman 1978; Hartshorn 1978). The "Chablis" may be divided into the crown gap and the epicenter (the area of the fallen crown of the tree). Different locations in the Chablis will have different above- and below-ground environments. Mineral soil is exposed by tip ups, which form sites of colonization for certain plant species, e.g., mosses in temperate forests. Soil nitrogen and other nutrients are expected to be low in these sites, higher in the area under and near the trunk, and highest in the epicenter (Fig. 2). Disturbance size and heterogeneity interact in determining size and identity of plant populations. In oldfield succession heterogeneity generated by erosion and by clumping of the vegetation by clonal growth of some shrubs and trees generates high species diversity (Bazzaz 1975).

The level of heterogeneity in a disturbed site is in a dynamic state. The species populations whose invasion of the site is influenced by heterogeneity may themselves change it. Different species may occupy different locations within a disturbed site. Clearing of a site for agricultural purposes and its continued use in monoculture with uniform plowing, fertilizing, and biocide application reduces patchiness. After abandonment, patchiness may be created by random or non-random distribution of propagules as result of direction of seed source, path taken by animal dispersers, etc. The occupation of different locations by colonists after abandonment itself creates different patches by means of differential nutrient uptake and release (e.g., Whittaker and Woodwell 1968), different rhizosphere chemistry (e.g., Smith 1976; Waldendorp 1978), difference in the amount of stemflow (Peterson 1980), differences in the accumulation of debris under herbs (Watt

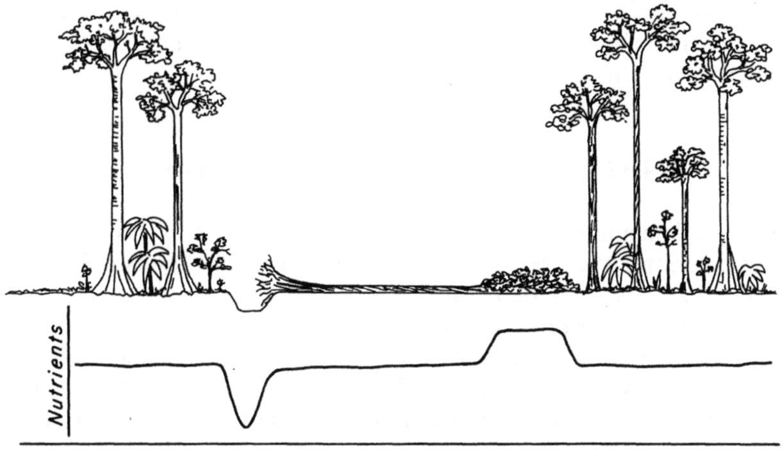

Fig. 2. Diagrammatic representation of nutrient heterogeneity in surface soil, created by a tree fall. Nutrients are lowest in the tip up and highest under the crown of the fallen tree. Light, temperature, and humidity profiles will also be different for different locations in the gap

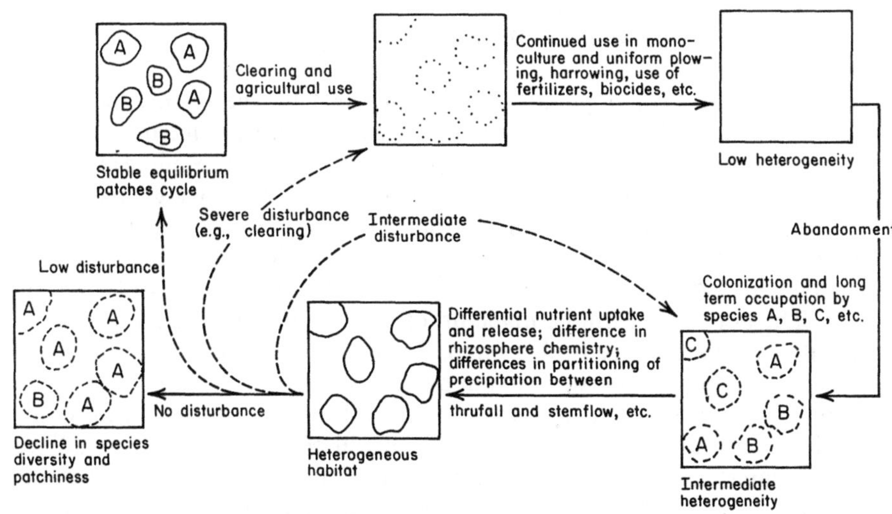

Fig. 3. Dynamics of environmental heterogeneity under different disturbance regimes

1981), etc. This patchiness in turn influences the pattern of later invasion, and the patch types may change location depending on random processes and the nature of species replacement. For example, in low diversity deciduous forests, species usually exchange locations with A replacing B and B replacing A, but in high diversity forest species replacement is more random (Barden 1980; Runkle 1981). Another clearing or disturbance initiates another cycle of heterogeneity change (Fig. 3). Size and level of heterogeneity in disturbed sites have indirect effects on the plant life history features and population characteristics as well. Various herbivores, predators, pollinators, dispersers, and pathogens may be attracted to different patches in a disturbed site and their interaction with the plants may change the level of heterogeneity in that site.

5.4.3.6 Nature of the Biological Neighborhood

The biological neighborhood of a disturbance, as a source of colonization, may influence population structure of the site. Diverse neighborhoods offer a variety of potential colonizers while depauperate neighborhoods do not. The degree of patchiness of the neighborhood plays a role as well, in that patchy colonization sources may offer many species with different colonization abilities and requirements. This patchiness itself is determined by size and frequency of disturbance and therefore there is an interconnected, integrated system of patch dynamics where all the attributes of disturbance are dependent on each other.

5.4.4 Population Characteristics Responsive to Disturbance

5.4.4.1 Density, Dispersion, and Age Structure

Several attributes of populations are influenced by the disturbance regime (Table 1). Density and dispersion are quite sensitive to size and intensity of disturbance. Growth rates are influenced by the level and pattern of resource availability after disturbance. Age structure may be a function of intensity and frequency as well. In severe disturbance age class distribution of the colonizers may be narrow; in less severe disturbance, where some individuals of different ages resprout and new individuals are recruited from the seed bank, the age class distribution may be wide. The population will be made up of several cohorts instead of one.

Intensity of disturbance is also important in selecting for certain life history adaptations and population structure in nature. Severe fires destroy both live and dead biomass, reduce the chance of sprouting, kill seeds in soil, and lead to much change in the resource base of a site. Invasion will be by dispersal from adjacent areas and change in populations and attainment of stable communities takes much time. Light fire, in contrast, may be followed by populations derived from the seed bank in situ and much resprouting. Stable communities are reestablished quickly and there may be less change in the pattern of distribution of individuals of sprouting species; they maintain their original location.

Frequency of disturbance may regulate age structure of the population. Fire frequency in the boreal forests of Jasper Park in the Canadian Rockies has determined the age structure of lodgepole pine communities (Tande 1979). Disturbance caused by man and by oak wilt disease has structured the age class distribution of *Prunus serotina* in southern Wisconsin (Auclair and Cottam 1971). Age structure of several species in northern hardwood forests in New Hampshire, U.S.A. has been used by Henry and Swan (1974) to reconstruct disturbance history of the area as to timing of occurrence, frequency, kind and intensity of the disturbance.

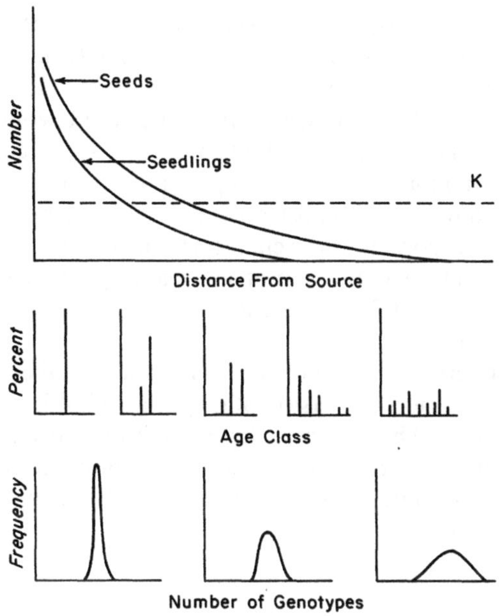

Fig. 4. Relationship of seed and seedling population densities, age class distributions, and frequency of plant genotypes to distance from source (reflecting disturbance size). *K* site's carrying capacity for the species

The size of disturbance and the nature of the biotic neighborhood influence age structure of the populations that occupy the site after disturbance as well. The number of seeds dispersed into the site from the edge of the undisturbed vegetation decreases with distance from the source (Fig. 4). The density of the resulting seedlings from one dispersal episode may exceed the carrying capacity of the site up to a certain distance from the source but is increasingly less at greater distances. Thus repeated invasion episodes may contribute to the seedling population away from the source. In very large disturbances, the filling of space by that particular species may occur only after the few scattered individuals that arrived at different times mature and themselves begin to disperse seeds. As a result, different age class distributions occur at different distances within the disturbed site. Near the edge populations will be even-aged and away from it they will be mixed-aged (Fig. 4). Density-dependent mortality will prevail near the edge while density-independent mortality prevails away from the edge.

5.4.4.2 Genotypic Variability in Populations

In small disturbances, and near the edge of intact vegetation in large disturbances, the plant population structure is not only more-or-less even-aged and even-sized but the degree of relatedness between the individuals may be high. This is especially true if the intact vegetation has high species diversity and only one or a few individuals of a species act as seed source for colonization of the disturbed site. The resulting seedling population would have low genotypic variability near the source and higher variability away from the source (Fig. 4). Conspecific competition would be intense and the thinning process in these locally crowded populations would be governed by chance events. The result would be clumped distributions in which neighboring individuals are likely to be genotypically similar but different from individuals in other clumps. In the event of further disturbance, these clumped and related individuals would experience somewhat similar changes in resource levels and respond to these changes in the same manner. Thus these disturbances may result in local extinction of some genotypes. Whether or not certain genotypes will be eliminated from the entire area would depend on the type and extent of that disturbance.

Large scale, long term disturbance may have additional consequences as well. For example, after glaciation reinvasion of deglaciated habitats may involve many localized founder effects. This is especially likely if reinvasion is by long distance transport of propagules. Furthermore, reinvading populations may have lower genotypic diversity, as some genotypes may have been disproportionately destroyed during glaciation.

5.4.4.3 Interactions Between Species

Competitive and coevolutionary interactions between species may be modified by severe disturbance because of the formation of new species assemblages. Some species may lag behind and never reinvade the site after these disturbances. Examples of species that do not reinvade after glaciation along with other members of the former assemblage are found in European flora. Conversely, species may greatly increase their importance and geographic extent after severe disturbance. In the North American grassland, *Agropyron smithii* experienced competitive release during the great drought of the 1930s. It occupied sites usually dominated by *Andropogon* before the drought. The species' phenology, physiology, and reproductive strategies suit it well for expansion. It grows vigorously in the spring when soil moisture is available, producing large numbers of seeds and rhizomes. It has high transpiration rates and depletes the soil moisture, depressing the performance of later growing grasses (Weaver 1954).

5.4.5 Life History Characteristics and Disturbance

5.4.5.1 Life Span

There is much evidence to show that length of life cycle of species that predominate in a given habitat is closely related to the frequency of disturbance. Agricultural fields which are plowed annually are dominated by annuals. Herbaceous perennials are prominent in habitats that are frequently but not annually disturbed. In fire-prone habitats, e.g., California chaparral, the garrigue of southern France, the pine barrens of New Jersey, U.S.A., life cycle of the dominants coincides with

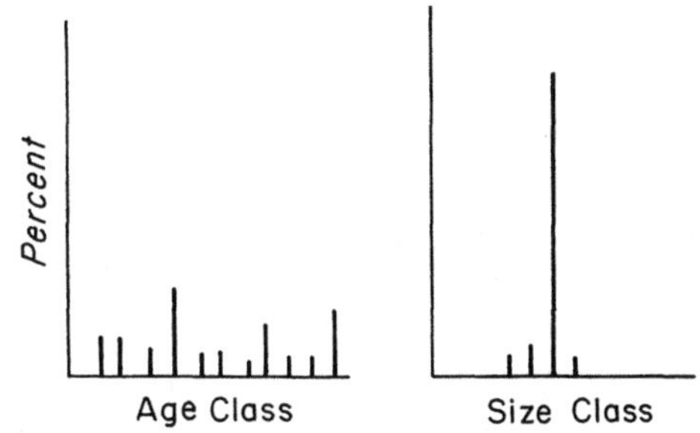

Fig. 5. Relationship between age class and size class distribution of seedlings and young individuals of species whose seeds germinate under the canopy. Released from suppression when a gap is created, individuals of different ages may reach maturity at the same time

fire frequency. Different fire frequencies select for dominance of different species within a given geographic area. Fire frequency in the garrigue of southern France regulates *Pinus halepensis* populations. If fires are infrequent (once in 100 years) the pines are maintained but under more frequent fires (50 year cycle) the pine forests are usually replaced by a garrigue dominated by *Rosmarinus officinalis* (Trabaud 1980). These relationships are not simply cause-effect, but rather complex in that life span itself may regulate fire frequency as discussed earlier. Giant sequoia may establish on glacial outwash and the individual life span may be close to the average length of the interglacial period!

5.4.5.2 Reproductive Strategies

Age at first reproduction in iteroparous plants may also be influenced by disturbance frequency. In habitats that burn frequently and especially unpredictably, early reproduction is advantageous and may be necessary, especially for plants that do not sprout after fire. Fire frequency is especially important for species that reproduce only sexually. They will be eliminated by successive fires which occur more frequently than the time span required for the plant to reach sexual maturity.

In plant populations, age-structure may be less important than size-structure in determining the response of the population to its environment. Size rather than age may determine time of reproduction in some herbaceous plant species (e.g., Werner 1975; Regehr and Bazzaz 1979). In tropical forests (Schulz 1960; Poore 1968) seeds of some light-demanding species germinate under intact canopy soon after they are shed. They grow very slowly as seedlings and their size may be the same despite differences in age (Fig. 5). When the canopy opens by disturbance the seedlings grow rather quickly, compete intensely and some die. Those that reach flowering and fruiting size simultaneously may be of different ages. In the boreal forests of North America, *Abies balsamea* forms seedling populations which accumulate over several years. When the canopy is damaged, e.g., by the spruce budworm, the seedlings are released and a complex age structure-size structure results (Ghent 1958). Some small individuals are older than some larger individuals.

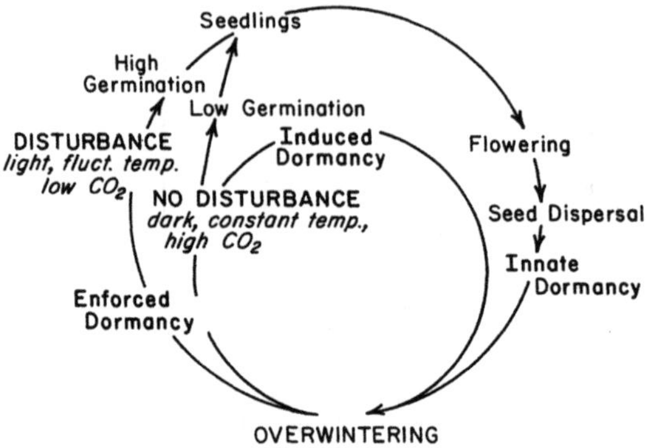

Fig. 6. Relationship between germination of *Ambrosia artemisiifolia* and disturbance. (Bazzaz 1979)

Again individuals of different ages may reach maturity and reproduction simultaneously.

Reproductive allocation may be higher in species that colonize soon after disturbance and may be lower for species that invade later. Reproductive allocation was highest in populations of *Polygonum canadense* in highly disturbed habitats and declined with increased habitat stability (Hickman 1977). A similar situation is found in several species of *Solidago* (Abrahamson and Gadgil 1973) and in *Taraxacum* (Gadgil and Solbrig 1972).

5.4.5.3 Germination, Growth and Response Breadth

The physical environment of and the level of resources available in a site are significantly modified by disturbance. Light levels and temperature increase, relative humidity decreases, and nutrients may become abundantly available initially but decline later. Wind, temperature, and CO_2 profiles change. These changes are dependent on disturbance characteristics. Nevertheless, daily and seasonal environmental variability increase (Bazzaz 1979). The colonist after disturbance must possess the appropriate life history features for establishment, growth and reproduction in these habitats. Differences in intensity, periodicity and spatial distribution of disturbance have provided the major selective force for the evolution of regeneration strategies (Grime 1979). Many of the pioneers have seed germination behavior that is keyed to disturbance. These seeds usually have efficient dispersal in space and/or in time. Disturbance provides the seeds with a suitable environment for germination including appropriate light, temperature fluctuations, oxygen concentrations, and nutrient relations (Fig. 6). Fires simultaneously remove competitors and open serotinous cones of many colonizing conifers. Physiological characteristics of seed germination, seedling growth and growth of mature individuals of plants that colonize after disturbance have been extensively studied and documented in both temperate and tropical habitats (see reviews by Bazzaz 1979; Bazzaz and Pickett 1980, and literature therein). Seedlings and mature individuals of colonizers have photosynthetic characteristics and allocation patterns promot-

Table 2. Mean niche breadth of species from early successional (disturbed) and late successional (less disturbed) communities

Niche parameter	Early succession (disturbed)	Late succession (less disturbed)
Position of roots in soil (a)	0.68	0.43
Nutrients		
Herbs (b)	0.77	0.70
Trees (c)	0.91	0.81
Moisture		
Herbs (d)	0.81	
Trees (c)	0.89	0.65
Pollinators (e)	0.31	0.19

Data from (a) Parrish and Bazzaz (1976), (b) Parrish and Bazzaz (1982a), (c) Parrish and Bazzaz (1982b), (d) Pickett and Bazzaz (1978), (e) Parrish and Bazzaz (1979)

ing rapid growth that is generally associated with high competitive abilities, contrary to r- and K-selection theory which assumes competitive inferiority for opportunistic species. Nevertheless, these species do have many of the life history features associated with the r-type strategies of MacArthur and Wilson (1967) and the R (ruderal) strategy of Grime (1977).

The nature of competitive interactions between individuals is thought to be different for species of disturbed habitats than for species of less disturbed ones. First, the probability of conspecific competition between individuals of disturbed habitats is higher than for individuals of less disturbed ones; the reverse is thought to be true for interspecific competition (Odum 1969). Second, the role of interspecific competitive interactions has been less important in the evolution of life history strategies of the colonists than for species of less disturbed communities. Third, interspecific competitive interactions should therefore be stronger between individuals in disturbed habitats than between individuals in less disturbed ones. Ecological theory also predicts that the species of disturbed habitats have broad, highly overlapping niches which are necessary for coping with the high level of variability and unpredictability of disturbed environments. Experimental work on response breadths and overlap of several species on a number of resource axes (e.g., Parrish and Bazzaz 1976, 1979, 1982a; Werner and Platt 1976; Pickett and Bazzaz 1978) (Table 2) supports this prediction. Several of the species of disturbed habitats have been shown to experience more biomass reduction in the presence of heterospecific neighbors than do species of less disturbed communities. The former species produce relatively less biomass from a given resource base than do species of less disturbed habitats (Parrish and Bazzaz 1982b). The species of disturbed habitats also are morphologically (Bradshaw 1965) and physiologically plastic (Bazzaz 1979). They are capable of responding quickly to resource release, e.g., light, nutrients, and water (e.g., Marks 1974; Vitousek 1977). They accumulate high concentrations of nutrients when available (Fig. 7).

The interactions of population and life history characteristics with disturbance generate much of the pattern in vegetation. The role of natural disturbance in the

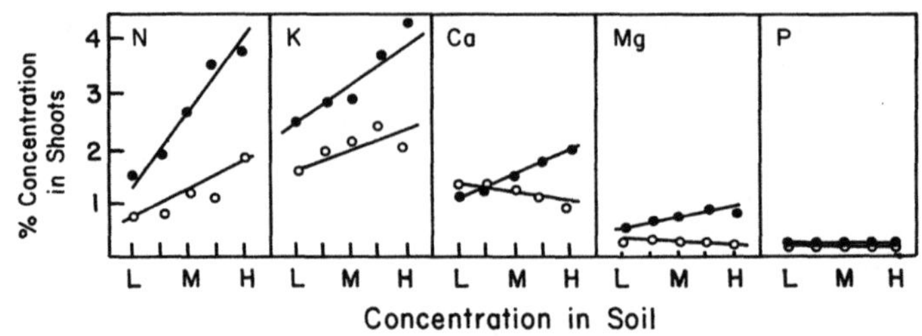

Fig. 7. Pattern of accumulation of the nutrients N, K, Ca, Mg, and P in shoots of plants of early successional ●---● (disturbed), and late successional ○---○ (less disturbed) habitats. (Data from Parrish and Bazzaz 1982a)

generation and maintenance of species diversity in many vegetation types is now well-appreciated (e.g., Connell 1978; Huston 1979; Pickett 1980; Denslow 1980). Differential regeneration characteristics contribute substantially to coexistence of plant populations (Grubb 1977). Specialization of different species on different gap sizes, distributions, and frequencies in tropical forests has been proposed as a factor in the resource partitioning between species populations (Denslow 1980). Experimental studies with annuals show that indeed different species perform differently in similar gaps and that different species require different minimum gap sizes to survive (Bazzaz and Grubb unpubl.). Disturbance frequency and size have selected species of different physiological and other adaptations to different canopy gaps (Bazzaz and Pickett 1980). Thus in high diversity communities there exists a larger number of life history designs than in low diversity communities. However, I suggest that in high diversity forests with frequent disturbances of small scale, high turnover rates, and close proximity there may be a blending of life history strategies of the component species. In these situations, the progeny of any individual likely would experience several gap types of different resource availability patterns; thus, there would be selection for performance in many rather than in one gap type resulting in blurring of differences among species life history strategies.

5.4.6 Conclusions

Disturbance seems to occur in all ecosystems. It plays a major role in the generation and maintenance of species diversity and related aspects of community organization. Disturbance is an important selective force in the evolution of plant life history designs and the resulting population characteristics. Thus there may be different life history designs for different disturbance regimes. The size, intensity, frequency, and regularity of occurrence of disturbance vary considerably and these interact dynamically with the species that experience this variation. There are many agents of disturbance in nature. However irrespective of its characters and causes, disturbance may be defined as *a sudden change in the resource base of a unit of the landscape that is expressed as a readily detectable change in population response.*

Résumé

Les perturbations affectent *tous* les systèmes écologiques, mais l'Homme est à l'origine de perturbations particulièrement intenses et brutales.

Une perturbation est un changement soudain des ressources dans une unité de paysage, qui se traduit par un changement perceptible dans les «réponses» des populations. Les caractères principaux d'une perturbation sont son étendue, sa fréquence, son intensité, la saison où elle survient, l'hétérogénéité qu'elle crée (Fig. 3), et la nature de l'environnement à l'intérieur duquel elle intervient.

Les «réponses» des populations dépendent de leur densité, de leur mode de répartition spatiale, de leur âge, de leur structure démographique (Fig. 4 et 5), de leur variabilité génétique, mais aussi de leurs modes de relations interspécifiques, de leur longévité, des modalités de leur production, de leur germination (Fig. 6), de leur croissance et de leur plasticité.

References

Abrahamson WG, Gadgil M (1973) Growth form and reproductive effort in goldenrods (*Solidago*, Compositae). Am Nat 107:651–660

Auclair AN, Cottam G (1971) Dynamics of black cherry (*Prunus serotina* Erhr.) in southern Wisconsin forests. Ecol Monogr 41:153–177

Barden LS (1980) Tree replacement in a cone hardwood forest of the southern Appalachians. Oikos 35:16–19

Bazzaz FA (1975) Plant species diversity in old-field successional ecosystems in Southern Illinois. Ecology 56:485–488

Bazzaz FA (1979) The physiological ecology of plant succession. Annu Rev Ecol Syst 10:351–371

Bazzaz FA, Pickett STA (1980) Physiological ecology of tropical succession: A comparative review. Annu Rev Ecol Syst 11:287–310

Bradshaw AD (1965) Evolutionary significance of phenotypic plasticity in plants. Adv Genet 13:115–155

Connell JH (1978) Diversity in tropical rain forests and coral reefs. Science 199:1302–1310

Connell JH, Slatyer RO (1977) Mechanisms of succession in natural communities and their role in community stability and organization. Am Nat 111:1119–1144

Denslow JS (1980) Gap partitioning among tropical rainforest trees in tropical succession. Biotropica 12:47–55

Ellenberg H (1963) Vegetation Mitteleuropas mit den Alpen. In: Walter H (Hrsg) Einführung in die Phytologie, vol IV. Ulmer, Stuttgart, S 1–943

Forman RTT, Boerner RE (1981) Fire frequency and the Pine Barrens of New Jersey. Bull Torrey Bot Club 108:34–50

Gadgil M, Solbrig OT (1972) The concept of r and K selection: evidence from wild flowers and some theoretical considerations. Am Nat 106:14–31

Garwood NC, Janos DP, Brokaw N (1979) Earthquake caused landslides: A major disturbance to tropical forests. Science 205:997–999

Ghent AW (1958) Studies of regeneration in forest stands devastated by spruce budworm. II. Age, height, growth, and related studies of balsam fir seedlings. For Sci 4:135–146

Gomez-Pompa A (1971) Possible papel del la vegetacion secundaria en la evolucion de la flora tropical. Biotropica 3:125–135

Greig-Smith P (1979) Pattern in vegetation. J Ecol 67:755–779

Grime JP (1977) Evidence for the existence of three primary strategies in plants and its relevance to ecological and evolutionary theory. Am Nat 111:1169–1194

Grime PJ (1979) Plant strategies and vegetation processes. Wiley, New York, p 222

Grubb PJ (1977) The maintenance of species richness in plant communities: The importance of the regeneration niche. Biol Rev 52:107–145

Hartshorn GS (1978) Tree falls and tropical forest dynamics. In: Tomlinson PB, Zimmerman MH (eds) Tropical trees as living systems. Cambridge Univ Press, Cambridge, pp 617–638
Heinselman ML (1973) Fire and the virgin forests of the Boundary Waters Canoe Area, Minnesota. J Quater Res 3:329–382
Henry JD, Swan JMA (1974) Reconstructing forest history from live and dead plant material. An approach to the study of forest succession in southwest New Hampshire. Ecology 55:772–783
Hickman JC (1977) Energy allocation and niche differentiation in four co-existing annual species of *Polygonum* in western North America. J Ecol 65:317–326
Horn HS (1974) The ecology of secondary succession. Annu Rev Ecol Syst 5:25–37
Huston M (1979) A general hypothesis of species diversity. Am Nat 113:81–101
Johnson EA, Rowe JS (1975) Fire in the subarctic wintering ground of the Beverley Caribou herd. Am Nat 94:1–4
Levin SA, Paine RT (1974) Disturbance, patch formation, and community structure. Proc Natl Acad Sci USA 71:2744–2747
Loucks OL (1970) Evolution of diversity, efficiency, and community stability. Am Zool 10:17–25
MacArthur RH, Wilson EO (1967) The theory of island biogeography. Princeton Univ Press, Princeton, p 203
Marks PL (1974) The role of pin cherry (*Prunus pensylvanica* L.) in the maintenance of stability in northern hardwood ecosystems. Ecol Monogr 44:73–88
McNaughton SJ (1979) Grazing as an optimization process: grass-ungulate relationships in the Serengeti. Am Nat 113:691–703
Mooney HA, Bonnicksen TM, Christensen NL, Lotan JE, Reiners WA (eds) (1981) Fire regimes and ecosystem properties. USDA For Serv Gen Tech Rep WO-26:594
Muller CH, Hanawalt RB, McPherson JK (1968) Allelopathic control of herb growth in the fire cycle of California chaparral. Bull Torrey Bot Club 95:225–231
Noble IR, Slatyer RO (1980) The use of vital attributes to predict successional changes in plant communities subject to recurrent disturbances. Vegetatio 43:5–21
Odum EP (1969) The strategy of ecosystem development. Science 164:262–270
Oldeman RAA (1978) Architecture and energy exchange. In: Tomlinson PB, Zimmerman MH (eds) Tropical trees as living systems. Cambridge Univ Press, New York, pp 535–560
Oliver CD, Stephens EP (1977) Reconstruction of mixed-species forest in central New England. Ecology 58:562–572
Parrish JAD, Bazzaz FA (1976) Underground niche separation in successional plants. Ecology 57:1281–1288
Parrish JAD, Bazzaz FA (1979) Difference in pollination niche relationships in early and late successional plant communities. Ecology 60:597–610
Parrish JAD, Bazzaz FA (1982a) Responses of plants from three successional communities to a nutrient gradient. J Ecol 70:233–248
Parrish JAD, Bazzaz FA (1982b) Competitive interactions in plant communities of different successional ages. Ecology 63:314–320
Perozzi RE, Bazzaz FA (1978) The response of an early successional community to shortened growing season. Oikos 31:89–93
Peterson DL (1980) Nutrient dynamics of forest communities in central Illinois. PhD Thesis, Univ Illinois, Urbana, p 138
Pickett STA (1976) Succession: An evolutionary interpretation. Am Nat 110:107–119
Pickett STA (1980) Non-equilibrium coexistence of plants. Bull Torrey Bot Club 107:238–248
Pickett STA, Bazzaz FA (1978) Organization of an assemblage of early successional species on a soil moisture gradient. Ecology 59:1248–1255
Pickett STA, Thompson JN (1978) Patch dynamics and the design of nature reserves. Biol Conserv 13:27–37
Poore MED (1968) Studies in Malaysian rain forest. I. The forest on triassic sediments in Jengka Forest Reserve. J Ecol 56:143–196
Regehr DL, Bazzaz FA (1979) The population dynamics of *Erigeron canadensis*, a successional winter annual. J Ecol 67:923–933
Runkle JR (1981) Gap regeneration in some old-growth forests of the eastern United States. Ecology 62:1041–1051

Schulz JP (1960) Ecological studies on rainforest in Northern Surinam. Elsevier, North-Holland Amsterdam New York, p 267
Smith WH (1976) Character and significance of forest tree root exudates. Ecology 57:324–331
Spurr SH, Barnes BV (1973) Forest ecology, 2nd edn. Ronald Press, New York, p 571
Stephens GR (1981) Defoliation and mortality in Connecticut Forests. Bulletin 796. Conn Agric Exp Stn, New Haven
Tande GF (1979) Fire history and vegetation pattern of coniferous forests in Jasper National Park, Alberta. Can J Bot 57:1912–1931
Trabaud L (1980) Impact biologique et écologique des feux de végétation sur l'organisation, la structure et l'évolution de la végétation des zones de garrigues du Bas-Languedoc. Theses, Acad Montpellier, p 288
Vitousek PM (1977) The regulation of element concentrations in mountain streams in the northeastern United States. Ecol Monogr 47:65–87
Waldendorp JW (1978) The rhizosphere as part of the plant-soil system "Structure and functioning of plant populations." Verh K Ned Akad Wet, Afd Natuurkd Reeks 70:237–276
Watt AS (1947) Pattern and process in the plant community. J Ecol 43:490–506
Watt AS (1981) Further observations on the effect of excluding rabbits from Grassland A in East Anglian Breckland: the pattern of change and factors affecting it (1936–1973). J Ecol 69:509–536
Weaver JE (1954) North American prairie. Johnson, Lincoln Nebr
Webb LJ (1958) Cyclones as an ecological factor in tropical lowland forest, North Queensland. Aust J Bot 6:220–228
Werner PA (1975) Predictions of fate from rosette size in Teasel (*Dipsacus fullonum* L.). Oecologia 20:197–201
Werner PA, Platt WJ (1976) Ecological relationships of co-occurring goldenrods (*Solidago:* Compositae). Am Nat 110:959–971
White PS (1979) Pattern, process, and natural disturbance in vegetation. Bot Rev 45:229–299
Whitmore TC (1975) Tropical rain forests of the Far East. Claredon Press, Oxford, p 278
Whittaker RH, Woodwell GM (1968) Dimension and production relations of trees and shrubs in the Brookhaven Forest, New York. J Ecol 56:1–25
Woods KD (1979) Reciprocal replacement and the maintenance of codominance in a beech-maple forest. Oikos 33:31–39

Subject Index

MIX
Papier aus verantwortungsvollen Quellen
Paper from responsible sources
FSC® C105338

If you have any concerns about our products,
you can contact us on
ProductSafety@springernature.com

In case Publisher is established outside the EU,
the EU authorized representative is:
Springer Nature Customer Service Center GmbH
Europaplatz 3, 69115 Heidelberg, Germany

Printed by Libri Plureos GmbH
in Hamburg, Germany